AF540135

Sustainable *Rural Development* through Agriculture

The Editors

Dr. Shobhana Gupta, currently serving as Deputy Director (Extension) in RVSKVV, Gwalior (MP), did her M.Sc. in Home Science Extension Education and Ph.D. in Agricultural Extension. Recipient of University Gold Medal, Dr. Gupta had a continuous brilliant career in education. She has a vast experience of nearly 18 years of working in KVK system. Being a promising, young and dynamic women scientist, she has received many awards on various forums for her enthusiasm towards working for the betterment of rural community. She has to her credit 5 books, 18 booklets, 25 research papers, 65 popular scientific articles, 8 technical bulletins, 40 radio and 10 TV programmes etc. She is life member of many scientific societies and editorial board of various journals of academic interest. She writes from a background of long and varied practical experience as extension worker, teacher and administrator.

Dr. S.S. Tomar is working as Dean, College of Agriculture, Gwalior and Director Extension Services, RVSKVV, Gwalior. He has an excellent academic record throughout his career. An alumnus from JNKVV, Jabalpur, during last 34 years, he has shown multi-functional skills in Research, Teaching Extension and extracurricular activities. He is guided 36 M.Sc. (Ag.) and Ph.D. Thesis. He has to his credit 105 papers including lead papers presented in National and International seminars and conferences, 45 research papers published in reputed NAAS rated journals, 27 technical bulletins, many radio and T.V. programmes and more than 25 articles. He is also authored five books and two manuals. He is life and founder member of many societies. Recipient of national and state level 12 awards, Dr. Tomar writes from the background of an organizer, trainer, teacher and an administrator.

Sustainable *Rural Development* through Agriculture

— Editors —

Dr. Shobhana Gupta
Dr. S.S. Tomar

2014

BIOTECH BOOKS®

ISBN 978-81-7622-327-0

Published by: **BIOTECH BOOKS®**
4762-63/23, Ansari Road, Darya Ganj,
New Delhi - 110 002
Phone: +91-011-23262132
E-mail: biotechbooks@yahoo.co.in

Printed at: **Chawla Offset Printers**
Delhi - 110 052

PRINTED IN INDIA

राजमाता विजयाराजे सिंधिया कृषि विश्वविद्यालय,
रेस कोर्स रोड ग्वालियर (म.प्र.) – ४७४००२
Rajmata Vijayaraje Scindia Krishi Vishwa Vidyalaya,
Race Course Road, Gwalior (M.P.) - 474002
Tel : 0751-2467673, Fax : 0751-2464141, E-mail: vcrvsaugwa@mp.gov.in

प्रो. अनिल कुमार सिंह
कुलपति
Prof. Anil Kumar Singh
Vice - Chancellor

No./VC/2013-14/3608

Date: 29/03/2014

Foreword

Promoting sustainable agriculture and rural development is a matter of global concern as its major objective is to increase food production and enhance food security in an environmentally sound manner contributing to sustainable natural resource management. This will involve education related initiatives, utilization of economic incentives and development of new as well as appropriate technologies, thus, ensuring stable supplies of nutritionally adequate food and provide access to those supplies by vulnerable groups; production for markets; employment and income generation to alleviate poverty; and natural resource management and environmental protection. The available literature highlights the linkages between the economic, social and environmental objectives of sustainable agriculture. It was also noticed that food security-although a policy priority for all countries-remains an unfulfilled goal. Agriculture has a special and important place in society and helps to sustain rural life and land. Rural Development is, therefore, indicated as one of the thematic areas along with agriculture.

Rural development generally refers to the process of improving the quality of life and economic well-being of people living in relatively less accessible and under developed areas. It has traditionally centered on the exploitation of land-intensive natural resources such as agriculture and forestry. However, changes in global production networks and increased urbanization have changed the character of rural areas. Increasingly tourism, niche manufacturers, and recreation have replaced resource extraction and agriculture as dominant economic drivers. The need for rural communities to approach development from a wider perspective has created more focus on a broad range of development goals rather than merely creating an environment for agricultural or resource based businesses. Education, entrepreneurship, physical infrastructure, and social infrastructure all play a vital role in developing rural regions. Rural development is also characterized by its emphasis on locally produced economic development strategies. Rural development programmes run by Government organizations and NGOs generally focus on finding out ways to improve rural lives with participation of the rural people themselves to meet the required local needs. An outsider may not comprehend the setting, culture, language and other things prevalent in the local area. Consequently there is a plethora of rural development approaches used globally. There is, therefore, an urgent need to focus on these approaches for sustainable agricultural development.

This publication entitled "*Sustainable Rural Development through Agriculture*" edited by Dr. Shobhana Gupta and Dr. S.S. Tomar is reailly very timely as they have documented the perspective and research based articles portraying the role of agriculture in overall rural development and offering strategies to overcome current and perceived problems. I congratulate the editors for their efforts in bringing out this publication. I am sure this book will be useful for students, planners, policy makers, professionals and other stake holders.

(A.K.Singh)

Preface

By the year 2025, 83 per cent of the expected global population of 8.5 billion will be living in developing countries. Yet the capacity of available resources and technologies to satisfy the demands of this growing population for food and other agricultural commodities remains uncertain. Agriculture has to meet this challenge, mainly by increasing production on land already in use and by avoiding further encroachment on land that is only marginally suitable for cultivation. Major adjustments are needed in agricultural, environmental and macroeconomic policy, at both national and international levels, in developed as well as developing countries, to create the conditions for sustainable agriculture and rural development with focus on increasing food production in a sustainable way and enhance food security. This will involve education initiatives, utilization of economic incentives and the development of appropriate and new technologies, thus ensuring stable supplies of nutritionally adequate food, access to those supplies by vulnerable groups, and production for markets; employment and income generation to alleviate poverty; and natural resource management and environmental protection.

The priority must be on maintaining and improving the capacity of the higher potential agricultural lands to support an expanding population. However, conserving and rehabilitating the natural resources on lower potential lands in order to maintain sustainable man to land ratios is also necessary. Sustainable rural development can be achieved through policy and agrarian reform, crop diversification, land conservation and improved management of available inputs as well as the support and participation of rural people, Governments, the private sector and international cooperation.

Both the purposive development of agriculture and rural economy and society has been pursued for many decades, but often separately rather than together. Indeed for a long time rural development was conducted on a sector-by-sector basis. However, the concept of integrated rural development has gradually emerged, whereby interrelations between agriculture and other sectors of economy and society form the basis for more holistic systems and network approaches to development. Sustainability, as another integrative concept, has served to underpin this hybrid approach *i.e.* natural and human systems taken together to the promotion of rural development.

The meaning of sustainability for agriculture and rural development contains three widely recognized dimensions: environment, economy, and society. In more detail, the main environmental dimension includes utilization of natural capital, such as soil (land), water, and mineral resources, so that their use is reproducible over succeeding generations; the improvement of biodiversity; and recycling of wastes and nutrients that does not cause pollution of the biosphere, especially water resources. In the economic dimension, emphasis is given to maintaining agricultural raw materials and services to the nonfarm population by means that provide satisfactory economic returns to land, labor, and capital, even though the definition of satisfactory is contested and is socially determined.

The maintenance of economically viable employment opportunities is extended to other nonfarm, land-based industries (*e.g.*, forestry, mineral extraction, and fishing), manufacturing, and services (*e.g.*, tourism) located in rural regions. With regard to the social dimension, sustainable development includes the long-term retention of an optimum level of population, the maintenance of an acceptable quality of life, the equitable distribution of material benefits from economic growth, and the building of capacity in the community to participate in the development process, including the use of knowledge to create new choices and options over time. In the promotion of sustainable agriculture and rural development, these interrelated environmental, economic, and social dimensions are pursued simultaneously rather than separately.

Views and data presented in the collected articles are those of the authors of the chapters and not necessarily of editors or of their organizations. Again, responsibility of discrepancy, if there is in any article, lies on authors, not on editors and publisher.

We are extremely indebted to Prof. A.K.Singh, Hon'ble Vice Chancellor, RVSKVV, Gwalior for his empowering leadership and encouragement by permitting us to bring out the book on "Sustainable Rural Development through Agriculture". We record our heartfelt thanks to Directors, Deans and Officers of the University for their guidance and valuable suggestions. We are thankful to all contributors of chapters in the book which have resulted in a comprehensive coverage.

It is hoped that this compilation will be useful for the extension professionals, students and other stake holders.

Dr. Shobhana Gupta
Dr. S.S. Tomar

Contents

Foreword v

Preface vii

List of Contributors xiii

— Theme I –
Agriculture: Key Challenges

1. **Food Security in India: Its Challenges and Solution** 3
U.N. Shukla, G.N. Parihar, Smita Singh, Rakesh Kumar and Narendra Kumawat

2. **Climate Change a Threat to Agriculture in 21st Century** 29
D.S. Tomar, S.K. Kaushik, Rekha Tiwari and Arvind Saxena

3. **Climate Change and Vegetable Crop Production** 42
Udit Kumar and Birendra Prasad

4. **Dryland Farming in India** 62
Narendra Kumawat, P.S. Shekhawat, Rakesh Kumar and U.N. Shukla

5. **Shifting Cultivation: A Perspectives in North Eastern Hilly States** 81
Rakesh Kumar, U.N. Shukla and Narendra Kumawat

— Theme II –
Responding to Challenges Through Agriculture

6. **Exploring Indigenous Technical Knowledge** 103
Jitendra Chauhan and Shobhana Gupta

7. **Mitigation and Adaption Strategies in Agriculture against the Changing Climate** **113**
Sarju Narain, O.P. Maurya, Vikas Kumar and R.R. Kushwaha

8. **Strategies for Drought Mitigation in Western Rajasthan** **127**
Lokesh Kumar Jain

9. **Sustainable Crop Production Techniques in Western Rajasthan** **135**
Lokesh Kumar Jain and P.D. Kumawat

10. **Sustainable Use of Agricultural Biodiversity in Western Rajasthan** **141**
Lokesh Kumar Jain, R.K. Rathore and P.C. Meena

11. **Combating Agricultural Risks through Knowledge Management in Rice Farming** **149**
R. Sendilkumar

12. **Factor's for Improving Rice Quality** **159**
Smita Singh, A.K. Singh, U.N. Shukla, L.B. Singh, Anchal Sharma, Deepak Srivastawa, Satish Singh Baghel, and M. Jerman

13. **To Conserve Resource Base in Rice through System of Rice Intensification** **168**
Smita Singh, L.B. Singh, Deepak Srivastava and Satish Singh Baghel

14. **Genetic Purity and Varietal Maintenance** **181**
Birendra Prasad and Shambhoo Prasad

15. **Orchard Management and Varietal Narrative of Mango** **188**
Abhishek Bahuguna, Sandhya Bahuguna and Birendra Prasad

— *Theme III* –
New Initiatives

16. **Intensifying Smallholders' Income through Profitable Enterprises by KVKs: Case Studies** **213**
S.R.K. Singh, Anupam Mishra, Prem Chand and A.P. Dwivedi

17. **Strengthening Extension System: Issue and Options** **231**
Sarju Narain, Vikas Kumar, Sudhir Kumar Rawat and R.R. Kushwah

18. **Farmer Field School: A Platform for Transformative Learning in Rural India** **247**
Sudhir Kumar Rawat, M.K. Awasthi, S.C. Singh, Sarju Narain and Shobhana Gupta

19. Rural Development through Agro-based Industries in India 262
G.L. Meena, S.S. Burark, V.P. Chahal and Prem Chand

20. Agri-Business in India: An Overview 276
G.L. Meena, D.C. Pant, V.P. Chahal and Prem Chand

21. Sustaining Vegetables Production through Farmers' Own Seed: An Integrated Approach 285
Krishan Pal Singh, Beena Nair, Prem Chand, S.K. Sharma and A.S. Bhati

22. Scope of Utilization of Under Utilized Feeds for Enhancing Animal Production 307
Pankaj Lavania and Shobhana Gupta

23. Enhancing Marketing Efficiency in Domestic Trade of Milk and Milk Products 317
B.S. Chandel, Prem Chand and J.P. Dhaka

24. Accelerated Rural Finance for Rapid Agricultural Growth: Constraints and Solutions 338
Sarju Narain, Vikas Kumar, O.P. Maurya and R.R. Kushwaha

25. Alternate Use of Land for Better Livelihood in Western Rajasthan 345
Lokesh Kumar Jain, Neeraj Hada and S.S. Bhadauria

26. Seed Production Technology of Wheat (*Triticum aestivum* L.) 350
Vipin Chandra Joshi, M.K. Nautiyal and Birendra Prasad

27. Women and Livestock: A Sustainable Management 361
Sudhir Kumar Rawat, S.C. Singh, M.K. Awasthi and Sarju Narain

Index 379

List of Contributors

Sl.No.	Name	Designation
1.	**U.N. Shukla**	Assistant Professor (Agro), Agriculture University, Jodhpur (Raj.)
2.	**G.N. Parihar**	Director Research, Agriculture University, Jodhpur (Raj.)
3.	**Smita Singh**	Technical Assistant (Agro.), College of Agriculture, Rewa, (M.P.)
4.	**Rakesh Kumar**	Scientist (Agro), ICAR RC NEH Region, Nagaland Centre Jharnapani, Medziphema
5.	**Narendra Kumawat**	SRF (Agro), ARS, SKRAU, Bikaner
6.	**D.S. Tomar**	Scientist (Agro), KVK Ujjain
7.	**S.K. Kaushik**	Scientist (PBG), KVK Ujjain
8.	**Rekha Tiwari**	Scientist (H.Sc.), KVK Ujjain
9.	**Arvind Saxena**	Associate Professor (Ext Edu), College of Agriculture, Ganj Basoda
10.	**Udit Kumar**	Assistant Professor (Horti) RAU, Pusa, Bihar
11.	**Birendra Prasad**	Senior Research Officer, Department of Genetics and Plant Breeding, College of Agri., GBPUAT, Pantnagar
12.	**P.S. Shekhawat**	Professor (Agro) ARS, SKRAU, Bikaner (Raj.)
13.	**Jitendra Chauhan**	Joint Director (KVK) and Head RBS College Bichpuri, Agra (U.P.)
14.	**Shobhana Gupta**	Deputy Director Extension, RVSKVV, Gwalior (M.P.)

Sl.No.	Name	Designation
15.	**Sarju Narain**	Assistant Professor (Agri. Ext.), BN (PG) College, Raath, Hamirpur
16.	**O.P. Maurya**	Assistant Professor (Agri. Eco.), RSM (PG) college, Dhampur, Bijnor
17.	**Vikas Kumar**	Scientist, Division of Social Science, IGFRI, Jhansi
18.	**R.R. Kushwaha**	Assistant Professor (Agri. Eco.) NDAUT, Faizabad
19.	**Lokesh Jain**	Assistant Professor (Agro.) College of Agriculture, Sumerpur Pali Agriculture University, Jodhpur (Raj.)
20.	**R.K. Rathore**	Assistant Professor (PB.) College of Agriculture, Sumerpur Pali Agriculture University, Jodhpur (Raj.)
21.	**P.C. Meena**	Assistant Professor (Agri. Eco.) College of Agriculture, Sumerpur Pali Agriculture University, Jodhpur (Raj.)
22.	**R. Sendilkumar**	Associate Professor (Agi. Ext.) KAU, Thrissur
23.	**A.K. Singh**	Principal Scientist (PBG), JNKVV, Jabalpur (M.P.)
24.	**L.B. Singh**	Field Extension Officer, Department of Horticulture, College of Agriculture, Rewa, (M.P.)
25.	**Anchal Sharma**	Research Associate (MPWSRP), Department of Horticulture, College of Agriculture, Rewa, (M.P.)
26.	**Deepak Srivastava**	PG. Student (Agro.), College of Agriculture, Rewa, (M.P.)
27.	**Satish Singh Baghel**	Technical Assistant (Agro), College of Agriculture, Rewa, (M.P.)
28.	**M. Jerman**	Technical Assistant (PBG), College of Agriculture, Rewa, (M.P.)
29.	**Shambhoo Prasad**	Ph.D. Scholar (Seed Science and Tech.), GBPUAT, Pantnagar
30.	**Abhishek Bahuguna**	Institute of Medicinal and aromatic Plants, Uttrakhand Uni. of Horti. and Forestry, Uttarakhand
31.	**Sandhya Bahuguna**	Institute of Medicinal and aromatic Plants, Uttrakhand Uni. of Horti. and Forestry, Uttarakhand
32.	**S.R.K. Singh**	Sr. Scientist (Agri Ext.)ZPD, Zone-VII, Jabalpur (M.P.)
33.	**Anupam Mishra**	Zonal Project Director, ZPD, Zone-VII, Jabalpur (M.P.)
34.	**Prem Chand**	Scientist (Agri. Eco.), ZPD, Zone-VII, Jabalpur (M.P.)
35.	**A.P. Dwivedi**	Sr. Scientist (Agro.),ZPD, Zone-VII, Jabalpur (M.P.)
36.	**Sudhir Kumar Rawat**	SMS (AH and D)KVK Mohaba (U.P.)
37.	**S.C. Singh**	KVK Lalitpur (U.P.)

Sl.No.	Name	Designation
38.	**G.L. Meena**	Assistant Professor, Rajasthan College of Agriculture, MPUAT, Udaipur
39.	**D.C. Pant**	Professor, Rajasthan College of Agriculture, MPUAT, Udaipur
40.	**Krishan Pal Singh**	SMS (Horti.) KVK Banasthali, Distt. Tonk (Raj)
41.	**Beena Nair**	Assistant Professor, K.N. Modi University, Tonk (Raj).
42.	**S.K. Sharma**	Programme Coordinator, KVK Banasthali, Distt. Tonk (Raj)
43.	**A.S. Bhati**	Farm Manager, KVK Banasthali, Distt. Tonk (Raj)
44.	**Pankaj Lavania**	Assistant Professor (AH), MPUAT, KVK Sirohi,
45.	**B.S. Chandel**	Principal Scientist DES and M Division, NDRI, Karnal (Haryana)
46.	**J.P. Dhaka**	Principal Scientist (Retd.) DES and M Division, NDRI, Karnal (Haryana)
47.	**Vipin Chandra Joshi**	Ph.D. Scholar and SRF, Department of Genetics and Plant Breeding, College of Agri., GBPUAT, Pantnagar
48.	**M.K. Nautiyal**	Department of Genetics and Plant Breeding, College of Agri., GBPUAT, Pantnagar
49.	**M.K. Awasthi**	Animal Breeding Centre, Raibareli (U.P.)
50.	**V. P. Chahal**	Principal Scientist (Agri. Ext. Division), ICAR Headquarter, New Delhi
51.	**S.S. Burark**	Professor and Head, Department of Agr. Eco., RCA, MPUAT, Udaipur
52.	**Neeraj Hada**	SMS (Crop Science), Directorate of Extension Services, RVSKVV, Gwalior
53.	**S.S. Bhadauria**	Technical Officer to Director, Directorate of Extension Services, RVSKVV, Gwalior
54.	**P.D. Kumawat**	Junagadh Agriculture University, Junagadh, Gujarat

— Theme I –

Agriculture: Key Challenges

Chapter 1

Food Security in India: Its Challenges and Solution

U.N. Shukla, G.N. Parihar, Smita Singh, Rakesh Kumar and Narendra Kumawat

ABSTRACT

To address the challenge of sustainable development, future agricultural planning in the developing world has to ensure sufficient food production, employment and rural income generation while conserving natural resources. This has to be achieved in context which agriculture especially in developing countries is undergoing a transformation due to changing demands, markets and agricultural technologies. The pace of these changes is expected to increase rapidly in coming years and whole system of agricultural production may become quite different in next 10–20 years. Understanding the interactions between food security and global environmental change is highly challenging. This is nevertheless increasingly important as 50 per cent more food will be needed by 2030 and there are concerns that risk of food insecurity will likely grow. These concerns are compounded by simultaneous need to reduce negative environmental feedbacks from ways we meet these demands. Addressing climate change is, therefore, critical for future food security and attainment of the Millennium Development Goals, especially the goal of poverty alleviation. Developing nations and regions will need to implement strategies link with their development plans to enhance their adaptive capacity. Several options are available today that can reduce vulnerability. "No-regrets" adaptation strategies that promote sustainable development and are also effective in reducing negative impacts of climate change need to be identified and implemented for securing food production.

1.1 Introduction

Food security refers to the availability of food and one's access to it. A household is considered food-secure when its occupants do not live in hunger or fear of starvation. According to the World Resources Institute, global per capita food production has been increasing substantially for the past several decades. The FAO estimates that 1.02 billion people are suffering from chronic hunger in the world, mostly in Africa and south Asia, meaning that almost one sixth of humanity is suffering from hunger. Despite good economic performance with over 200 million people who are food insecure, India is home to the largest number of hungry people in the world. International Food Policy Research Institute sheds renewed light on acute Indian hunger situation. The Global Hunger Index (GHI) 2011 ranks India at the bottom with 67th position (out of 81 countries) with a GHI of 23.7, which the index characterizes as "alarming" food security situation. India has been ranked 66 in the list of 105 countries -much lower than neighboring China (ranked 39) and somewhat lower than Sri Lanka (62) in the 2012 Global Food Security Index released by the American chemical company DuPont (Figure 1.1). The Index has been developed by Economist Intelligence Unit (EIU) and is sponsored by DuPont. Founded in 1946 as an in-house research unit for The Economist newspaper, Economist Intelligence Unit is part of the Economist Group, which is the leading source of analysis on international business and world affairs. With India expected to become the most populous country in the world by 2025, feeding the population is likely to be one of the serious challenges that the country will face in the coming decades. India scores somewhat higher in category of 'availability' than in the other two 'affordability' and 'quality and safety' categories. In comparison, India is better off than Pakistan (75) and Bangladesh (81), according to the index calculations. High level of poverty, lower income, less public spending on farm research, poor infrastructure, sluggish supply of quality protein are some of the key challenges that India needs to address, it noted. On the positive side, however, presence of food safety net programs and access to farm credit has helped the country achieve some level of food security. According to EIU regional Director Pratiba Thaker, "Apart from the challenges of availability and accessibility as reflected in chronic household food insecurity, India also faces a nutrition challenge.

Much of the food security debate understandably centres on aspects of food production and this has long been subject of major research investment. Increasing production has always been an important strategy to help alleviate food insecurity, and it still is today. There is hence still a strong sentiment that producing more food will satisfy society's needs, and theoretically this is of course case produce enough and all will be fed. However, despite the fact that more than enough food is currently produced per capita to adequately feed the global population, about 925 million people remained food insecure in 2010 (FAO, 2010). For many, this gap in production vs. need is more related to political economy of interventions and political inertia in funding decisions than to technical ignorance.

According to a 2004 article from the BBC, China, the world's most populous country, is suffering from an obesity epidemic. In India, the second-most populous

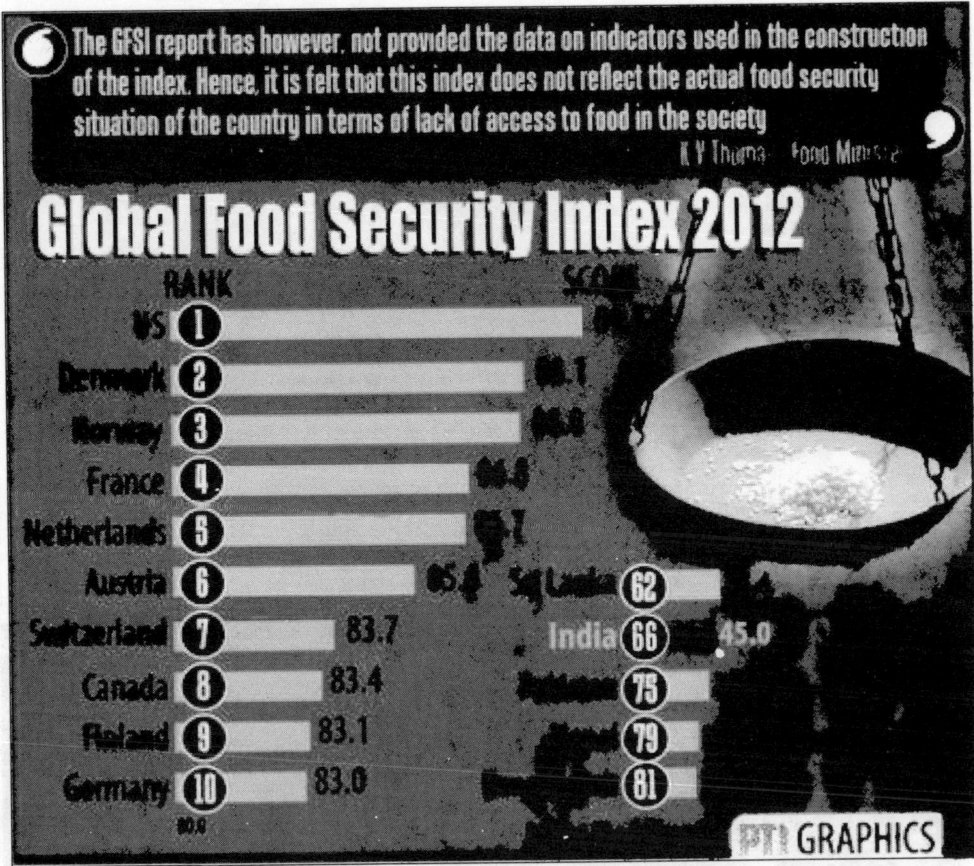

Figure 1.1: Global Food Security Index.

country in the world, 30 million people have been added to the ranks of the hungry since the mid 1990s and 46 per cent of children are underweight. Worldwide around 852 million people are chronically hungry due to extreme poverty, while up to 2 billion people lack food security intermittently due to varying degrees of poverty. Six million children die of hunger every year-17,000 every day. As of late 2007, export restrictions and panic buying, US Dollar Depreciation, increased farming for use in bio-fuels, world oil prices at more than $100 a barrel, global population growth, climate change, loss of agricultural land to residential and industrial development, and growing consumer demand in China and India are claimed to have pushed up price of grain. However, role of some of these factors is under debate. Some argue the role of bio-fuel has been overplayed as grain prices have come down to the levels of 2006. Nonetheless, food riots have recently taken place in many countries across the world. The ongoing global credit crisis has affected farm credits, despite a boom in commodity prices. Food security is a complex topic, standing at intersection of many disciplines In India approximately, 320 Indians go to bed without food every night and recent data is very much alarming and situation is going even worse. Food riots

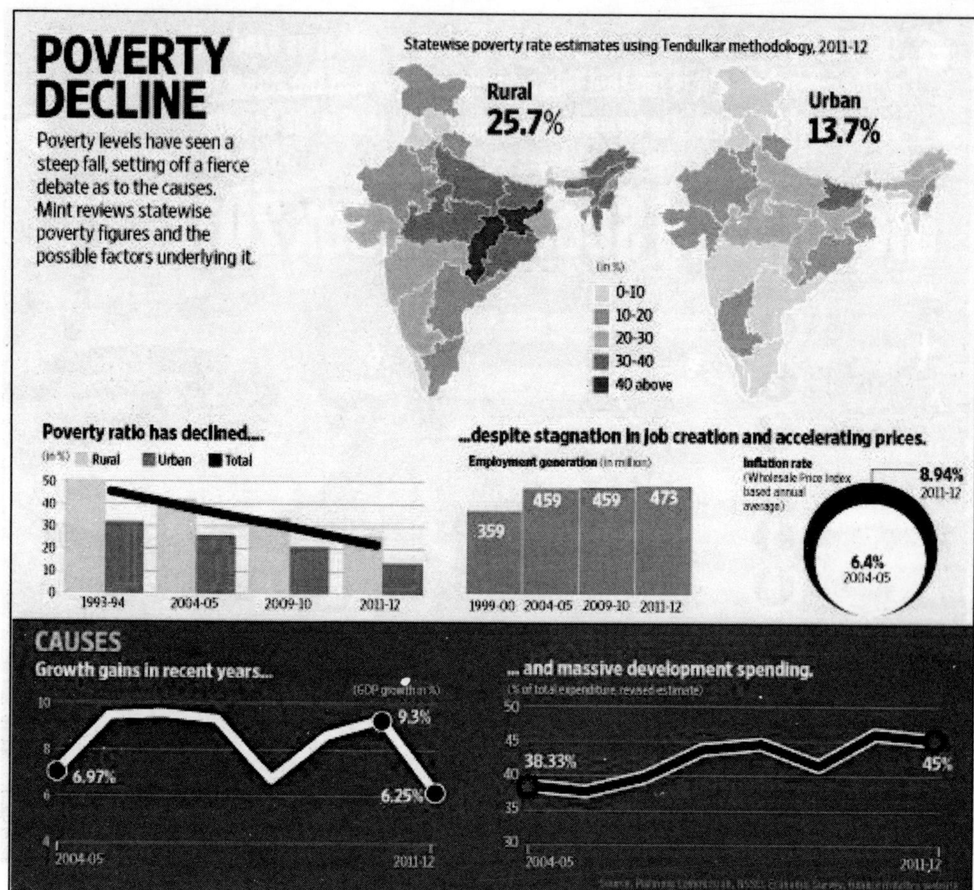

Figure 1.2: India's Poverty Decline Year by Year Due to Gross Domestic Increment and Allocation of Budget Toward Food Security (Planning commission of India, 2012).

have taken in many countries of the world. It's becoming very difficult to maintain food security. In rural context, agriculture development for small and marginal farmer is the most important dimension of food security. The diversification of agriculture for food *e.g.* cereals, pulses, edible oil yielding, vegetable, fuel and timber yielding plants, medicinal and fodder crops are necessary to meet the food and augment income to farmers to meet the food security. Natural vagaries like excessive rainfall, drought, and availability of water for irrigation, undulating topography, soil erosion and soil type such as degraded soil, acidic and alkaline soil affect the food security. Understanding the interactions between food security and global environmental change is highly challenging. This is nevertheless increasingly important as 50 per cent more food will be needed by 2030 (Godfray *et al.*, 2010) and there are concerns that the risk of food insecurity will likely grow. These concerns are compounded by the simultaneous need to reduce negative environmental feedbacks from the ways we meet these demands. The income levels of farmer families govern the access to food affordability. Public Distribution System (PDS) is not satisfactorily functioning. The families very poor within adequate income cannot escape food crisis. Globalization may and may not help food security. However, there are people who feel that globalization will definitely help food security due to trade but it matter of debate. We will have to aim at food security in developing countries through increased and stabilized food production on an economically and environmentally sustainable technologies/methods. Diversification in agriculture is highly required. We cannot afford to huger, malnutrition and famine. Accordingly, all of us including Govt. policies seriously need to redesign in order to meet the present demand and fill the gap of exiting system in order to ensure food security to every citizen of the country.

1.2 Definition

Before I write about food security we have to understand what food is and the requirement for its security. ***Food is something that gives you the energy to function and keeps you alive.*** This is true for both plant and animal kingdoms. Plants process their food directly from sunlight, water and soil while animals, including human beings depend upon plants and other animals. From time immemorial the method of taking energy (food) from a source has not changed. From the definition given by Food and Agricultural Organization (FAO) and United States Department of Agriculture (USDA) **Food security** can be explained that it is the **availability** of **enough** food **to all** the people **at all the times** where people have **physical** and **economic access** to **sufficient**, **safe** and **nutritious** food to meet the **dietary needs** and **food preferences** for **survival** and **active** and **healthy life**. For ensuring food security three important events have to take place, they are **production, distribution** and **consumption** which author would like to call it as a **food security cycle.**

Originally, the term "food security" was used to describe whether a country had access to enough food to meet dietary energy requirements. National food security was used by some to mean self-sufficiency *i.e.* the country produces the food it needs or that which its population demands. Again, it was seldom made clear whether self-sufficiency meant that all citizens had access to enough food to meet energy and nutritional requirements or whether meeting economic demand from domestic production was enough to claim self-sufficiency. Using the latter definition, all countries could claim self-sufficiency simply by leaving the domestic market to equate demand and supply at whatever price would result. National food sovereignty was and still is used to measure the extent to which a country has the means to make available to its people the food needed or demanded, irrespective of whether the food is domestically produced or imported. A country that does not produce the food it needs or its population is prepared to buy and does not have the hard currency to import what is missing, would not be food sovereign. Much of the food security debate understandably centres on aspects of food production and this has long been the subject of major research investment. Increasing production has always been an important strategy to help alleviate food insecurity, and it still is today. There is hence still a strong sentiment that producing more food will satisfy society's needs, and theoretically this is of course the case: produce enough and all will be fed.

However, despite the fact that more than enough food is currently produced per capita to adequately feed the global population, about 925 million people remained food insecure in 2010 (FAO, 2010). For many, this gap in production vs. need is more related to the political economy of interventions and political inertia in funding decisions than to technical ignorance (see quotation above). Given that food prices are again high (in March 2011, the food index remained 36 per cent above its level a year earlier, World Bank 2011) there is a strong likelihood that this number will again rise. This link between food prices and numbers of food insecure people underscores the importance of the affordability of food in relation to food security. This is reflected in the commonly-used definition stemming from the 1996 World Food Summit (FAO, 1996) which states that food security is met when "***all people, at all times, have physical and economic access to sufficient, safe, and nutritious food to meet their dietary needs and food preferences for an active and healthy life***". This definition puts the notion of access to food centre stage. Further, not only does it bring in a wide range of issues related to a fuller understanding of food security, but some key words such as "food production" and "agriculture" which might have been expected in such a definition is not included; the emphasis changed from increasing food production to increasing access to food for all. This definition also integrates notions of food availability and food utilization. The majority of the more recent (*i.e.* since 1990s) definitions have the notion of access to food central and are now manifestly very valuable in raising the pı ofile of access to food *vis à vis* producing food.

1.3 Poverty in India

The percentage of the population living below the poverty line in India decreased to 22 per cent in 2011-12 from 37 per cent in 2004-05, according to data released by the Planning Commission in July 2013. This section presents data on recent poverty estimates and goes on to provide a brief history of poverty estimation in the country.

1.3.1 National and State-wise Poverty Estimates

The Planning Commission estimates levels of poverty in the country on the basis of consumer expenditure surveys conducted by the National Sample Survey Office (NSSO) of the Ministry of Statistics and Programme Implementation. The current methodology for poverty estimation is based on the recommendations of an Expert Group to Review the Methodology for Estimation of Poverty (Tendulkar Committee) established in 2005. The Committee calculated poverty levels for the year 2004- 05. Poverty levels for subsequent years were calculated on the basis of the same methodology, after adjusting for the difference in prices due to inflation. Table 1.1 shows national poverty levels for the last twenty years, using methodology suggested by the Tendulkar Committee. According to these estimates, poverty declined at an average rate of 0.74 percentage points per year between 1993-94 and 2004-05, and at 2.18 percentage points per year between 2004-05 and 2011-12.

State-wise data is also released by the NSSO. Table 1.2 shows state-wise poverty estimates for 2004-05 and 2011-12. It shows that while there is a decrease in poverty for almost all states, there are wide inter-state disparities in the percentage of poor below the poverty line and the rate at which poverty levels are declining.

Table 1.1: National Poverty Estimates (per cent below poverty line) (1993–2012)

Year	Rural	Urban	Total
1993 – 94	50.1	31.8	45.3
2004 – 05	41.8	25.7	37.2
2009 – 10	33.8	20.9	29.8
2011 – 12	25.7	13.7	21.9

Source: Press Note on Poverty Estimates, 2011-12, Planning Commission; Report of the Expert Group to Review the Methodology for Estimation of Poverty (2009) Planning Commission; PRS.

Table 1.2: State-wise Poverty Estimates (per cent below poverty line) (2004-05, 2011-12)

State	2004-05	2011-12	Decrease
Andhra Pradesh	29.9	9.2	20.7
Arunachal Pradesh	31.1	34.7	–3.6
Assam	34.4	32	2.4
Bihar	54.4	33.7	20.7
Chhattisgarh	49.4	39.9	9.5
Delhi	13.1	9.9	3.2
Goa	25	5.1	19.9
Gujarat	31.8	16.6	15.2
Haryana	24.1	11.2	12.9
Himachal Pradesh	22.9	8.1	14.8
Jammu and Kashmir	13.2	10.4	2.8
Jharkhand	45.3	37	8.3
Karnataka	33.4	20.9	12.5
Kerala	19.7	7.1	12.6
Madhya Pradesh	48.6	31.7	16.9
Maharashtra	38.1	17.4	20.7
Manipur	38	36.9	1.1
Meghalaya	16.1	11.9	4.2
Mizoram	15.3	20.4	–5.1
Nagaland	9	18.9	–9.9
Odisha	57.2	32.6	24.6
Puducherry	14.1	9.7	4.4
Punjab	20.9	8.3	12.6
Rajasthan	34.4	14.7	19.7
Sikkim	31.1	8.2	22.9

Contd...

Table 1.2–Contd...

State	2004-05	2011-12	Decrease
Tamil Nadu	28.9	11.3	17.6
Tripura	40.6	14.1	26.5
Uttar Pradesh	40.9	29.4	11.5
Uttarakhand	32.7	11.3	21.4
West Bengal	34.3	20	14.3
All India	**37.2**	**21.9**	**15.3**

Source: Review of Expert Group to Review the Methodology for Estimation of Poverty (2009) Planning Commission, Government of India; Press Note on Poverty Estimates, 2011-12 (2013) Planning Commission, Government of India; PRS.

However, the concept of poverty and estimates of its magnitude and profile are quite relevant in the context of policy formulation, its process and outcome evaluation. It matters most in a country low-income developing country like India, which has been pursuing development strategies and policy programmes for 'Growth with Poverty Reduction'. Consistent with this policy concern, the concept of poverty and norm for its definition has evolved over time depending upon information availability, prevailing exigencies, policy imperatives and priorities. One important norm used consistently for defining poverty line relates to nutrition, the energy intake criterion in particular. The Government of India (GoI) has been using a minimum dietary energy requirement norm of 2400 kcal per person per day for the rural sector and 2100 kcal for the urban sector while the Food Agricultural Orgnisaion norm for India as a whole for 2003-05 is 1770 kcal.

Table 1.3: Percentage of Population Below Poverty Line Calculated by the Lakdawala Committee and the Tendulkar Committee for the year 2004-05

Committee	Rural	Urban	Total
Lakdawala Committee	28.3	25.7	27.5
Tendulkar Committee	41.8	27.5	37.2

Source: Report of the Expert Group on Estimation of Proportion and Number of Poor, 1993, Perspective Planning Division, Planning Commission; Report of the Expert Group to Review the Methodology for Estimation of Poverty, 2009, Planning Commission; PRS

1.4 Food Grain Production, Procurement and Distribution

1.4.1 Food Grain Production

While the agriculture sector may not be the focus of government attention due to its preoccupation with the so called economic reforms (liberalization and westernization of economy) dictated by Western lending agencies, food grain production growth has been steady (if not strong) since mid 2000s. The widespread farmers' suicide due to indebtedness is price nation has to pay in order to become an industrialized society -following western footsteps. The compound annual rate of growth of food grain production, which stood at just 0.8 per cent during 2000-01 to 2005-06, accelerated to 2.9 per cent per annum during 2005-06 to 2012-13. As a result, production in agricultural year 2012-13 is estimated to have touched a record 255.4 million tonnes (Figure 1.3). However, the per capita production (after adjusting for seed, feed and wastage) has been just 164.9 kg in 2011, which had been exceeded as far back as in 1984, and was below recent peak in 2008 of 171 kg. Thus, there was no growth in per capita food grain production. Yet interestingly, the government has been burdened with a rising stock of food grain-unable to distribute the amount it procures (Figure 1.4).

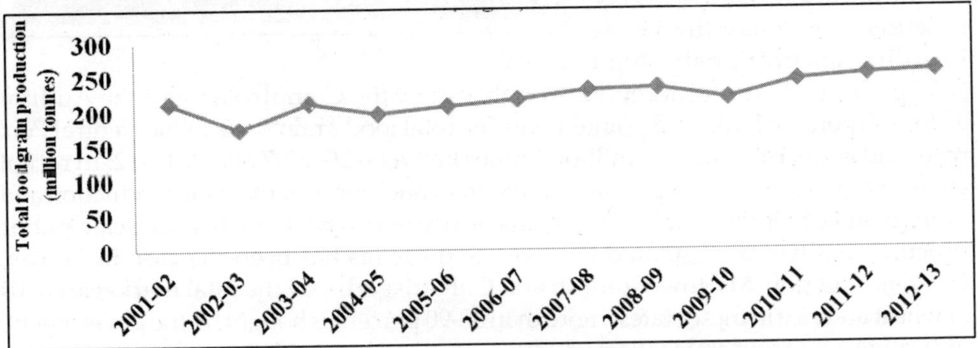

Figure 1.3: India: Foodgrain Production Trends.
***Source*: Ministry of Agriculture, 2013.**

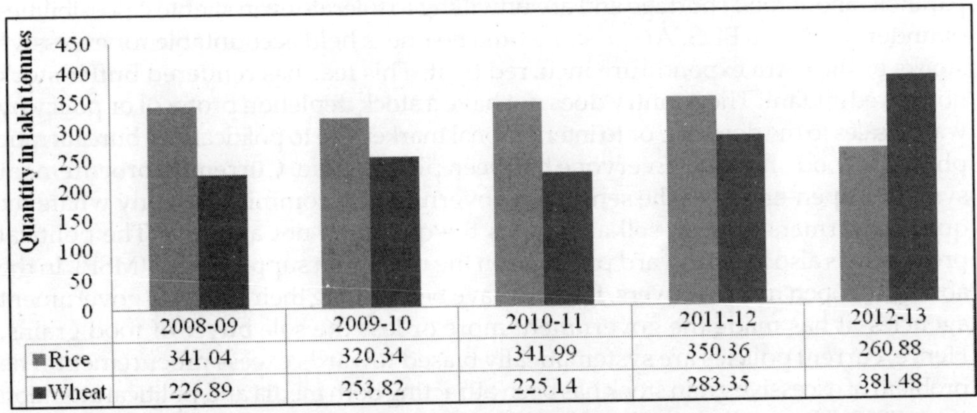

	2008-09	2009-10	2010-11	2011-12	2012-13
■Rice	341.04	320.34	341.99	350.36	260.88
■Wheat	226.89	253.82	225.14	283.35	381.48

Figure 1.4: Procurement of Rice and Wheat for Central Pool (in lakh tonnes).
***Source*: Ministry of Agriculture, 2013.**

1.5 Procurement and PDS Distribution

As shown in the chart here, until 1989-90 the procurement and distribution matched. Since then government has been buying more than it distributed year by year; the gap became only wider since early 2000s. It has led to the accumulation of food grain stock way beyond what is needed. Clearly, buffer stock norms have been consistently exceeded early 2000s. It is nothing but a reflection mismanagement of procurement and distribution policy. Beyond buffer stock norms and strategic reserves, averaged difference between procurement and subsidized distribution should be close to zero. No wonder, due to accumulated errors of over procurements today the FCI is hoarding an irrationally high food grain stock. A recent report brought out by the Comptroller and the Auditor (CAG) General of India also pointed out-The total food grain stock in the Central Pool recorded an increase of 45.8 million tonnes between 2006-07 and 2011-12. In recent years, the government has procured more than one-thirds of the total production and more than half of the marketed surplus of rice and wheat. Such large scale public procurement has strangulated the private trade (as has been the case in Punjab, Haryana and now Madhya Pradesh and Chhattisgarh). Of the total market arrivals of wheat and rice in these states, more than 80-90 percent is bought by the government, indicating a de-facto state takeover of grain trade (DFPD, 2013).

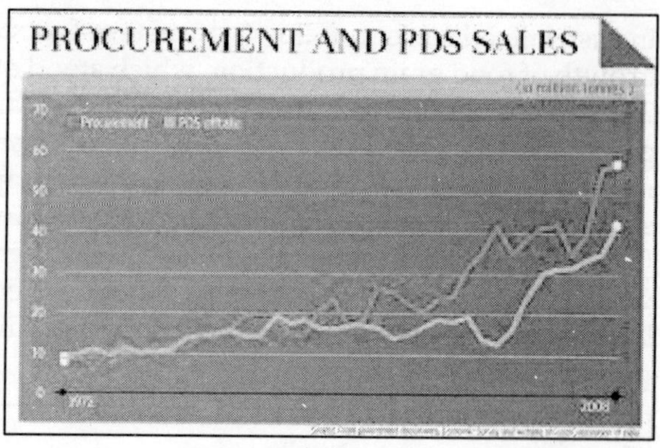

1.5.1 Current Procurement Policies Promote Accumulation

The explanation lies in fact that government and the bureaucrats are highly paranoid about food shortage and are unwilling to tolerate even slightest possibilities of under-supply to PDS. At the same time no one is held accountable for excessive stocks or the extra expenditure incurred by it. This fear has rendered buffer stock norms redundant. The country does not have a stock depletion protocol or policy by way of sales to the domestic or to international market due to political and bureaucratic phobia of food shortage. Everyone has been playing safe. Currently, procurement system is open-ended in the sense that government is committed to buy whatever quantity farmers wish to sell as long as the quality is not an issue. The political pressure has also kept upward pressure on the minimum support price (MSP). In the absence of open market buyers, farmers have been selling their grains to government agencies- it has made the government more or less the sole buyer of food grains. Hence, current policies are systematically biased towards excess procurement. The problem of excessive grain stock has been attracting both media and political attention since 2012. The media has often highlighted the issue of widespread rotting grain stock in makeshift storage facilities of the FCI while millions of poor go to bed hungry. The issue found attention of the Parliament also. In September 2010, hearing right to food public interest petition, the Supreme Court asked the government to distribute to poor the food grains that would otherwise rot. But unfortunately, situation has only worsened since then. The excessive stock with FCI serves no useful purpose. Liquidating it will ease inflation, reduce food subsidy and bring rationalization in the grain market (DFPD, 2013).

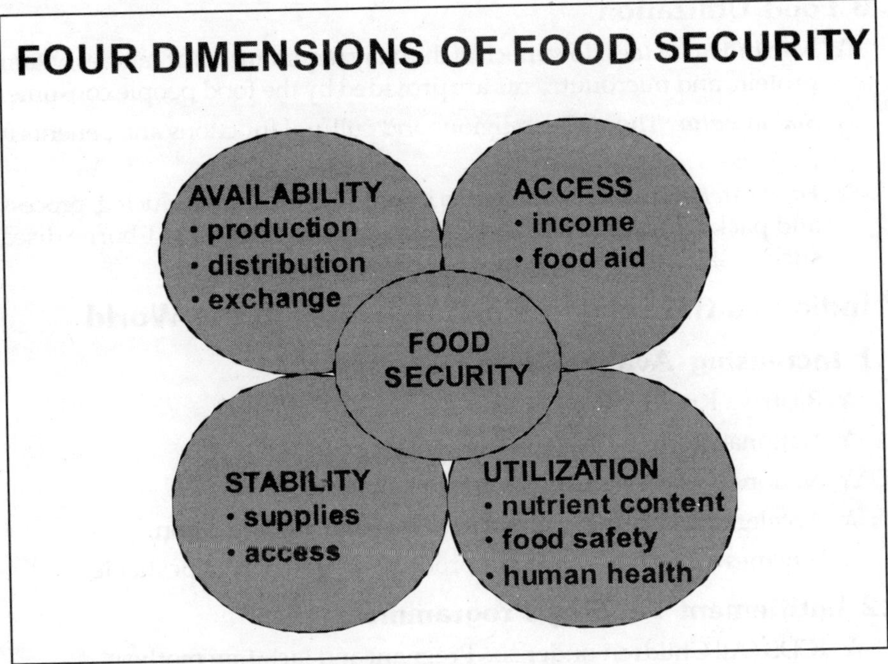

1.6 Dimension of Food Security

Food security outcomes are grouped into three components (Availability, Access and Utilization), each of which comprises three elements.

1.6.1 Food Availability

- ☆ *Production:* How much and which types of food are available through local production.
- ☆ *Distribution:* How food is made available (physically moved), in what form, when and to whom.
- ☆ *Exchange:* How much of available food is obtained through exchange mechanisms such as barter, trade, purchase, or loans.

1.6.2 Access to Food

- ☆ *Affordability*: The purchasing power of households or communities relative to price of food.
- ☆ *Allocation:* The economic, social and political mechanisms governing when, where and how food can be accessed by consumers.
- ☆ *Preference:* Social, religious or cultural norms and values that influence consumer demand for certain types of food.

1.6.3 Food Utilization

- *Nutritional value:* How much of the daily requirements of calories, vitamins, protein, and micronutrients are provided by the food peopie consume.
- *Social value:* The social, religious and cultural functions and benefits food provides.
- *Food safety:* Toxic contamination introduced during producing, processing and packaging, distribution or marketing food; and food-borne diseases such as salmonella and CJD.

1.7 India has the Largest Food Schemes in the World

1.7.1 Increasing Availability

- Rastriya Krishi Vikas Yojana
- National Horticulture Mission
- National Food Security Mission
- Accelerated efforts in rainwater harvesting and irrigation
- Schemes for Enhancing factor productivity (fertilizers pesticide, etc.)

1.7.2 Entitlement Feeding Programmes

- ICDS (All Children under six, Pregnant and lactating mother)
- MDMS (All Primary School children)

1.7.3 Food Subsidy Programmes

- Targeted Public Distribution System (35 kg/month of subsidised food grains)
- Annapurna (10 kg of free food grain for destitute poor)

1.7.4 Employment Programmes

- Mahatma Gandhi National Rural Employment Scheme (100 days of employment at minimum wages)

1.7.5 Social Safety Net Programmes

- National Old Age Pension Scheme (Monthly pension to BPL)
- National Family Benefit Scheme (Compensation in case of death of bread winner to BPL families)

1.7.6 Improving Access

- Mahatma Gandhi National Rural Employment Guarantee Act
- Integrated Child Development Services
- School Noon Meal Programme
- Annapoorna
- Universal and Targeted Public Distribution Systems
- Food Security Act (proposed)

1.7.7 Improving Absorption

- ☆ Rajiv Gandhi Drinking Water Mission
- ☆ Total Sanitation Programme
- ☆ National Rural Health Mission
- ☆ National Urban Health Mission

1.8 The Food Security Outcomes and their Elements

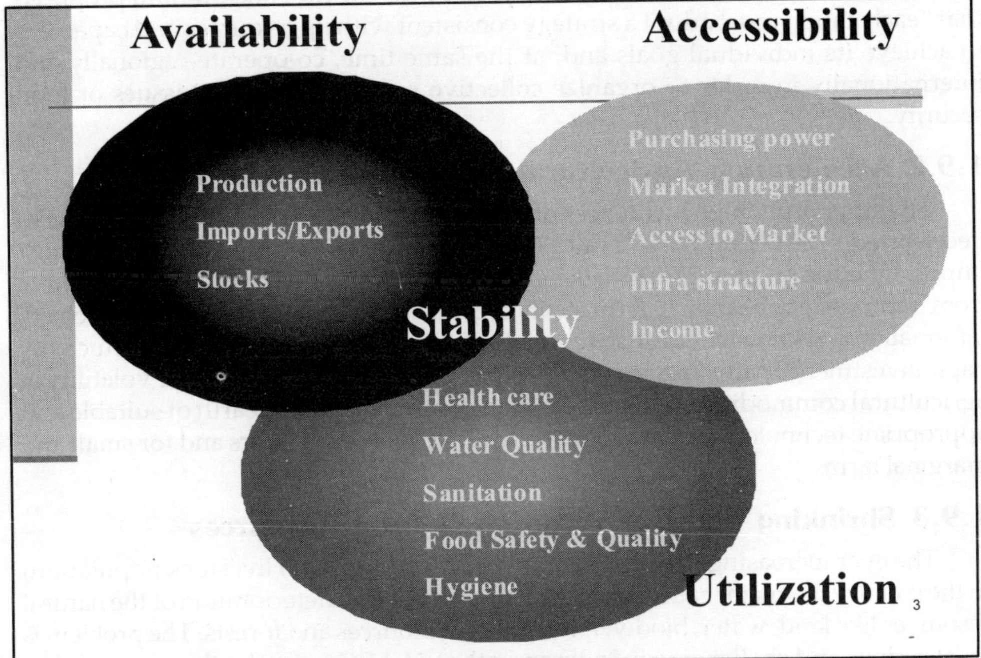

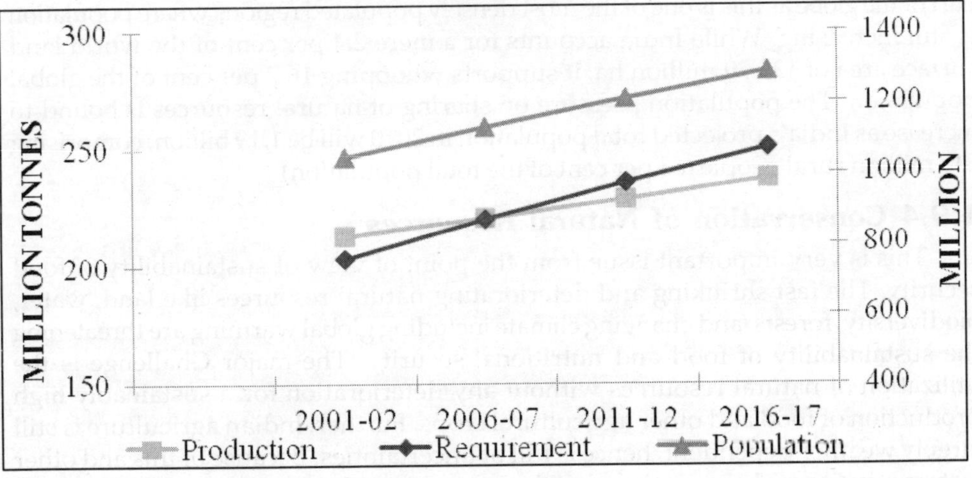

1.9 Issues and Challenges

1.9.1 Fighting Hunger and Poverty

The issue of food security has been identified as a major objective to be pursued by global community by Rome Declaration on World Food Security and the World Food Summit Plan of Action in 1996. The Rome Declaration took into consideration the multi-faceted character of food security and emphasised that "concerted national action, and effective international efforts" were needed to "supplement and reinforce national action." The Plan of Action adopted by the World Food Summit proposed that "each nation must adopt a strategy consistent with its resources and capacities to achieve its individual goals and, at the same time, co-operate regionally and internationally in order to organize collective solutions to global issues of food security."

1.9.2 Accelerating Agricultural Production Growth Rate

The stubbornly high incidences of under nutrition and poverty are ascribed to decelerated production growth rates in recent years, and must be seen as the most important issue and foremost challenge before the nation. Often the system suffers from improper application of farm technology, deficiencies in the input supply chain, information and knowledge capacity with the farmers and huge yield and productivity gaps, investment apathy in agriculture sector, poor infrastructure, price volatility in agricultural commodities, collapse of extension services and dearth of suitable and appropriate technology and support systems for women farmers and for small and marginal farms.

1.9.3 Shrinking and Deteriorating Natural Resources

The ever increasing pressure of human population (also livestock population) in the country is a major cause of shrinking as well as fast deterioration of the natural resources like land, water, biodiversity, genetic resources and forests. The problem is most serious and challenging in India and other SAARC countries than in any other part of the globe as this is one of the most densely populated regions where population is still increasing. While India accounts for a mere 2.4 per cent of the world land surface area of 135.79 million ha, it supports whopping 16.7 per cent of the global population. The population pressure on sharing of natural resources is bound to increase as India's projected total population in 2010 will be 1.19 billion, comprising 815 million rural people (68 per cent of the total population).

1.9.4 Conservation of Natural Resources

This is very important issue from the point of view of sustainability of food security. The fast shrinking and deteriorating natural resources like land, water, biodiversity, forests, and changing climate including global warming are threatening the sustainability of food and nutritional security. The major Challenge is the utilization of natural resources without any deterioration for a sustainably high production of food and other agricultural crops. Further, Indian agriculture is still largely weather dependent, hence is full of uncertainties. Because of this and other factors, new issues keep coming, and there are no standard solutions for them. Only way is to overcome them by experience, collective wisdom, and of course thorough appropriate research based technology application. As regards water, the situation is precarious as all the known sources of fresh water, whether surface or ground water, are dwindling and drying. Himalayan glaciers that are a major source of water to Afghanistan, Bangladesh, Bhutan, India Nepal and Pakistan (more than 3/4th of SAARC region) are melting and fading away. Great rivers are at the verge of extinction, water in the remaining ones is highly polluted. Water reservoirs, big or small, in country side have silted and under greed or population push, ground water table has gone so deep that it is rendered useless for common man. Further, rainfall is indicating a receding trend in several parts of India and making the situation worse.

1.9.5 Unstable Food Self-Sufficiency and Sub-optimal Buffer Stocking

Given huge and increasing India's population size, high incidence of hunger and poverty, heavy dependence on agriculture for livelihood security, limited impact of the "trickle down" effect, the little or negative impact of the market forces and liberalization on the wellbeing of the majority poor and increasing agricultural unemployment, on one hand, and the international market price volatility, uncertainty of timely and cost effective availability of adequate quantities of bio safe and desired quality food grains in the international market, on the other hand, demand that India should largely remain food self-sufficient. Moreover, the upheavals that India's entry as an importer in the global grain market creates in the prices and the resultant hardship to poorer and smaller importing countries also suggest that India should remain food self-sufficient. This approach will protect both its poor producers and consumers. However, for other crops, we may focus more on crops with comparative advantage and boost our export earnings which could help fund our imports, if needed. Thus, the main challenge is to produce enough food within our resources not only for meeting current macro level per capita food requirement, but more essentially to ensure at micro level sufficient nutritious food for each household member. Along with, peoples purchasing power ought to be increased to enable them to buy the needed food from the market or the PDS. Moreover, sufficient food stock must be maintained as buffer for any contingency. Considering the recent uncertainties and controversies in buffer stocking, the size, irregularities and role of the PDS, which is generally dependent on food stocks held by the public sector, the prospect of the food grains futures markets and trend of international prices, a detailed policy on buffer stocking is called for.

1.9.6 Continued High Dependence on Agriculture, Growing Rural-Urban Divide and Farmers' Indebtedness

Agriculture is expected to be the engine of India's growth and prosperity contributing 18 per cent of the national GDP and 57 per cent of total employment and 73 per cent of India's total rural employment (NCF Reports, 2006). Of the 77 per cent of India's population living on less than Rs. 20 a day, 85 per cent of them are from rural areas and most of them are marginal/sub marginal farmers and landless agricultural workers. And, it is these people who comprise the bulk of the undernourished people in the country. Further, Nearly 85 per cent of Indian agriculture is conducted on small/marginal farms and about 50 per cent of the farmers are under debt and paradoxically most of them belong to agriculturally advanced States, Andhra Pradesh being 82 per cent indebted, while in Tamil Nadu and Punjab this percentage is 75 and 65, respectively. The worst part is that 84 per cent of farmer households spend (₹2770 month^{-1}) more than they earn (₹2115). The input-risk-output imbalance has worsened for the farmers. The per capita income disparity between the farm and non-farm sectors has increased from 1:3 to 1:5 during the last 30 years (Agricultural Statistics at a Glance, 2006).

1.9.7 Rainfed Areas and Resource-Poor Farmers Bypassed by Green Revolution

Green Revolution technologies are scale neutral but not resource neutral, since inputs are needed 27 for output. Hence they have generally bypassed the vast rainfed areas and resource-poor farmers. A large number of rainfed crops and cropping systems were not touched by the green revolution process.

1.9.8 Technology Fatigue

The faulty use of inputs in Green Revolution areas has not only reduced the production efficiency, but also caused environmental and economic losses. For instance, the excessive drawl of underground water for irrigation has resulted in drastic drop of water table and uncontrolled flooding of fields uncoupled with drainage has caused serious water logging and salinity problems. Likewise, unbalanced use of fertilizers has adversely impacted soil health and lowered fertilizer productivity. The problem of technological fatigue is further compounded with huge technology transfer gaps at various levels. The gaps between potential and realizable and between realizable and average realized yields in country are generally around 50 to 100 percent, respectively.

1.9.9 Inadequate Harnessing of Cutting Edge Technologies

Gene revolution and ICT revolution are sweeping the world, especially fast expanding knowledge-rich and knowledge-based economies. The first Green Revolution in India has failed to connect itself with these revolutions. The powers of biotechnology and ICT have not been internalized in India's agricultural transformation process, although there are some "islands" of successes, *viz* Bt cotton hybrids and e-chaupal and sporadic Village Knowledge Centres. The national biotechnology policy is still in the making while area under "illegal" biotech varieties (Bt cotton hybrids) has been expanding (along with "legal" varieties).

1.9.10 Huge Post-Harvest Losses

On an average, post-harvest losses of the tune of 12 to 15 per cent occur, whereas in case of fruits and vegetables the average losses are about 30 per cent. Rural agro-processing, strengthening market infrastructures and value addition are extremely limited.

1.9.11 Climate Change and Mounting Risks

Our farm and fisher families are being increasingly subjected to the fury of nature in the form of drought, unseasonal and heavy rains and floods, and climate change.

Temperatures in the Ganga basin and similar other regions are expected to rise by 0.1° to 0.3°C by 2010, and by 1.0° to 3.5°C in next 90 years of this century (Perry *et al.*, 2007). When this happens, it will shrink Polar Regions, enhance the temperate climate Zone of the world, and alter the pattern of rainfall in different parts. India and other South Asian countries are expected to be most adversely affected. Mechanisms for risk mitigation are poor or absent. Hardly 10 per cent of farmers are covered by crop insurance.

1.9.12 Poor Farmer-Market Links

High levels of food production in limited areas, with a widespread distribution system, have undoubtedly affected market prices in food grains. In the absence of price support systems in large parts of the country, increases in productivity have led to localized gluts, with local prices crashing, and thereby retarding the incentives to increase production. While this may not have had very much of an effect on the overall demanded-supply position, it has affected the pace of improvement in the livelihoods of people in non Green Revolution areas. Warehouse system, including in rural areas, and farmers linkage with futures markets could help farmers in realizing optimize prices and also send signal for undertaking crop production.

1.9.13 Agricultural Product Price Insecurity

Higher prices when a farmer becomes a buyer and a lower price at the time he/she becomes a seller puts producers at disadvantaged positions. As long as there are resources, or entitlements in Sen's notion, to buy food from the market, food security would not constitute a major problem. But, average income of a farmer is one-fifth of that of a non-farmer, his economic access to livelihood is inherently low.

1.9.14 Investment Apathy in Agriculture

Despite the centrality of agriculture in the national socio-economic, environmental and political milieu, investment, and capital formation in agriculture have steadily declined and has reached all time low in relative terms in recent years resulting in poor infrastructure and performance. In order to obviate low gross capital formation, a suitable climate ought to be created for private sector investment by ensuring the market and export reforms. The investment intensity in research at 0.34 per cent of agricultural (GDP) is only half of the overall average for all developing countries (0.6 per cent). Further, there is considerable interstate variability in intensity of state funding ranging from 0.08 per cent in Uttar Pradesh to 1.4 per cent in Himachal Pradesh (NCF, 2006).

1.9.15 Inadequate Institutional Support

Rising capital intensity, particularly in the high-growth sectors of agriculture, has set in motion a new set of forces leading to biased knowledge, technological and market developments and thus exacerbating the problems of poor and small farmers. Declining growth in public investments and eroding institutional infrastructure are other disturbing features of the current trend. World agriculture, particularly trade, places high premium on quality, and public health, food safety and overall agricultural bio-security concerns have become central themes of global regulatory negotiations.

The access to farm credit, insurance and subsidies on part of the marginal farmers and the poor is rather inequitable.

1.9.16 Redefining the Role of the State

Agricultural growth in recent years has thrown new sectors and regions into prominence. Livestock, fisheries, horticulture, specialty enterprises (spices, medicinal, aromatic, organic) and value added products illustrate this trend. Fortunately, the ownership of livestock is more egalitarian. Enhancing small farm productivity and increasing small farm income through crop-livestock integrated production systems and multiple livelihood opportunities through agro- processing and biomass utilization are essential both to meet food production targets and for reducing hunger, poverty and rural unemployment. Market driven diversification in a global perspective has become the new paradigm driving future agricultural growth. The most profound shift pertains to rapid privatization in all domains-production, consumption, investment, technology etc. and concomitant decline in State control. Alternative instruments and approaches are evolving to transform agriculture and a very important part of this 'learning' phase is a redefinition of the role of the State. Public goods, welfare imperatives, other regulatory needs, and other areas of market failure will continue to need government intervention.

1.10 Challenges Ahead for Food Security in India through FCI and PDS System

India's food security policy has a laudable objective to ensure availability of food grains to common people at an affordable price and it has enabled the poor to have access to food where none existed. The policy has focused essentially on growth in agriculture production (once India used to import food grains) and on support price for procurement and maintenance of rice and wheat stocks. The responsibility for procuring and stocking of food grains lies with the FCI and for distribution with the public distribution system (PDS).

1.10.1 Minimum Support Price

The FCI procures food grains from the farmers at government announced minimum support price (MSP). The MSP should ideally be at a level where procurement by FCI and off take from it are balanced. However, under continuous pressure from powerful farmers lobby, government has been raising the MSP and it has now become higher than what the market offers to the farmers. Also, with quality norms in procured grains not strictly observed, farmers pressurise FCI to procure grains beyond its procurement target and carrying capacity. The MSP has now become more of a procurement price rather than being a support price to ensure minimum production. The rich farmers and traders have cornered most of benefits under the support price policy. The small farmers lack access to FCI and being steeped in poverty resort to distress selling. Constricted warehousing facility has further aggravated their miseries. At times, the same farmers later pay more to buy it from PDS.

1.10.2 Input Subsidies

Over years, to keep food grain prices at affordable levels for poor, government has been imposing restrictions on free trade in food grains. This has suppressed food grain prices in local market, where farmers sell a part of their produce and as compensation, they are provided subsidies on agriculture inputs such as fertilizers, power and water. These subsidies have now reached unsustainable levels and also led to large scale inefficiencies in use of these scarce inputs. Overuse of fertilizer and water has led to water logging, salinity, depletion of vital micronutrients in the soil and reduced fertility. The high subsidies have come at the expense of public investments in the critical agriculture infrastructure, thereby reducing agriculture productivity. Besides the high MSP, input subsidies and committed FCI purchases have distorted the cropping pattern with wheat and paddy crops being grown more for the MSP they fetch, despite there being relatively less demand for them. Punjab and Haryana are classic examples here. This has also led to a serious imbalance in inter-crop parities despite no significant increase in the yield of wheat and paddy.

1.10.3 Issue Price

The people are divided into two categories: below poverty line (BPL) and above poverty line (APL), with issue price being different for each category. However, this categorisation is imperfect and a number of deserving poor have been excluded from the BPL fold. Moreover, some of the so called APL slip back to BPL, say with failure of even one crop and it is administratively difficult to accommodate such shifts.

1.10.4 Fiscal Deficit

The government has sought to curtail the food subsidy bill by raising the issue price of food grains and linking it to the economic cost at which the FCI supplies food grains to PDS. The economic cost comprises cost of procurement, that is, MSP, storage, transportation and administration and is high mainly because of the artificially inflated MSP and also due to the operational inefficiencies of the FCI. This has pushed the issue price to APL category higher than the market rates and to BPL category beyond their purchasing power, resulting in plummeting of off take from the PDS.

1.10.5 High-Priced, Low-Quality Food Grains

Indian rice and wheat find little place in international market. Recently, two Indian consignments were rejected even by Iraq on quality considerations. The result is bulging stocks with FCI amidst widespread starvation. Also, low quality of PDS grains and the poor service at PDS shops have forced many people to switchover to market, which offers better quality grains, allows purchase on credit, ensures flexibility to purchase in small quantities

1.10.6 Market Demand

The PDS entitlement meets only around 25 per cent of the total food grain requirement of a BPL family and it has to depend more on market for meeting its needs. Also with APL families essentially opting for market purchases, market demand has risen. However, massive FCI procurement has crowded out market supplies, resulting in a relative rise in rates. The poor are the most hurt in this bargain.

1.10.7 Food for Work Scheme

The government is running food-for-work scheme to give purchasing power to the poor who get paid for their labour in cash and food grains. The scheme is, however, not successful, since Central Government is required to meet only the food grain component and cash strapped States are expected to meet the cash component.

1.11 Recommendations for Food Security

1.11.1 Technical and Institutional Reforms

There is a need to shift from the existing expensive, inefficient and corruption ridden institutional arrangements to those that will ensure cheap delivery of requisite quality grains in a transparent manner and are self-targeting.

1.11.1.1 Futures Market and Free Trade

The present system marked by input subsidies and high MSP should be phased out. To avoid wide fluctuations in prices and prevent distress selling by small farmers, futures market can be encouraged. Improved communication systems through use of information technology may help farmers get better deal for their produce. Crop insurance schemes can be promoted with government meeting a major part of insurance premium to protect farmers against natural calamities. To start with, all restrictions on food grains regarding inter-State movement, stocking, exports and institutional credit and trade financing should be renounced. Free trade will help make-up the difference between production and consumption needs, reduce supply variability, increase efficiency in resource-use and permit production in regions more suited to it.

1.11.1.2 Food-for-Education Programme

To achieve cent per cent literacy, the food security need can be productively linked to increased enrolment in schools. With the phasing out of PDS, food coupons may be issued to poor people depending on their entitlement.

1.11.1.3 Modified Food-for-Work Scheme/Direct Subsidies

With rationalisation of input subsidies and MSP, Central Government will be left with sufficient funds, which may be given as grants to each State depending on the number of poor. The State government will in turn distribute the grants to the village bodies, which can decide on the list of essential infrastructure, work the village needs and allow every needy villager to contribute through his labour and get paid in food coupons and cash.

1.11.1.4 Community Grain Storage Banks

The FCI can be gradually dismantled and procurement decentralised through the creation of food grain banks in each block village of district, from which people may get subsidised food grains against food coupons. The food coupons can be numbered serially to avoid frauds. The grain storage facilities can be created within two years under existing rural development schemes and the initial lot of grains can come from existing FCI stocks. If culturally acceptable, possibility of relatively cheap coarse grains, like bajra and ragi and nutritional grains like millets and pulses meeting the nutritional needs of the people can also be explored. This will not only enlarge the food basket but also prevent such locally adapted grains from becoming extinct. The community can be authorised to manage the food banks. This decentralised management will improve delivery of entitlements, reduce handling and transport costs and eliminate corruption, thereby bringing down the issue price substantially. To enforce efficiency in grain banks operation, people can also be given an option to obtain food grains against food coupons from open market, if rates in the grain banks are higher, quality is poor or services are deficient. A fund can be set up to reimburse the food retailers for the presented coupons. This competition will lead to constant improvement and lower prices. It must also be mandatory to maintain a small buffer stock at the State level, to deal with exigencies.

1.11.1.5 Enhancing Agriculture Productivity

The government, through investments in vital agriculture infrastructure, credit linkages and encouraging use of latest techniques, motivate each district/block to achieve local self-sufficiency in food grain production. However, instead of concentrating only on rice or wheat, food crop with a potential in area must be encouraged. Creation of necessary infrastructure like irrigation facilities will also simulate private investments in agriculture. The focus on accelerated food grains production on a sustainable basis and free trade in grains would help create massive employment and reduce the incidence of poverty in rural areas. This will lead to faster economic growth and give purchasing power to the people.

Note: A five-year transitory period may be allowed while implementing these. Thus, India can achieve food security in the real sense and in a realistic timeframe.

1.11.2 Adaptation Strategies

Any perturbation in agriculture can considerably affect food systems and thus increase the vulnerability of a large portion of the resource-poor population. Therefore, it is important to understand possible adaptation and coping strategies of producers in response to global climatic change. Adaptation strategies need to take a number of factors into consideration, including globalization and population and income growth and resulting changes in food preferences and demand, as well as the socioeconomic and environmental consequences of alternative adaptation options (Aggarwal, 2008).Therefore, there is a need to identify "no-regrets" adaptation strategies that are beneficial for the sustainable development of agriculture under multiple alternative scenarios. These adaptations should occur at multiple scales, including the individual, society, farm, village, watershed, and national levels.

1.11.2.1 Sustainable Agriculture

Although the first Green Revolution (1960s and 70s) increased global yield, Revolution came at a price: per capita hunger also increased, as small farmers were forced out of subsistence agriculture and into urban slums, often due to high cost of GR seeds and inputs required to grow them (fertilizers, pesticides, and machinery). The second wave of Green Revolution focuses on genetically modified organisms (GMOs) as central way in to feed the world's growing population; however, this second wave of the GR may be worse for small farmers, as large corporations own the patents to seed. In addition, in this second wave of the Green Revolution, the focus is not on sustainable agriculture, as high amounts of inputs (*i.e.*, fertilizers, pesticides and intensive irrigation) are required (Pretty *et al.*,2006). Because industrial agricultural inputs and infrastructure are expensive, rely on fossil fuels, and degrade the environment in numerous ways, many experts agree that relying upon unsustainable agriculture will, in the long term, increase global food insecurity. Studies involving small farms have indicated that sustainable agricultural practices can actually increase yield (Pretty *et al.*, 2006).

1.11.2.2 Altered Agronomy of Crops

Small changes in climatic parameters can often be managed by altering planting dates, spacing and input management. Use of alternate crops or cultivars more adapted to changing environmental conditions can further ease the pressure on food systems. For example, for the case of wheat, early planting or use of longer-duration cultivars may offset most of loss associated with increased temperatures in South Asia. Available germplasm of various crops needs to be evaluated for heat and drought tolerance.

1.11.2.3 Development of Resource-Conserving Technologies

Recent research has shown that surface seeding or zero-tillage establishment of upland crops after rice produces similar yields to when planted under conventional tillage over a diverse set of soil conditions. This reduces costs of production, allows earlier planting and thus higher yields, results in less weed growth, reduces the use of natural resources, such as fuel and steel for tractor parts and shows improvements in efficiency of water and fertilizers. In addition, such resource-conserving technologies restrict the release of soil carbon thus mitigating the accumulation of CO_2 in the atmosphere. It is estimated that zero tillage saves at least 30 litres of diesel as compared to conventional tillage. This leads to 80 kg ha^{-1} year^{-1} reduction in CO_2 production. If these savings could be translated even partially to large arable areas, substantial carbon dioxide emissions to the atmosphere could be reduced.

1.11.2.4 Augmenting Production and its Sustainability

The yield potential of many crops is much higher given climate conditions than amount that is currently harvested in many parts of the world. For example, potential yields of rice and wheat are calculated to be more than 6 mt/ha while their actual yields range between 2 and 3 million tonnes ha^{-1}, on average (Aggarwal *et al.*, 2000). Yield gaps are very large in eastern India, for example, and, hence, this region could increase the food security of India while mitigating the adverse impacts of climate change, given efforts to close this yield gap. Institutional support in form of improved extension services, markets, and infrastructure needs to be provided in such regions to increase the stability of agricultural production and bridge yield gaps.

1.11.2.5 Improved Land Use and Natural Resource Management

Policies and Institutions Strategies to adapt to climate change also involve natural resource management policies and institutions such as crop insurance, subsidies and pricing policies related to water and energy. Necessary provisions need to be included in development plans in order to address twin objectives of managing environmental change and improving resource-use productivity. Rational pricing of surface and groundwater for example can arrest its excessive and injudicious use. Availability of assured prices and infrastructure would promote better utilization of groundwater. Policies, such as financial incentives for green manuring would encourage farmers to enrich organic matter in the soil and thus improve soil health.

1.11.2.6 Improving Agricultural Biodiversity

Improving agricultural biodiversity through sustainable agricultural practices may also alleviate food insecurity (Thrupp, 2000). Industrial agriculture relies upon mono-cropping in which one genetic type of crop is planted on large tracts of land, while sustainable farms frequently plant a genetically diverse array of both crop type and species. Mono-cropping increases crop susceptibility to both pests and diseases several historical famines and crop decimations were due to a pest or disease devastating mono-cropped agricultural plantings (Thrupp, 2000). With mono-cropping also comes an increased need for chemical fertilizers and pesticides, which can erode soil biodiversity and in turn negatively affect yields over time. Enhancing biodiversity through the use of sustainable agricultural practices can protect communities from food insecurity associated with both crop loss and decreased yield (Bennett *et al.*, 2012).

1.11.2.7 Improved Risk Management through Early Warning Systems and Crop Insurance

The increasing probability of floods and droughts, and other climate uncertainties may seriously increase vulnerability of eastern India and of resource-poor farmers to global climate change. Crop insurance can provide protection to farmers in event that their farm production is reduced due to natural calamities. In view of climate change and uncertainty regarding future agricultural technologies and trade scenarios, it would be very useful to have an early warning system in place to alert the public to the spatial and temporal distribution of extreme events. Such a system would help identify potential food-insecure areas and communities and type of risk. Modern information technologies would greatly facilitate development of such early warning systems.

1.11.2.8 Recycling Wastewater and Solid Wastes in Agriculture

Since freshwater supplies are limited, agriculture sector should evaluate the potential for using industrial wastewater and sewage. Such effluents, once properly treated, can be a source of nutrients for crops. Since water serves multiple uses and users, effective inter-departmental coordination in government is needed to develop location-specific framework of sustainable water management and optimum recycling of water.

1.12 National Food Security Mission (NFSM)

- ☆ The growth in food grain production has stagnated during recent past while consumption need of growing population is increasing.
- ☆ To meet growing food grain demands, National Development Council in its 53rd meeting adopted a resolution to enhance the production of rice, wheat and pulses by 10, 8 and 2 million tons respectively by 2011.

- ☆ The proposed Centrally Sponsored Scheme 'National Food Security Mission (NFSM) is to operationalize the resolution of NDC and enhance the production of rice, wheat and pulses.

1.12.1 Key Features of NFSM

- ☆ The scheme to be implemented in a mission mode through a farmer centric approach
- ☆ All the Stakeholders to be actively associated at the District levels for achieving the set goal.
- ☆ The scheme aims to target the select districts by making available the improved technologies to the farmers through a series of planned interventions.
- ☆ A close monitoring mechanism proposed to ensure that interventions reach to the targeted beneficiaries.

1.12.2 Objectives of the Mission

- ☆ Increasing production of rice, wheat and pulses through area expansion and productivity enhancement in a sustainable manner;
- ☆ Restoring soil fertility and productivity at individual farm level;
- ☆ Enhancing farm level economy to restore confidence of farmers of targeted districts

1.12.3 Strategies

1. Expansion of area of Pulses and Wheat, No expansion of area in rice
2. Bridging the yield gap between potential and present level of productivity through
   - Acceleration of seed production
   - Integrated Nutrient Management and Integrated Pest Management
   - Promotion of new production technologies like hybrid rice, timely planting of wheat and promotion of new improved variety of Pulses
   - Supply of input ensuring their timely availability
   - Farmers Training and Visits

1.12.4 Components of NFSM

Rice

- ☆ Demonstration of improved technology including hybrid and SRI
- ☆ Incentive for quality seeds of HYV/hybrids
- ☆ Popularization of new varieties through seed mini kits
- ☆ Promotion of micro nutrients, lime and gypsum
- ☆ Promotion of mechanical weeders and other farm implements
- ☆ Integrated pest management

- ☆ Extension, training and mass media campaign
- ☆ Awards for best performing district in each State.
- ☆ Assistance for innovative interventions at local level

Wheat

- ☆ Demonstration of improved technology
- ☆ Incentive for quality seeds of HYVs to raise the SRR
- ☆ Promotion of micronutrient use in deficient areas
- ☆ Incentive for promotion of application of Gypsum
- ☆ Popularization of Zero till Machines and rotavator
- ☆ Providing subsidy on diesel pumpsets and community generators for irrigation
- ☆ Extension, training and mass media campaign awards for best performing districts
- ☆ Assistance for innovative interventions at local level

Pulses

- ☆ Increasing seed replacement rate to 25 per cent from present level of 7-8 per cent
- ☆ Promotion of improved production technologies
- ☆ Integrated Nutrient Management (INM)
- ☆ Integrated Pest Management (IPM)
- ☆ Promotion of micronutrients/gypsum/bio-fertilizers
- ☆ Promotion of sprinkler irrigation
- ☆ Pilot Project on tackling the menace of blue bull
- ☆ Extension, training and mass media campaign
- ☆ Awards for best performing districts
- ☆ Pilot project on demonstration ICRISAT Technologies.

1.13 Conclusion

There are strong, direct relationships between agricultural productivity, hunger, and poverty which would reflect in malnourishment. Therefore for food security improvements in agricultural productivity, maintaining or increasing the area of cultivable land, encourage the farmers to use modern technology and multi-crop cultivation with adequate financial support are the needs of the hour. In addition the government has to take a bold step to include water as a central subject and link all the rivers which should be made as the top priority of any government which comes to power.

References

Aggarwal PK 2008. Global climate change and Indian agriculture: impacts, adaptation and mitigation. Indian J Agric Sci 78: 911–919.

Aggarwal PK, Talukdar KK and Mall RK 2000. Potential yields of rice-wheat system in the Indo- Gangetic plains of India. Rice-Wheat Consortium Paper Series 10. New Delhi, India. RWCIGP, CIMMYT 16p.

Aggarwal, PK and Singh A K 2010. Implications of Global Climatic Change on Water and Food Security.In: C. Ringler *et al.* (eds), Global Change: Impacts on Water and Food Security, 49 DOI: 10.1007/978-3-642-04615-5_3, © Springer-Verlag Berlin Heidelberg 2010.

Bennett A, Bending G, Chandler D, Hilton S, and Mills P 2012. Meeting the demand for crop production: the challenge of yield decline in crops grown in short rotations. Biological Reviews 87: 52–71.

Department of Food and Public Distribution 2013. Annual report, Ministry of consumer affairs, food and public distribution, Government of public distribution, pp 25-30.

FAO 2010. Crop Prospects and Food Situation. No. 2 (May), Rome.

Godfray HCJ, Beddington JR, Crute IR, Haddad L, Lawrence D and Muir JF 2010. Food security: The challenge of feeding 9 billion people. Science 327: 812-818.

Pretty JN, Noble AD, Bossio D, Dixon J, Hine RE, Penning de Vries FWT and Morison JIL 2006. Resource-Conserving Agriculture Increases Yields in Developing Countries. Environ. Sci. Technol 40: 1114–1119.

Thrupp LA 2000. Linking Agricultural Biodiversity and Food Security: The Valuable Role of Sustainable Agriculture. Royal Institute of International Affairs 76, 265-281.

2014, Sustainable Rural Development through Agriculture
Pages 29–41
Editors: Dr. Shobhana Gupta and Dr. S.S. Tomar
Published by: BIOTECH BOOKS, NEW DELHI

Chapter 2

Climate Change a Threat to Agriculture in 21st Century

D.S. Tomar, S.K. Kaushik, Rekha Tiwari and Arvind Saxena

Introduction

What is in Store for the World?

The predicted pattern of climate change globally as seen presently is a matter of concern in many areas of socio- economic activities, such as agriculture, forestry, etc., and is a major threat for biodiversity and ecosystem function. Climate change is a result from emission of greenhouse gases (*e.g.* CO_2, CH_4, and N_2O, etc.) in the past century that will cause atmospheric warming (IPCC, 2007). The effects have become particularly obvious over the last 30 years in the natural environment and it will affect all level of life, from the individual, population species community and ecosystem to the eco-region level (Lepetz *et al.*, 2009). With the present efforts and pace at which this threat is being addressed, the world is likely to warm by more than 3°C above the preindustrial climate. On the contrary if mitigation commitments and pledges are fully implemented, there is roughly a 20 percent chance of exceeding 4°C by 2100. If they are not met, a warming of 4°C could occur as early as the 2060s. Such a warming level and associated sea-level rise of 0.5 to 1 meter, or more, by 2100 would not be the end point: a further warming to levels over 6°C, with several meters of sea-level rise, would likely occur over the following centuries. Thus, while the global community has committed itself to holding warming below 2°C to prevent "dangerous" climate change, the sum total of current policies—in place and pledged—will very likely lead to warming far in excess of these levels. Indeed, present emission trends put the world plausibly on a path toward 4°C warming within the century. A world in which warming reaches 4°C above preindustrial levels would be one of unprecedented heat waves, severe drought, and major floods in many regions, with serious impacts on human systems, ecosystems, and associated sectors.

Why to be Worried About

Every one today who can foresee the dangerous changes marching fast is worried because by 2050, the world's population is expected to reach 9 billion. Most of these people are expected to live in developing countries and have higher incomes than currently is the case, which will result in increased demand for food. In the best of circumstances, the challenge of meeting this demand in a sustainable manner will be enormous. When one takes into account the effects of climate change (higher temperatures, shifting seasons, more frequent and extreme weather events, flooding, and drought) on food production, that challenge grows even more daunting. Slow increases in world food production and declining rates of yield growth in main food crops threaten world food security. Land and water constraints, underinvestment in rural infrastructure and agricultural innovation, lack of access to agricultural inputs, and disruptions are impairing productivity growth and the needed production response. These factors, combined with sharp increases in food prices in recent years, have added to concerns about the food and nutrition situation of people around the world, especially the poor in developing countries. Nearly every agricultural commodity—including dairy products, meat, poultry, oil seeds and cassava—has been part of the rising price trend. Several factors have contributed to these unprecedented food price increases: climate change, rising energy prices and subsidized bio-fuel production, income and population growth, globalization, and urbanization. The livelihoods of many small holders and rural people depend directly on their ability to produce and market agricultural products. Therefore, agricultural growth in developing regions remains fundamental for poverty reduction and food security. If poverty and hunger are to be eradicated in the longer term, substantial investments must be made in agricultural research and innovation.

1. Plant Responses to Climate change

Plants are grouped in to 'C3', 'C4' and 'CAM' plants according to their photosynthetic metabolic pathways. Around 95 per cent of the world plant biomass grouped in 'C3' plant species (*e.g.* wheat, rice, fruits and vegetables), C4 (*e.g.* maize or corn, sugarcane and sorghum) and CAM (*e.g.* Pineapple). These divisions into groups largely based on the enzymes involved in photosynthetic fixation of CO_2, namely Rubisco, PEP carboxylase and to some extent carbonic anhydrase, which are significantly different in their response to CO_2 enrichment. CO_2 together with other minerals can activate Rubisco by binding at a non-catalytic site on the enzyme protein. The process of photorespiration rate is high in C3 plants and the relative proportion of CO_2 and O2 inside the leaf determines the rate of photorespiration. In contrast, PEP carboxylase in C4 plants not inhibited by O2 and thus photorespiration is negligible. PEP carboxylase also has a higher effective affinity for CO_2 than Rubisco in the absence of O2. Therefore, we would expect higher atmospheric CO_2 concentrations to increased photosynthesis and growth of C3 plants but not to the same extent, if any, in C4 plants (Bolin, 1989). The result from experiments done on wild grass species shows that under elevated CO2 condition both C3 and C4 species show increase in the total plant biomass of 44 per cent and 33 per cent respectively, the increased in C3 species was greater in tiller formation whereas in C4 was greater in leaf area. Net CO_2 assimilation rates (*A*), that means (flux of CO_2 between leaf and atmosphere through photosynthesis) increased in both C3 and C4 species with 33 per cent and 25 per cent respectively, while, stomatal conductance (*Gs*)(the ability of CO_2 entering, or water vapour exiting through the stomata) decreased for C3 and C4 species by 24 per cent and 29 per cent, respectively. Many simulation results showed that increased biomass production were observed in both C3 and C4 plants under elevated [CO_2]; although the enhancement of shoot production by elevated CO_2 varied with temperature and precipitation. In C3 species, the response of NPP to increased temperatures was negative under dry and ambient CO_2 condition, but was positive under wet and doubled CO_2 condition; whereas, the responses of NPP of C4 species to elevated CO_2 was positive under all temperature and precipitation levels particularly at high precipitation level (Chen *et al.*, 1996). Plant growth in elevated atmospheric CO_2 has shown to be less vulnerable to drought, maintaining higher growth rate on drought condition than plants under lower CO_2. Elevated CO_2 also enhances plant resistance to heat, frost stresses, likely reflecting greater concentrations of membrane stabilizing sugars in the tissues and it induces greater nutrient deficiency, and as observed in several studies it leads to accumulation, of secondary carbon rich chemicals such as tannins (Niinemets, 2010).

1.1 Responses of Field Crops to Climate Change

Higher CO_2 levels leads plants to produce a larger number of mesophyll cell, chloroplasts, longer stems and extended length, diameter and number of large roots, forming good lateral root production with different branching patterns. In some crops this, results in the increase in root to shoot ratios (Qaderi and Reid, 2009). The potential of crop productivity increased under an increased in local average temperature range of 1-3°C, but it decreased above this range (IPCC, 2007), probably the reason could be low vernalization,shortened phenological phases decrease in photosynthesis rate, and increased transpiration. (Qaderi and Reid, 2009). Elevated CO_2 have a positive effect on some annual C3 field crops, such as soybean, peanut, and rice cultivars, etc. Growth and development accelerated throughout the vegetative phase, and before flowering stage started seven days earlier, which contributed to the higher grain yield and change in the chemical composition of the rice grain (Uprety *et al.*, 2010). Some studies also show that a reduction in maize (C4 species) yield occurred under elevated [CO_2] condition due to shortened growing period and a yield reduction also recorded in some experiment on winter wheat (C3 species) due to an effect on vernalization period (Alexadrov and Hoogenboom, 2000). Whereas an increase in the yield of spring wheat with 8-10 per cent was observed when water was no limiting; similarly, a cotton crop exposed to free–air CO_2 enrichment (FACE) was stimulated and show increased about 48 per cent of harvestable yield and 37 per cent of biomass under elevated (550 ppm) [CO2] level (Easterling and Apps, 2005). The difference in responses in different ecosystems to elevated CO_2 might be due to difference in water, soil, nutrient availability and temperature variation.

1.2 Responses of Forest Trees to Climate Change

Different processes in plants or forest ecosystems and their interaction with climate variability is complex, due to different response of physical, biological, and chemical processes. An increase in the ambient CO_2 concentration could reduce the opening of stomata required to allow a given amount of CO_2 to enter in the plant that might reduce transpiration of the trees. These could increase the efficiency of water use by forest plants and increase productivity to some extent (Bolin *et al.*, 1989). Trees are capable of adjusting to a warmer climate, although the response expected from species are different and the effect on photoinhibition and photorespiration are more difficult to generalize (Saxe *et al.*, 2001). As forest trees are characterized by the C3 photosynthetic path way their productivity and demand for nutrient is highly affected by atmospheric CO_2 concentration and temperature. The total productivity expected from trees (especially from trees with indeterminate growth) growing under elevated CO_2 is larger than estimated in crops (Lukac *et al.*, 2010). Estimated increased production from trees is higher than crops only achieved especially if the combination of absorption and increased nutrient use efficiency is attained.However, the long-term response of forest to rising level of [CO_2] is still uncertain. The current over all response of trees is positive and results from a review of 49 papers on effects of elevated CO_2 on different tree species shows that net primary production (NPP, photosynthesis minus plant respiration) on average increased with 23 per cent at an elevated CO_2 concentration of 550 ppm as compared with 370 ppm (Norby *et al.*, 2005). Whereas, enhanced in temperatures can lead to heat and more water logging stress in bogs and cause more severe heat, drought and photo-inhibition stress periods in temperate bog and forest ecosystems (Niinemets, 2010).

1.3 Photosynthesis and Plant Respiration Processes

Respiration can be highly affected by temperature, and its rate is determined by status of carbohydrate and supply of adenylate (enzyme catalyzing the conversion processes). The sucrose content of the tissue can govern the capacity of mitochondrial respiration, and mitochondrial respiration plays a great role in growth and survival of plants (Atkin *et al.*, 2005). One would expect at least a short period increases in respiration rate from parts of plants those show increased growth and assimilation due to elevated [CO_2], that is source leaves, individual sink tissue (fruit, seed, steam, root etc.) and total sink tissue. Nevertheless, a few reports concluded that long-term treatment with increased concentration of CO_2 resulted in declined whole-plant respiration. Whereas, result of a few other experiments show that a short-term increase in temperature on plants growing in cold climate areas such as Arctic have resulted in greater potential impact on plant respiration than in plants growing in warmer areas. One of the reasons might be that tropical plants more acclimate to higher temperatures than the Arctic cold area plants.

The response of plant respiration to long-term change in temperature is dependent on the level of effects of temperature on plant development, and on other direct and interactive effects of temperature and abiotic factors (*e.g.* Irradiance, nutrient availability and drought). Evidence shows that the response of respiration to temperature is dynamic, with plant respiration often acclimating to long-term changes in temperature. In addition, both degree of acclimation and value of Q10 (proportional change in respiration with a 10°C increase in temperature) vary in response to the surrounding environment and/or the metabolic condition of the plants. There is variability in Q10 as day and night time temperature varies (*e.g.* nights are increasing to a higher extent than daytime). The Q10 of leaf *R* is often not always reduced in the light compared with the Q10 of leaf *R* in dark, and Q10 values are often lower in water-stressed plants than in their fully- watered counterparts, root and leaves also differ in their Q10 values as upper and lower canopy leaves. Q10 of both root and leaf *R* dark generally decreased as temperature increased. Rise in [CO_2] does not show a predictable, systematic effect on Q10 of dark *R* of stems root or leaves. Different studies show a variation in the effect of rise atmospheric [CO_2] on Q10 value of *R* above ground plant parts in dark condition, but the overall result indicate that elevated [CO_2] has little impact on the average Q10 values (Atkim *et al.*, 2005). However, Q10 dark *R* is greater in some plants grown under elevated [CO_2] for *e.g.* According to a study both Q10 light and Q10 dark increased in leaves *R* of *Xanthium strumarium*. However, there has been a variation in different literatures results; Overall, higher [CO2] does not on average alter the temperature sensitivity of dark *R* in roots, leaves or shoots. In most plants as temperature increase with optimal range, the rate of respiration as well as the rate of metabolism increased, because increased respiration results with higher energy available, that means as long as nutrients are available the metabolism processes within the plant will also increase. Following increased temperature to a certain level, the rate of photosynthesis is also increases but not as much as respiration. That indicates the amount of CO_2 produced from increased respiration is faster than the amount of O2 released from increased photosynthesis.

Temperature affect photosynthesis through altering the activities of enzymes, electron transport and leaf temperature (leaf-to air vapor pressure difference) can influence the stomatal conductance. As evaporation increases, stomata tends to close to reduce water loss through transpiration, following this stomata closure reduction in CO_2 assimilation rate occur due to less rate of CO_2 supply to chloroplast, this is indirect temperature response. Temperature also affects photosynthetic metabolism directly showing a change in the activity of ribulose-1,5- carboxylase oxygenase (Rubisco) processes associated with the regeneration of Rubisco's substrate,rubulose-1,5- bisphosphate (RuBP) through the Calvin cycle.

2. Agriculture and Climate Change

Agriculture contributes to climate GHGs emission and highly affected by change in climate parameters. In an intensive farming, we expect high greenhouse gases emission because of using high amount of inputs and chemicals, due to these changes of human activity natural divers and climate change impacts varies accordingly in different part of the world. Vulnerability to climate change depends not only on physical and biological responses but also on socio economic characteristics. Low-income population especially those who cultivate crops under rain fed and non-irrigated agriculture systems in drylands, arid and semi-arid areas highly affected by severe hard ship due to climate change.

2.1 Effect of Climate Variables on Productivity

According to the report of IPCC, (2007) The overall impacts of higher temperatures on crop responses at the plot level, without considering changes in the frequency of extreme events, moderate warming. (During first half of this 21st century) may benefit in crop and pasture productivity in temperate regions, while it may reduce productivity in tropical and semi-arid regions. Modelling studies indicate small beneficial effect in temperate corresponds to local mean temperature increases in 1-3°C with association of an increased in CO_2 and rainfall changes. In contrast, models show that tropical regions show a negative yield impacts for major crops with moderate rise in temperature (1-2°C), but further warming projected in all regions in the end of twenty-first century results in the increased on negative impacts. Climate plays a major role in determining the yield level by increasing or reducing in global perspective from temperate to tropics. Many experiments show that CO_2 is a limiting factor, in which higher concentration of CO_2 enhances photosynthesis and crop growth, modifying water and nutrient cycles, these responses found to hold even for plants grown under different stressful conditions.When other environmental factors such as water shortage, less light, shortage or excess of minerals, very high or very low temperature limit yield, then CO_2 concentration will have little or no effect. Nevertheless, in certain stressful environments the relative photosynthetic response of plants to CO_2 enrichment is actually increased (Bolin *et al.*, 1989).

For most crops, growing under elevated CO_2 conditions both quality and total yield shows improved (more ear of plant per m^2). The increased CO_2 induce and makes an increase in the grain weight and according to the observation, it was greater under average phosphorus treatment compared to higher phosphorus level. This influence of CO_2 and phosphorus supply was attributed to increase in the number of cells within endosperm, which is the result of enhanced rate of cell division during grain development or by greater amount of grain filling during ripening phase (Uprety *et al.*, 2010). However, it has also been shown that elevated CO_2 concentrations may have negative effects on the grain quality from wheat in terms of protein content, it alters wheat grain lipids and doubled the number of mitochondria in wheat leaves, lower seed nitrogen concentration and decreases grain and flower protein.

3. Adapting to a Changing Climate

There could be many definitions of climate change adaptation but the one which suits most from the point of view of natural resource conservation given by Glick *et al.*, 2009 is "Climate change adaptation for natural systems is a management strategy that involves identification, preparation for, and responses to expected climate change in order to promote ecological resilience, maintain ecological function, and provide necessary elements to support biodiversity and sustainable ecosystem services". In general the term climate change adaptation can be concluded by the following phrases; "climate change safe guards", "preparing for warming world", "protecting wild life and natural resource from warm" and "coping mechanisms". Ecological responses to climate changes are already visible. The impact of global warming is discernable in both animal and plant biology. For example, averaged across several hundred temperate-zone species (including plants and animals), the shift in spring phenology, such as breeding or blooming, is about 5 d earlier in each decade during the past half century. Satellite imagery has shown that the beginning of summers is 8 ± 4 d earlier in North and the duration of the active growing season decreased by 12 ± 5 d. "The goals of adaptation strategies are to improve the knowledge and skills of farmers, to encourage adoption of new technologies, and to expand the array of options available to farmers" However, because of the given uncertainties and serious consequences of potentially inaccurate assessments of climate change and the required adaptation strategies, recommended aggressive study and research into how best to limit and mitigate the impacts of climate change on agriculture. Slight complacency is very risky and vigorous effort toward understanding and preparing for potentially serious impacts on agriculture by developing adaptation strategies is needed.

Practical Approach for Wheat Growing Farmers of Central India: Screening Varieties for wider Adaptability under Climatic Change Scenario

Methodology

Ujjain district of *Malwa* plateau was purposively selected for the study, as it has more varietal diversification during both crop season and also crop diversification started according to climatic change. KVK Ujjain has adopted village clusters based upon the agro-ecological situation for the conducting the screening of wheat cultivars. Three clusters of 20-20 villages were selected; from three block *i.e.* Ujjain Block, Ghatiya block and Mehidpur block. One village from each cluster was selected for the collection of data on the basis of observation, PRA and feedback was also collected during the interaction with the farmers. KVK also demonstrated some special intervention for covering the climatic changes as a mitigation programme during 2007-08 to 2010-11). The study was simultaneously conducted at research farm and at cultivators field with the selected varieties and after three years varieties were grouped according to performance and farmers likings.

The annual rainfall of the study region varies from 890 to 1000 mm. summer and winter temperatures are not as extreme as in the arid west but the summer temperature may reach around 45° C. In winters, minimum may reach 3.5° C. The groundwater table varies from 45 to 125 meters but the annual fluctuations are high, especially in the years when the south-west monsoon fails and the yearly replenishments rates are low. Surface water sources are scarce and so harnessing of ground water resources has been going on at an accelerated pace. A very common feature of the area is drought, which appears in regular cycles (3 or 5 years) and continues to exert a toll on production. To cater the problem of existing soybean-wheat crop sequence, identification of suitable wheat varieties for different traits was done. The experiment was laid out in Randomized block design with three replications, with a plot size of 6m×4m, inter-plot distance of 0.75m and block distance of 1m. Sixteen wheat varieties, off which eleven were *Triticum aestivum* L, four were of *Triticum durum* and two were

Triticum aestivum L designated as *sharbati* identified for central zone along with one most popular variety LOK-1 was included as local check. The experiment was sown in the third week of November in all the three years as this period coincides with the most ideal period for sowing (average temperature < 23° C). Seeds were sown in rows 25cm apart at the rate of 120 kg ha$^{-1,}$ with recommended dose of fertilizer 120-60-20-20 kg ha^{-1} N,P_2O_5, K_2O and S was applied in the form of Urea, Single Super phosphate and muriate of potash and bentonite sulphur, respectively. One come up irrigation followed by three irrigations at CRI, maximum tillering and grain filling stages was applied. Number of days taken to achieve the various phenological stages were noted when 50 per cent of the plants in all replication reached the respective stages *i.e.*, emergence, CRI, tillering, ear emergence, grain filling, milk stage, dough stage and maturity. The daily meteorological data were collected from MPWSRP weather station at the research farm. The various measurements of accumulated heat units were calculated according to the formulae of Rajput (1980).

1. Growing degree days (GDD) = $\Sigma [(T_{max} + T_{min})/2 - T_b]$.
2. Heat use efficiency (HTU) = Grain yield (kg ha^{-1}) ÷ GDD.
3. Phenothermal index (PTI) = GDD ÷ Growth days.

Outcome

Farmers Feedback

Discussions held during the Participatory Rural Appraisal (PRA) revealed that the drought has a very serious implication on the rural livelihood. The level of migration is highest in the small and marginal segment of farmers amounting to almost 55 and 38.5 percent. Similarly all the parameters undertaken show that small and marginal farmers are the worst affected. Even the large farmers are compelled to invest in infrastructural development, reduced consumption of fertilizers and other inputs and dependence on dry farming practices. These changes in strategy only helps to sustain the immediate needs of the family but on the future scale all the farmers have to bear the wrath of drought.

Technological Interventions

Phenological development is usually analyzed in terms of progress toward stages such as germination, seedling emergence, and initiation of floral primordia, flowering, and physiological maturity. Progress is estimated by integrating a developmental rate over the interval from one stage to the next. The rate usually is a function of temperature and photoperiod. The result of the experiment suggests that to attain a particular growth stage the GDD varied from species to species and variety to variety. Based on three years performance among the aestivum group GW-366, GW-322, HI-1418, DL-803-3, Raj-3777, GW-273, HI-1544 and JW-3020 gave higher yield over the local check LOK-1 by a margin of as low as 3 per cent and as high as 26 per cent. Whereas among the durum HI-8498 and HI-8381 gave higher yield over the check. Among the sharbati group though the yields of both the varieties was less than the check but the heat use efficiency was higher than the durum which suggests that these varieties can perform well under rainfed situations. Over years the yields were

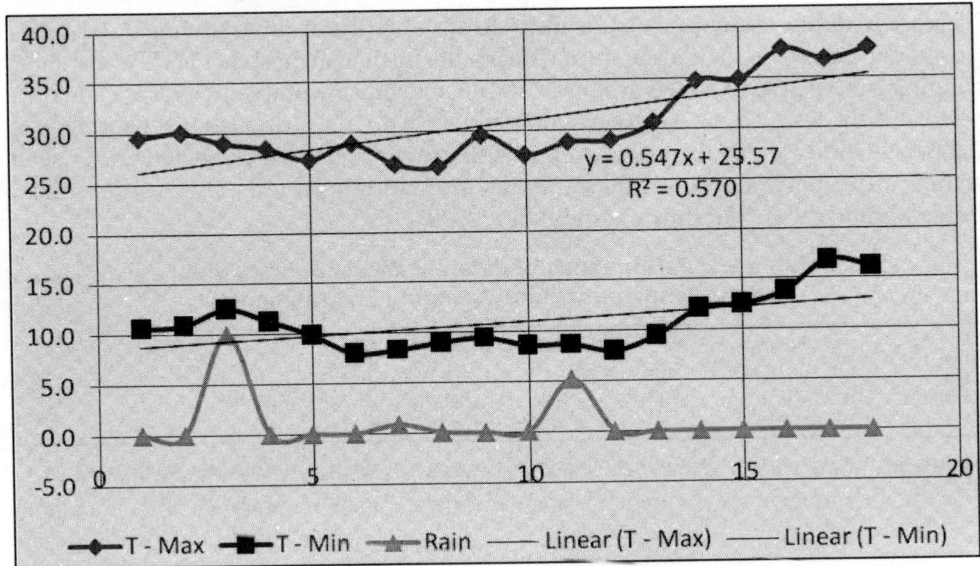

Figure 2.1: Mean Maximum Minimum and Rainfall Pattern of Three Years (2007-2009).

affected by rise in minimum temperature more than the rise in maximum temperature. The change in minimum temperature over the years was 0.9° C per annum as against 0.45° C in case of maximum temperature. Kalra *et al.*, 2007 found that the optimal sowing dates in northwest India for winter crops, especially wheat, are relatively early compared to the eastern part of the country. The model output indicates that

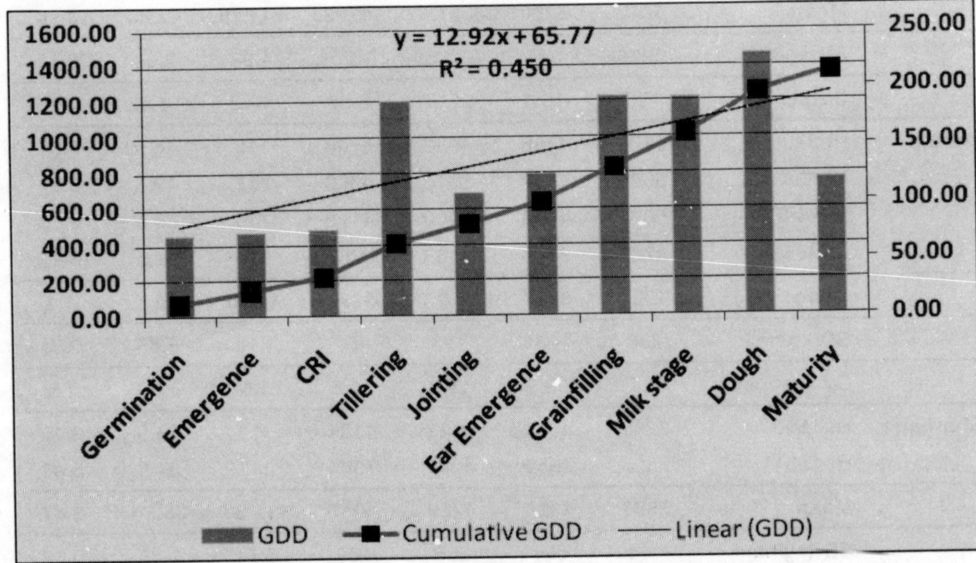

Figure 2.2: Degree Day Requirement of Wheat at different Phonological Stages.

increase in temperature by 1–3°C is likely to advance the optimal sowing dates by 5–8 days per degree rise in temperature. The results further suggest that HTU of varieties ranging between 4.2 and 4.8 are more stable and thermo stable as evidenced from yield and the test weight. The regression equation obtained and the trends in Figure 2.2 clearly indicate that the yield levels of different varieties were affected with rise in temperature between post milking stage and dough period. The results are in agreement to the earlier findings of Sikder, 2009.

Table 2.1: Performance of different Wheat Species and their Genotypes Under Changing Environment

Wheat	*Years* / *Variety Species*	*2007-08* *Yield*	*2008-09* *Yield (Kg/ha)*	*2009-10* *Yield (Kg/ha)*	*Mean Yield q/ha (Kg/ha)*	*Days to Maturity*	*Test Weight*	*HTU*
Aestivum	GW-366	5702	6037	5208	5649	108	57.5	4.81
	GW-322	5231	5662	4792	5228	106	48.5	4.54
	HI-1418	5248	5874	4021	5048	116	43.2	4.01
	DL 803-3	5616	4748	4583	4982	119	45.0	3.85
	RAJ-3777	4925	5637	4167	4910	109	34.5	4.15
	GW-273	6197	4753	3333	4761	104	43.5	4.21
	HI-1544	4478	5103	4208	4596	111	44.0	3.81
	JW 3020	4968	5537	3188	4564	118	32.2	3.56
	LOK-1(LC)	4057	4479	4883	4473	109	46.5	3.78
	HI-1479	3921	4578	4325	4275	109	39.5	3.61
	MP-4010	5184	5347	2188	4240	114	37.0	3.42
	Mean	**5048**	**5250**	**4081.47**	**4793**	**111.18**	**42.85**	**3.98**
	SD(kg/ha)	690	547	877.58	425	4.92	7.04	0.42
	CV(per cent)	13.7	10.4	21.50	8.87	4.42	16.42	10.67
Durum	HI-8498	5776	4486	4042	4768	118	46.5	3.31
	HI-8381	5099	4449	3375	4308	117	43.5	3.34
	HD-4672	5072	4503	3000	4192	116	36.5	3.36
	HI-8663	4729	2410	3863	3667	114	52.5	3.42
	Mean	**4901**	**3457**	**3431.3**	**3930**	**115.0**	**44.5**	**3.4**
	SD(kg/ha)	438	1035	473	452	1.71	6.65	0.05
	CV (per cent)	8.9	29.9	13.8	11.5	1.5	14.9	1.5
Sharbati	HI-1500	3269	4686	4417	4124	113	38.5	3.45
	HI-1531	3897	3969	4042	3969	112	34.5	3.48
	Mean	**3583**	**4328**	**4229**	**4047**	**112.50**	**36.50**	**3.47**
	SD (kg/ha)	444	507	265	109	0.71	2.83	0.02
	CV (per cent)	**12.4**	**11.7**	**6.3**	**2.70**	**0.63**	**7.75**	**0.63**

Table 2.2: Phenothermal Index and Heat Use Efficiency of different Wheat Species and their Genotypes Under Changing Environment

Wheat Species	*Variety*	*Yield*	*Days to Maturity*	*PTI*	*HUE*
Aestivum	GW-366	5649	108	12.67	4.13
	GW-322	5228	106	12.91	3.82
	HI-1418	5048	116	11.80	3.69
	DL 803-3	4982	119	11.50	3.64
	RAJ-3777	4910	109	12.56	3.59
	GW-273	4761	104	13.16	3.48
	HI-1544	4596	111	12.33	3.36
	JW 3020	4564	118	11.60	3.33
	LOK-1(LC)	4473	109	12.56	3.27
	HI-1479	4275	109	12.56	3.12
	MP-4010	4240	114	12.01	3.10
	Mean		**111.18**	**12.33**	**3.50**
Durum	HI-8498	4768	118	11.60	3.48
	HI-8381	4308	117	11.70	3.15
	HD-4672	4192	116	11.80	3.06
	HI-8663	3667	114	12.01	2.68
	Mean		**115**	**11.78**	**3.09**
Sharbati	HI-1500	4124	113	12.11	3.01
	HI-1531	3969	112	12.22	2.90
	Mean		**112.5**	**12.17**	**2.96**

Phenothermal Index

Phenothermal indices of different genus of wheat and their varieties given in Table 2.2. clearly indicate that the PTI of aestivum group of varieties on an average was 12.33 with a yield range of 4240 to 5649 kg ha^{-1} followed by sharbati group with PTI of 12.17 and the least in durum group with 11.78. This clearly indicates that the heat tolerant varieties have a higher PTI as is reflected by their yield performance. The durum group is high yielders but is probably not good enough in tolerating heat stress and/or moisture stress and hence had lower PTI. Days to maturity have been found to be closely related to PTI. More the duration lower is the PTI due to combined effect of rising temperature (more heat) coupled with moisture stress.

Heat Use Efficiency

From the results (Table 2.2) it was observed that all the cultivars in aestivum used the heat more efficiently as compared to sharbati and durum group of wheat. Among the varieties GW-366 gave the highest HUE (4.13) and least by HI-8663 (2.68). This suggests that the varieties more suitable in the central plateau region with black soils are the ones with higher HUE ranging between 3.10 to 4.13. Similar results were also been reported by Rajput *et al.* (1987) and Sikder, 2009.

Conclusion

To tackle problems associated to climate change we have to focus on practicing activities of mitigation and potential adaptation measures. To protect the environment, mitigations should be proactive by preventing and minimizing emission from source before they are emitted to the atmosphere or by activities of conservation and substitution of energy, carbon sink, and using modern technologies etc. Through adaptation measures, we can protect marketable and nonmarketable ecosystem services from damage and our actions have to be ongoing and flexible. Identifying uncertainties in climate change can help us to reduce the challenges in performing adaptation practices and trying to protect and resist risks of climate events. Wheat is a highly thermo-sensitive crop, particularly during its reproductive phase. Rising trend of maximum temperature during critical stages hamper optimum production on sustained basis. Hence, selection of suitable varieties based on scientific parameter coupled with participatory varietal selection by the farming community would definitely help in coping with the vagaries of nature for a secured livelihood in our agrarian economy.

References

Aggrawal, P.K., Bandyopadhyay,S.K.,Pathak,H.,Kalra,N.,Chander,S.,Kumar,S., 2000. Analysis of yield trends of rice wheat system in north – western India. *Outlook on Agriculture29* (4), 259-268.

Alexandrov, V. A., and Hoogenboom, G., 2000. Vulnerability and Adaptation Assessments of Agricultural Crops under Climate Change in the Southeastern USA; theor. Appl. Climatol. **67**,45-63.

Atkin, O. K., Bruhn, D., Hurry, V. M., and Tjoelker, M. G., 2005. The hot and the cold: unraveling the variable response of plant respiration to temperature: Functional Plant Biology, **32**, 87-105.

Bolin B., Döös B. R., Jager J., and Warrick R. A.,eds., 1989. The Green House Effect climate change and Ecosystems: SCOPE 29 International Council of Scientific Union. Great Britain.

Easterling W., and Apps M., 2005. Assessing the consequences of climate change for food and forest resources: a view from the IPCC: Climate change **70**, 165-189.

Glick P., Staudt A., and Stein B., 2009. A New Era for Conservation: Review of climate change adaptation literature: National Wildlife Federation.

Kalra Naveen, Chakraborty D., Sharma Anil, Rai H. K., Monica Jolly, Chander Subhash,. Ramesh P. K, Bhadraray S., Barman D., Mittal R. B., Mohan Lal and Sehgal Mukesh.,2008. Effect of increasing temperature on yield of some winter crops in northwest India. CURRENT SCIENCE, (94) 1: 82-88

Lepetz V., Massot, M. and Schmeller, D.S., and Clobert, J., 2009. Biodiversity monitoring: some proposals to adequately study species' responses to climate change. Biodiversity and Conservation **18**, 3185- 3203.

Glick P., Staudt A., and Stein B., 2009. A New Era for Conservation: Review of climate change adaptation literature: National Wildlife Federation.

IPCC, 2007: Summary for Policymakers. In: Climate Change 2007: Mitigation. Contribution of Working Group III to the Fourth Assessment Report of the Intergovernmental Panel on Climate Change [B. Metz, O.R. Davidson, P.R. Bosch, R. Dave, L.A. Meyer (eds)], Cambridge University Press, Cambridge, United Kingdom and New York, NY, USA.

Niinemets, Ü., 2010. Responses of forest trees to single and multiple environmental stresses from seedlings to mature plants: past stress history, stress interactions, tolerance and acclimation: Forest Ecology and Management **260,** 1623-1639.

Rajput, R. P., Deshmukh, M. R. and Paradkar V. K. 1987. Accumulated heat unit and phenology relationships in wheat (*Triticum aestivum* L) as influenced by planting dates under late sown conditions. *J. Agron. Crop Sci.* 159, 345- 348.

Rajput, R. P. and Sastry, P. S. N. 1985. Comparative performance of three varieties of rain fed soybean in a monsoon climate. Response to thermal environment. *In*"Proc. Int. Symp. on CATAC". p. 88.

Rajput, R. P. 1980. Response of soybean crop to climate and soil environments. Ph. D Thesis. Indian Agricultural Research Institute, New Delhi.

Rodomiro Ortiz, Kenneth D. Sayre, Bram Govaerts, Raj Gupta, G.V. Subbarao c, Tomohiro Ban, David Hodson, John M. Dixon,J. Iva´n Ortiz-Monasterio, Matthew Reynolds.,2008. Climate change: Can wheat beat the heat? Agriculture, Ecosystems and Environment 126 : 46–58

Sikder S.2009. Accumulated Heat Unit and Phenology of Wheat Cultivars as Influenced by Late Sowing Heat Stress Condition. *J Agric Rural Dev* 7(*1 and* 2), *57-64.*

Tao,F.,Yokozawa,M.,Zhang,Z.,Hayashi,Y.,Grassl,H.,Fu,C.,2004. Variability in climatology and agricultural production in china in association with the East Asia summer monsoon and El Nino South Oscillation. *Climate Research.28, 23-30*

Uprety D. C., Sen S., and Dwivedi N., 2010. Rising atmospheric carbon dioxide on grain quality in crop plants: *physiol Mol Biol Plants.*

2014, Sustainable Rural Development through Agriculture *Pages* **42–61**
Editors: **Dr. Shobhana Gupta and Dr. S.S. Tomar**
Published by: **BIOTECH BOOKS, NEW DELHI**

Chapter 3
Climate Change and Vegetable Crop Production

Udit Kumar and Birendra Prasad

Climate change and global warming has brought serious challenges for natural and agro-ecosystem to sustain to support the living beings. The success and failure of agricultural enterprises depends on the whims of weather and extreme and uncertain weather events. The frequencies of such weather events are likely to increase manifold under climate change and global warming scenario. These, changes can have disastrous effects on agricultural production and thus food security of any region. Farmers would be affected severely because they subsist mostly on climate sensitive economic sectors. Climate change is one of the most important global environmental challenges in the history of mankind. The carbon dioxide, methane, nitrous oxide, sulphur dioxide, etc. form greenhouse gas (GHG) pools in the atmosphere. The change in climate is mainly caused by increasing concentration of these Green House Gases in the atmosphere. In 1980s, scientific evidences linking GHGs emission due to human activities causing global climate change, started to concern everybody. Subsequently, United Nations General Assembly in 1992 formed Intergovernmental Negotiating Committee for Framework Convention on Climate Change (UNFCCC) which finally adopted the framework for addressing climate change concerns. Climate is defined as the "average weather", or more precisely, as the statistical description of the weather in terms of the mean and variability of relevant quantities over periods of several decades (typically three decades as defined by World Meteorological organization). These quantities are most often surface variables such as temperature, precipitation, and wind, but in a wider sense the "climate" is the description of the state of the climate system. Climate change according to Inter Governmental Panel on Climate

Change (IPCC) refers to 'a change in the state of the climate that can be identified (using statistical tests) by changes in the mean and/or the variability of its properties that persist for an extended period, typically decades or longer. However, UNFCCC, in its Article 1, defines "climate change" as "a change of climate which is attributed directly or indirectly to human activity that alters the composition of the global atmosphere and which is in addition to natural climate variability observed over comparable time periods". The UNFCCC thus makes a distinction between "climate change" attributable to human activities altering the atmospheric composition, and "climate variability" attributable to natural causes.

The Impact of Climate Change on Vegetable Production

A significant change in climate on a global scale will impact agriculture and consequently affect the world's food supply. Climate change *per se* is not necessarily harmful; the problems arise from extreme events that are difficult to predict (FAO 2001). More erratic rainfall patterns and unpredictable high temperature spells will consequently reduce crop productivity. Developing countries in the tropics will be particularly vulnerable. Latitudinal and altitudinal shifts in ecological and agro-economic zones, land degradation, extreme geophysical events, reduced water availability, and rise in sea level and salinization are postulated (FAO 2004). Unless measures are undertaken to mitigate the effects of climate change, food security will be under threat.

Vegetable Production

Vegetables are the best resource for overcoming micronutrient deficiencies and provide smallholder farmers with much higher income and more jobs per hectare than staple crops (AVRDC 2006). The worldwide production of vegetables has doubled over the past quarter century and the value of global trade in vegetables now exceeds that of cereals. In Asia, vegetable production grew at an annual average rate of 3.4 per cent in the 1980s and early 1990s, from 144 million MT in 1980 to 218 million MT in 1993 (Ali 2000). In addition, the area under vegetable cultivation increased at an annual rate of 2.1 per cent, from 12 million hectares in 1980 to 16.3 million hectares in 1993 with most of the increase coming from increased production in China. Amongst vegetable crops, tomatoes are the most important horticultural crop worldwide and grown on over 4 million hectares of land area (FAO 2006, Brown *et al.*, 2005). Tomato, cabbage, onion, hot pepper and eggplant are particularly important in Asia and Sub-Saharan Africa (Brown *et al.*, 2005). Yields in Asia are highest in the east where the climate is mainly temperate and sub-temperate. Most vegetables prefer cooler temperatures, thus productivity is lowest in the hot and humid lowlands of Southeast Asia (Ali 2000). In Sub-Saharan Africa (excluding South Africa) and tropical Asia, average tomato yields are only about 10 - 12 MT/hectare, well below the yields in temperate regions (FAO 2006). Vegetables are generally sensitive to environmental extremes, and thus high temperatures and limited soil moisture are the major causes of low yields in the tropics and will be further magnified by climate change.

Environmental Constraints Limiting Vegetable Productivity

1. Temperatures
2. Drought
3. Salinity
4. Flooding

Environmental stress is the primary cause of crop losses worldwide, reducing average yields for most major crops by more than 50 per cent (Boyer 1982, Bray *et al.*, 2000). The tropical vegetable production environment is a mixture of conditions that varies with season and region. Climatic changes will influence the severity of environmental stress imposed on vegetable crops. Moreover, increasing temperatures, reduced irrigation water availability, flooding, and salinity will be major limiting factors in sustaining and increasing vegetable productivity. Extreme climatic conditions will also negatively impact soil fertility and increase soil erosion. Thus, additional fertilizer application or improved nutrient-use efficiency of crops will be needed to maintain productivity or harness the potential for enhanced crop growth due to increased atmospheric CO_2. The response of plants to environmental stresses depends on the plant developmental stage and the length and severity of the stress (Bray, 2002). Plants may respond similarly to avoid one or more stresses through morphological or biochemical mechanisms (Capiati *et al.*, 2006). Environmental interactions may make the stress response of plants more complex or influence the degree of impact of climate change. Measures to adapt to these climate change-induced stresses are critical for sustainable vegetable production. Until now, the scientific information on the effect of environmental stresses on vegetables is overwhelmingly on tomato. There is a need to do more research on how other vegetable crops are affected by increased abiotic stresses as a direct potential threat from climate change.

1. Temperatures

Temperature limits the range and production of many crops. In the tropics, high temperature conditions are often prevalent during the growing season and, with a changing climate, crops in this area will be subjected to increased temperature stress. Analysis of climate trends in tomato-growing locations suggests that temperatures are rising and the severity and frequency of above-optimal temperature episodes will increase in the coming decades (Bell *et al.*, 2000). Vegetative and reproductive processes in tomatoes are strongly modified by temperature alone or in conjunction with other environmental factors (Abdalla and Verkerk 1968). High temperature stress disrupts the biochemical reactions fundamental for normal cell function in plants. It primarily affects the photosynthetic functions of higher plants (Weis and Berry 1988). High temperatures can cause significant losses in tomato productivity due to reduced fruit set, and smaller and lower quality fruits (Stevens and Rudich 1978). Pre-anthesis temperature stress is associated with developmental changes in the anthers, particularly irregularities in the epidermis and endothesium, lack of opening of the stromium, and poor pollen formation (Sato *et al.*, 2002). In pepper, high temperature exposure at the pre-anthesis stage did not affect pistil or stamen viability, but high post-pollination temperatures inhibited fruit set, suggesting that fertilization is sensitive to high temperature stress (Erickson and Markhart 2002). Hazra *et al.* (2007) summarized the symptoms causing fruit set failure at high temperatures in tomato; this includes bud drop, abnormal flower development, poor pollen production, dehiscence, and viability, ovule abortion and poor viability, reduced carbohydrate availability, and other reproductive abnormalities. In addition, significant inhibition of photosynthesis occurs at temperatures above optimum, resulting in considerable loss of potential productivity.

2. Drought

Unpredictable drought is the single most important factor affecting world food security and the catalyst of the great famines of the past (CGIAR 2003). The world's water supply is fixed, thus increasing population pressure and competition for water resources will make the effect of successive droughts more severe (McWilliam 1986). Inefficient water usage all over the world and inefficient distribution systems in developing countries further decreases water availability. Water availability is expected to be highly sensitive to climate change and severe water stress conditions will affect crop productivity, particularly that of vegetables. In combination with elevated temperatures, decreased precipitation could cause reduction of irrigation water availability and increase in evapotranspiration, leading to severe crop water-stress conditions (IPCC 2001). Vegetables, being succulent products by definition, generally consist of greater than 90 per cent water (AVRDC 1990). Thus, water greatly influences the yield and quality of vegetables; drought conditions drastically reduce vegetable productivity. Drought stress causes an increase of solute concentration in the environment (soil), leading to an osmotic flow of water out of plant cells. This leads to an increase of the solute concentration in plant cells, thereby lowering the water potential and disrupting membranes and cell processes such as photosynthesis. The timing, intensity, and duration of drought spells determine the magnitude of the effect of drought.

3. Salinity

Vegetable production is threatened by increasing soil salinity particularly in irrigated croplands which provide 40 per cent of the world's food (FAO 2002). Excessive soil salinity reduces productivity of many agricultural crops, including most vegetables which are particularly sensitive throughout the ontogeny of the plant. According to the United States Department of Agriculture (USDA), onions are sensitive to saline soils, while cucumbers, eggplants, peppers, and tomatoes, are moderately sensitive. In hot and dry environments, high evapotranspiration results in substantial water loss, thus leaving salt around the plant roots which interferes with the plant's ability to uptake water. Physiologically, salinity imposes an initial water deficit that results from the relatively high solute concentrations in the soil, causes ion-specific stresses resulting from altered K+/Na+ ratios, and leads to a build up in Na+ and Cl- concentrations that are detrimental to plants (Yamaguchi and Blumwald 2005). Plant sensitivity to salt stress is reflected in loss of turgor, growth reduction, wilting, leaf curling and epinasty, leaf abscission, decreased photosynthesis, respiratory changes, loss of cellular integrity, tissue necrosis, and potentially death of the plant (Jones 1986; Cheeseman 1988). Salinity also affects agriculture in coastal regions which are impacted by low-quality and high-saline irrigation water due to contamination of the groundwater and intrusion of saline water due to natural or man-made events. Salinity fluctuates with season, being generally high in the dry season and low during rainy season when freshwater flushing is prevalent. Furthermore, coastal areas are threatened by specific, saline natural disasters which can make agricultural lands unproductive, such as tsunamis which may inundate low-lying areas with seawater. Although the seawater rapidly recedes, the groundwater contamination and subsequent osmotic stress causes crop losses and affects soil fertility. In the inland areas, traditional water wells are commonly used for irrigation water in many countries. The bedrock deposit contains salts and the water from these wells are becoming more saline, thus affecting irrigated vegetable production in these areas.

4. Flooding

Vegetable production occurs in both dry and wet seasons in the tropics. However, production is often limited during the rainy season due to excessive moisture brought about by heavy rain. Most vegetables are highly sensitive to flooding and genetic variation with respect to this character is limited, particularly in tomato. In general, damage to vegetables by flooding is due to the reduction of oxygen in the root zone which inhibits aerobic processes. Flooded tomato plants accumulate endogenous ethylene that causes damage to the plants (Drew 1979). Low oxygen levels stimulate an increased production of an ethylene precursor, 1-aminocyclopropane-1-carboxylic acid (ACC), in the roots. The rapid development of epinastic growth of leaves is a characteristic response of tomatoes to water-logged conditions and the role of ethylene accumulation has been implicated (*Kawase 1981*). The severity of flooding symptoms increases with rising temperatures; rapid wilting and death of tomato plants is usually observed following a short period of flooding at high temperatures (Kuo *et al.*, 1982).

The Need for Adaptation to Climate Change

I. Enhancing Vegetable Production Systems

- ☆ Water-saving irrigation management
- ☆ Cultural practices that conserve water and protect crops
- ☆ Improved stress tolerance through grafting

II. Developing Climate-Resilient Vegetables

- ☆ Tolerance to high temperatures
- ☆ Drought tolerance and water-use efficiency
- ☆ Tolerance to saline soils and irrigation water

III. Climate-Proofing though Genomics and Biotechnology

- ☆ QTLs and gene discovery for tolerance to stresses.
- ☆ Engineering stress tolerance.

Potential impacts of climate change on agricultural production will depend not only on climate *per se*, but also on the internal dynamics of agricultural systems, including their ability to adapt to the changes (FAO 2001). Success in mitigating climate change depends on how well agricultural crops and systems adapt to the changes and concomitant environmental stresses of those changes on the current systems. Farmers in developing countries need tools to adapt and mitigate the adverse effects of climate change on agricultural productivity, and particularly on vegetable production, quality and yield. Current, and new, technologies being developed through plant stress physiology research can potentially contribute to mitigate threats from climate change on vegetable production. However, farmers in developing countries are usually small-holders, have fewer options and must rely heavily on resources available in their farms or within their communities. Thus, technologies that are simple, affordable, and accessible must be used to increase the resilience of farms in less developed countries. AVRDC – The World Vegetable Center has been working to address the effect of environmental stress on vegetable production. Germplasm of the major vegetable crops which are tolerant of high temperatures, flooding and drought has been identified and advanced breeding lines are being developed. Efforts are also underway to identify nitrogen-use efficient germplasm. In addition, development of production systems geared towards improved water-use efficiency and expected to mitigate the effects of hot and dry conditions in vegetable production systems are top research and development priorities.

I. Enhancing Vegetable Production Systems

Various management practices have the potential to raise the yield of vegetables grown under hot and wet conditions of the lowland tropics. AVRDC – The World Vegetable Center has developed technologies to alleviate production challenges such as limited irrigation water and flooding, to mitigate the effects of salinity, and also to ensure appropriate availability of nutrients to the plants. Strategies include modifying fertilizer application to enhance nutrient availability to plants, direct delivery of water to roots (drip irrigation), grafting to increase flood and disease tolerance, and use of soil amendments to improve soil fertility and enhance nutrient uptake by plants.

a) Water-Saving Irrigation Management

The quality and efficiency of water management determine the yield and quality of vegetable products. The optimum frequency and amount of applied water is a function of climate and weather conditions, crop species, variety, stage of growth and rooting characteristics, soil water retention capacity and texture, irrigation system and management factor (Phene 1989). Too much or too little water causes abnormal plant growth, predisposes plants to infection by pathogens, and causes nutritional disorders. If water is scarce and supplies are erratic or variable, then timely irrigation and conservation of soil moisture reserves are the most important agronomic interventions to maintain yields during drought stress. There are several methods of applying irrigation water and the choice depends on the crop, water supply, soil characteristics and topography. Application of irrigation water could be through overhead, surface, drip, or sub-irrigation systems. Surface irrigation methods are utilized in more than 80 per cent of the world's irrigated lands yet its field level application efficiency is often 40-50 per cent (von Westarp 2004). To generate income and alleviate poverty of the small-holder farmers in developing countries, institutions promote affordable, small-scale drip irrigation technologies developed by the International Development Enterprises (IDE). Drip irrigation delivers water directly to plants through small plastic tubes. IDE states that water losses due to run-off and deep percolation are minimized and water savings of 50-80 per cent are achieved when compared to most traditional surface irrigation methods. Crop production per unit of water consumed by plant evapo-transpiration is typically increased by 10-50 per cent. Thus, more plants can be irrigated per unit of water by drip irrigation, and with less labor. In Nepal, cauliflower yields using low-cost drip irrigation were not significantly different from those achieved by hand watering; however the long-term economic and labor benefits were greater using the low-cost drip irrigation (von Westarp 2004). The water-use efficiency by chili pepper was significantly higher in drip irrigation compared to furrow irrigation, with higher efficiencies observed with high delivery rate drip irrigation regimes (AVRDC 2005). For drought tolerant crop like watermelon, yield differences between furrow and drip irrigated crops were not significantly different; however, the incidence of Fusarium wilt was reduced when a lower drip irrigation rate was used. In general, the use of low-cost drip irrigation is cost effective, labor-saving, and allows more plants to be grown per unit of water, thereby both saving water and increasing farmers' incomes at the same time

b) Cultural Practices that Conserve Water and Protect Crops

Various crop management practices such as mulching and the use of shelters and raised beds help to conserve soil moisture, prevent soil degradation, and protect vegetables from heavy rains, high temperatures, and flooding. The use of organic and inorganic mulches is common in high-value vegetable production systems. These protective coverings help reduce evaporation, moderate soil temperature, reduce soil runoff and erosion, protect fruits from direct contact with soil and minimize weed growth. In addition, the use of organic materials as mulch can help enhance soil fertility, structure and other soil properties. Rice straw is abundant in rice-growing areas and generally recommended for summer tomato production. The benefits of rice straw mulch on fruit yield of tomato have been demonstrated in Taiwan (AVRDC 1981). In India, mulching improved the growth of eggplant, okra, bottle gourd, round melon, ridge gourd, and sponge gourd compared to the non-mulched controls (Pandita and Singh 1992). Yields were the highest when polythene and sarkanda (*Saccharum* spp. and *Canna* spp.) were used as mulching materials. In the lowland tropics where temperatures are high, dark-colored plastic mulch is recommended in combination with rice straw (AVRDC 1990). Dark plastic mulch prevents sunlight from reaching the soil surface and the rice straw insulates the plastic from direct sunlight thereby preventing the soil temperature rising too high during the day. During the hot rainy season, vegetables such as tomatoes suffer from yield losses caused by heavy rains. Simple, clear plastic rain shelters prevent water logging and rain impact damage on developing fruits, with consequent improvement in tomato yields (Midmore *et al.*, 1992). Fruit cracking and the number of unmarketable fruits are also reduced. Elimination of flooding and rain damage, as well as the reduced air temperature, was responsible for the higher yields of the crops grown under plastic shelters. Another form of shelter using shade cloth can be used to reduce temperature stress. Shade shelters also prevent damage from direct rain impact and intense sunlight. Planting vegetables in raised beds can ameliorate the effects of flooding during the rainy season (AVRDC 1979, 1981). Yields of tomatoes increased with bed height, most likely due to improved drainage and reduction of anoxic stress. Additive effects on yield were observed when tomatoes were planted in raised beds with rain shelters.

c) Improved Stress Tolerance through Grafting

Grafting vegetables originated in East Asia during the 20th century and is currently common practice in Japan, Korea and some European countries. Grafting, in this context, involves uniting of two living plant parts (rootstock and scion) to produce a single growing plant. It has been used primarily to control soil-borne diseases affecting the production of fruit vegetables such as tomato, eggplant, and cucurbits. However, it can provide tolerance to soil-related environmental stresses such as drought, salinity, low soil temperature and flooding if appropriate tolerant rootstocks are used. Grafting of eggplants was started in the 1950s, followed by grafting of cucumbers and tomatoes in the 1960s and 1970s (Edelstein 2004). Romero *et al.* (1997) reported that melons grafted onto hybrid squash rootstocks were more salt tolerant than the non-grafted melons. However, tolerance to salt by rootstocks varies greatly among species, such that rootstocks from *Cucurbita* spp. are more tolerant of salt than rootstocks from *Lagenaria siceraria* (Matsubara 1989). Grafted plants were also more able to tolerate low soil temperatures. *Solanum lycopersicum* x *S. habrochaites* rootstocks provide tolerance of low soil temperatures (10oC to 13oC) for their grafted tomato scions, while eggplants grafted onto *S. integrifolium x S. melongena* rootstocks grew better at lower temperatures (18oC to 21oC) than non-grafted plants (Okimura *et al.*, 1986). Vegetables generally are unable to tolerate excessive soil moisture. Tomatoes in particular are considered to be one of the vegetable crops most sensitive to excess water. In the tropics, heavy rainfall with poor drainage induces water-logged conditions that reduce oxygen availability in the soil thereby causing wilting, chlorosis, leaf epinasty, and ultimately death of the tomato plants. Genetic variability for tolerance of excess soil moisture is limited or inadequate to prevent losses. Research has shown that many accessions of eggplant are highly tolerant of flooding (Midmore *et al.*, 1997). Thus, the research institutes developed grafting techniques to improve the flood tolerance of tomato using eggplant rootstocks which were identified with good grafting compatibility with tomato and high tolerance to excess soil moisture. Tomato scions grafted onto eggplant rootstock grow well and produce acceptable yields during the rainy season (Midmore *et al.*, 1997). In addition to protection against flooding, some eggplant genotypes are drought tolerant and eggplant rootstocks can therefore provide protection against limited soil moisture stress.

II) Developing Climate-Resilient Vegetables

Improved, adapted vegetable germplasm is the most cost-effective option for farmers to meet the challenges of a changing climate. However, most modern cultivars represent a limited sampling of available genetic variability including tolerance to environmental stresses. Breeding new varieties, particularly for intensive, high input production systems in developed countries, under optimal growth conditions may have counter-selected for traits which would contribute to adaptation or tolerance to low input and less favorable environments. Superior varieties adapted to a wider range of climatic conditions could result from the discovery of novel genetic variation for tolerance to different biotic and abiotic stresses. Genotypes with improved attributes conditioned by superior combinations of alleles at multiple loci could be identified and advanced. Improved selection techniques are needed to identify these superior genotypes and associated traits, especially from wild, related species that grow in environments which do not support the growth of their domesticated relatives that are cultivated varieties. Plants native to climates with marked seasonality are able to acclimatize more easily to variable environmental conditions (Pereira and Chavez 1995) and provide opportunities to identify genes or gene combinations which confer such resilience.

a) Tolerance to High Temperatures

AVRDC - The World Vegetable Center has developed tomatoes and Chinese cabbage with general adaptation to hot and humid tropical environments and low-input cropping systems since the early 1970s. This has been achieved by developing heat-tolerant and disease-resistant breeding lines. The Center has made significant contributions to the development of heat-tolerant tomato and Chinese cabbage lines and the subsequent release of adapted, tropical varieties worldwide. Indeed, AVRDC - The World Vegetable Center's heat tolerant hybrids have resulted in the successful cultivation of Chinese cabbage in the lowland tropics. The key to achieving high yields with heat tolerant cultivars is the broadening of their genetic base through crosses between heat tolerant tropical lines and disease resistant temperate or winter varieties (Opena and Lo 1981). The heat tolerant tomato lines were developed using heat tolerant breeding lines and landraces from the Philippines (*e.g.* VC11-3-1-8, VC 11-2-5, Divisoria-2) and the United States (*e.g.* Tamu Chico III, PI289309) (Opena *et al.*, 1989). However, lower yields in the heat tolerant lines are still a concern.More heat tolerant varieties are required to meet the needs of a changing climate, and these must be able to match the yields of conventional, non-heat tolerant varieties under non-stress conditions. A wider range of genotypic variation must be explored to identify additional sources of heat tolerance. An AVRDC - The World Vegetable Center breeding line, CL5915, has demonstrated high levels of heat tolerance in Southeast Asia and the Pacific. The fruit set of CL5915 ranges from 15 per cent - 30 per cent while there is complete absence of fruit set in heat-sensitive lines in mean field temperatures of 35oC. Genetic studies at AVRDC - The World Vegetable Center indicated that heat tolerance in CL5915, based on fruit set and fruit number per cluster, is controlled by additive and dominant effects (Hanson *et al.*, 2002). The average heritability for fruit set is 0.26. However, bimodality of fruit set distribution and recovery of tolerant lines in early backcross generations suggests that only a few major genes and modifiers may control the heat tolerance trait (Opena *et al.*, 1990, 1992). Since then new breeding lines have been developed from CL5915 and other sources that exhibit increased heat tolerance. In Egypt, CL5915 lines were best combiners for percentage fruit set and total yield in hybrids developed from heat-tolerant and heat-sensitive lines (Metwally *et al.*, 1996).Germplasm evaluation for heat tolerance at AVRDC - The World Vegetable

Center conventionaliy relied upon field screening during the hot and humid season, with measurement of fruit set and yield. Generally less than 1 per cent of the screened lines or accessions exhibit a high level of heat tolerance (Villareal *et al.*, 1978). Although field screening is effective, the accuracy and speed of the process could be improved through the use of molecular markers.

b) Drought Tolerance and Water-Use Efficiency

Plants resist water or drought stress in many ways. In slowly developing water deficit, plants may escape drought stress by shortening their life cycle (Chaves and Oliveira 2004). However, the oxidative stress of rapid dehydration is very damaging to the photosynthetic processes, and the capacity for energy dissipation and metabolic protection against reactive oxygen species is the key to survival under drought conditions (Ort 2001, Chaves and Oliveira 2004). Tissue tolerance to severe dehydration is not common in crop plants but is found in species native to extremely dry environments (Ingram and Bartels 1996). Genetic variability for drought tolerance in *S. lycopersicum* is limited and inadequate. The best source of resistance is from other species in the genus *Solanum*. The Tomato Genetics Resource Center (TGRC) at the University of California, Davis has assembled a set of the putatively stress tolerant tomato germplasm that includes accessions of *S. cheesmanii, S. chilense, S. lycopersicum, S. lycopersicum* var. *cerasiforme, S. pennellii, S. peruvianum* and *S. pimpinellifolium. S. chilense* and *S. pennelli* are indigenous to arid and semi-arid environments of South America. Both species produce small green fruit and have an indeterminate growth habit. *S chilense* is adapted to desert areas of northern Chile and often found in areas where no other vegetation grows (Rick 1973, Maldonado *et al.*, 2003). *S. chilense* has finely divided leaves and well-developed root system (Sanchez-Pena 1999). *S. chilense* has a longer primary root and more extensive secondary root system than cultivated tomato (O'Connell *et al.*, 2007). Drought tests show that *S. chilense* is five times more tolerant of wilting than cultivated tomato. *S. pennellii* has the ability to increase its water use efficiency under drought conditions unlike the cultivated *S. lycopersicum* (O'Connell *et al.*, 2007). It has thick, round waxy leaves, is known to produce acyl-sugars in its trichomes, and its leaves are able to take up dew (Rick 1973). Transfer and utilization of genes from these drought resistant species will enhance tolerance of tomato cultivars to dry conditions, although wide crosses with *S. pennellii* produce fertile progenies, *S. chilense* is cross-incompatible with *S. lycopersicum* and embryo rescue through tissue culture is required to produce progeny plants. Research at institutions is in progress to identify the genetic factors underlying drought tolerance in *S. chilense* and *S. pennellii*, and to transfer these factors into cultivated tomatoes.

c) Tolerance to Saline Soils and Irrigation Water

Attempts to improve the salt tolerance of crops through conventional breeding programs have very limited success due to the genetic and physiologic complexity of this trait (Flowers 2004). In addition, tolerance to saline conditions is a developmentally regulated, stage-specific phenomenon; tolerance at one stage of plant development does not always correlate with tolerance at other stages (Foolad 2004). Success in breeding for salt tolerance requires effective screening methods, existence of genetic variability, and ability to transfer the genes to the species of interest. Screening for salt tolerance in the field is not a recommended practice because of the variable levels of salinity in field soils. Screening should be done in soil-less culture with nutrient solutions of known salt concentrations (Cuartero and Fernandez-Munoz 1999). Most commercial tomato cultivars are moderately sensitive to increased salinity and only limited variation exists in cultivated species. Genetic variation for salt tolerance during seed germination in tomato has been identified within cultivated and wild species. A cross between a salt-sensitive tomato line (UCT5) and a salt-tolerant *S. esculentum* accession (PI174263) showed that the ability of tomato seed to germinate rapidly under salt stress is genetically controlled with narrow-sense heritability (*h*2) of 0.75 (Foolad and Jones 1991). Several studies indicate that salt tolerance during seed germination in tomato is controlled by genes with additive effects and could be improved by directional phenotypic selection (Foolad 2004). In pepper, salt stress significantly decreases germination, shoot height, root length, fresh and dry weight, and yield. Yildirim and Guvenc (2006) reported that pepper genotypes Demre, Ilica 250, 11-B-14, Bagci Carliston, Mini Aci Sivri, Yalova Carliston, and Yaglik 28 can be useful as sources of genes to develop pepper cultivars with improved germination under salt stress. In Tunisia, the root system of salt tolerant cultivar 'Beldi' was unaffected by salt stress (van der Beek and Ltifi 1991). In addition, 'Beldi' significantly out-yielded other test cultivars at high salt treatments. Related wild tomato species have shown strong salinity tolerance and are sources of genes as coastal areas are common habitat of some wild species. Studies have identified potential sources of resistance in the wild tomato species *S. cheesmanii*, *S. peruvianum*, *S pennelii*, *S. pimpinellifolium*, and *S. habrochaites* (Flowers 2004, Foolad 2004, Cuartero *et al.*, 2006). Attempts to transfer quantitative trait loci (QTLs) and elucidate the genetics of salt tolerance have been conducted using populations involving wild species. Elucidation of mechanism of salt tolerance at different growth periods and the introgression of salinity tolerance genes into vegetables would accelerate development of varieties that are able to withstand high or variable levels of salinity compatible with different production environments.

III) Climate-Proofing though Genomics and Biotechnology

Increasing crop productivity in unfavorable environments will require advanced technologies to complement traditional methods which are often unable to prevent yield losses due to environmental stresses. In the past decade, genomics has developed from whole genome sequencing to the discovery of novel and high throughput genetic and molecular technologies. Genes have been discovered and gene functions understood. This has opened the way to genetic manipulation of genes associated with tolerance to environmental stresses. These tools promise more rapid, and potentially spectacular, returns but require high levels of investment. Many activities using these genetic and molecular tools are in place, with some successes. National and international institutes are re-tooling for plant molecular genetic research to enhance traditional plant breeding and benefit from the potential of genetic engineering to increase and sustain crop productivity.

a) QTLs and Gene Discovery for Tolerance to Stresses

Genetic enhancement using molecular technologies has revolutionized plant breeding. Advances in genetics and genomics have greatly improved our understanding of structural and functional aspects of plant genomes. Combining new knowledge from genomic research with traditional breeding methods enhances our ability to improve crop plants. The use of molecular markers as a selection tool provides the potential for increasing the efficiency of breeding programs by reducing environmental variability, facilitating earlier selection, and reducing subsequent population sizes for field testing. Molecular markers facilitate efficient introgression of superior alleles from wild species into the breeding programs and enable the pyramiding of genes controlling quantitative traits. Thus, enhancing and accelerating the development of stress tolerant and higher yielding cultivars for farmers in developing countries. Molecular marker analysis of stress tolerance in vegetables is limited but efforts are underway to identify QTLs underlying tolerance to stresses. QTLs for drought tolerance have been identified in tomato; Martin *et al.* (1989) identified three QTLs linked to water use efficiency in *S. pennellii* based on 13C composition. Three independent yield-promoting regions were identified in *S. pennellii* when grown in both wet and dry field conditions in Israel (Gur and Zamir 2004), while Foolad *et al.* (2003) identified four QTLs associated with seed germination drought tolerance, two of which were contributed by *S. pimpinellifolium*. *S. pimpinellifolium* is also commonly investigated as source of salt tolerance. QTL mapping indicates that salt tolerance is quantitatively inherited (Foolad 2004). Only a few major QTLs account for the majority of phenotypic variation indicating the potential for marker-assisted selection (MAS) for salt tolerance. Lin *et al.* (2006) identified random amplified polymorphic DNA (RAPD) markers linked to heat tolerance in AVRDC - The World Vegetable Center's tomato line CL5915. A follow-up study is currently underway at the Center to fully understand the genetic mechanism of heat tolerance of CL5915. Studies indicate that stress tolerance is quantitatively inherited and in some cases tolerance is dependent on the developmental stage of the plant. Consequently, multiple genes are predicted to be involved with the expression of stress tolerance. Studies on QTL analysis of stress tolerance is limited, and may reflect the limited variation of these traits. Furthermore, the environmental influence on the expression of stress tolerance traits is high and makes phenotyping difficult. Identification of environmentally-stable surrogate phenotypes or component traits is necessary to effectively evaluate genotypes and genetic populations. This is a critical tool to identify the QTLs underlying tolerance to environmental stresses. Many genes associated with stress tolerance have recently been determined using high throughput expression assays. Integration of QTL analysis with gene discovery and modeling of genetic networks will facilitate a comprehensive understanding of stress tolerance, permit the development of useful and effective markers for marker-assisted selection, and identify candidate genes for genetic engineering.

b) Engineering Stress Tolerance

Environmental stress tolerance is a complex trait and involves many genes (Wang *et al.*, 2003). In response to stresses, both RNA and protein expression profiles change. Approximately 130 drought-responsive genes have been identified using microarrays (Seki *et al.*, 2001, Reymond *et al.*, 2000). These genes are involved with transcription modulation, ion transport, transpiration control and carbohydrate metabolism. *DREB1A* and *CBF*, and *HSF* genes are transcription factors implicated in drought and heat response, respectively (Sung *et al.*, 2003, Sakuma *et al.*, 2002). Cell wall invertase (*INV*) and sucrose synthase (*SUSY*) play key roles in carbohydrate partitioning in plants (Déjardin *et al.*, 1999) and this regulation of carbohydrate metabolism in leaves may represent part of the general cellular response to acclimation and contribute to osmotic adjustment under stress. The *ERECTA* gene regulates plant transpiration efficiency in *Arabidopsis thaliana* (Masle *et al.*, 2005), and the *NHX* and *AVP1* genes are associated with ion transport (Zhang and Blumwald 2001). There are many more genes implicated with stress response and the current challenge is to identify the ones that confer a tolerant phenotype in the crop of interest. Although the function of these genes has been elucidated, particularly in *A. thaliana*, only a few genes have contributed to a tolerant phenotype when over-expressed in vegetables (Zhang *et al.*, 2004). Expression of *AVP1*, a vacuolar H+ pyrophosphatase from *A. thaliana*, in tomato resulted in enhanced performance under soil water deficit (Park *et al.*, 2005). The engineered tomato has a stronger, larger root system that allows the roots to make better use of limited water. The control plants suffered irreversible damage after five days without water as opposed to transgenic tomatoes which began to show water-stress damage only after 13 days but recovered completely as soon as water was supplied. The *CBF/DREB1* genes have been used successfully to engineer drought tolerance in tomato and other crops (Hsieh *et al.*, 2002). Orthologous genes of *CBF* have been found in most crop plants and functional tests indicate conservation of the pathway in these plant species. Constitutive over-expression of *CBF* genes results in salt, cold, or drought tolerance in several plant species. However, in addition to increased stress tolerance, the transgenic plants were dark-green and were stunted, with higher levels of soluble sugars and proline (Liu *et al.*, 1998). The use of stress-inducible promoters that have a low background expression of *CBF* under normal growth conditions can achieve increased stress tolerance without plant growth retardation (Lee *et al.*, 2003). Maintaining a low cytosolic Na+ concentration is essential to achieve salt tolerance and can be achieved by restricting inflow, increasing outflow, or increasing vacuole sequestration of Na+ (Zhang *et al.*, 2004). Increased expression of the *A. thaliana* tonoplast membrane Na+/H+ antiporter, *AtNHX1*, under a strong constitutive promoter, was reported to result in salt-tolerant tomatoes (Zhang and Blumwald 2001). The transgenic tomato plants grown in the presence of 200 mM NaCl were able to flower and set fruit. While the leaves accumulated high concentrations of sodium, the tomato fruits continued to contain only low concentrations of sodium. The *NHX1* system seems to be highly conserved between many different plant species and manipulation of this system in crop species is likely to result in improved salt tolerance. Research on the physiology of stress tolerance has demonstrated that tolerance to a specific stress is determined by several component traits and controlled by corresponding genes. A combination of a genome-wide scan of expression, using DNA arrays, and QTL analysis could provide important information in identifying the major genes association with stress tolerance.

Prioritizing Vegetable Research to Address Impact of Climate Change

It is unlikely that a single method to overcome the effects of environmental stresses on vegetables will be found. A systems approach, where all available options are considered in an integrated manner, will be the most effective and ultimately the most sustainable, particularly for developing countries in the tropics under a variable climate. This holistic strategy will need global integration of efforts; the resulting synergies will produce impact more quickly than the individual institutions working in isolation could accomplish. For this to succeed, adequate and long-term funding is necessary, scientific results have to be delivered, best approaches utilized and effective methods sustained to deliver global public goods for impact.

AVRDC - The World Vegetable Center, as the world's leading international center focused on vegetable research and development, has expanded its research to further address the potential challenges posed by climate change. The Center's success in its major objectives of reducing malnutrition and alleviating poverty in developing countries through improved production and consumption of safe vegetables will involve adaptation of current vegetable systems to the potential impact of climate change. Vegetable germplasm with tolerance to drought, high temperatures and other environmental stresses, and ability to maintain yield in marginal soils must be identified to serve as sources of these traits for both public and private vegetable breeding programs. This germplasm will include both cultivated and wild accessions possessing genetic variation unavailable in current, widely-grown cultivars. Genetic populations are being developed to introgress and identify genes conferring tolerance to stresses and at the same time generate tools for gene isolation, characterization, and genetic engineering. Furthermore, agronomic practices that conserve water and protect vegetable crops from sub-optimal environmental conditions must be continuously enhanced and made easily accessible to farmers in the developing world. AVRDC-The World Vegetable Center has developed technologies on integrated crop management and many of these approaches directly provide solutions to mitigate the effects of climate change. Stressing the importance of effective delivery methodology to transfer technology and disseminate knowledge, AVRDC knows that for a technology to be successfully adopted, an effective extension strategy must be in place that includes technical, socio-economic, and political considerations. Finally, capacity building and education are key components of a sustainable adaptation strategy to climate change.

Enhancing adaptation of tropical production systems to changing climatic conditions is a huge undertaking. It requires the combined efforts of many national and international institutions and an effective and efficient strategy to be able to deliver technologies that can mitigate the effects of climate change on the diverse crops and production systems. The scientific information and technologies developed through these initiatives must be readily accessible, consolidated and utilized in a strategic way. This can only be achieved through collaboration, complementarity, and coordinated objectives to address the consequences of climate change on the world's crop production.

Conclusions

The impacts of climate change can be at any stage of crop growth and development, thus influencing the quality and yield. Currently, the world agriculture especially the vegetable production is passing through a difficult situation and faced with the challenge of food/nutritional security to meet the requirement for ever growing population. We have to produce more and more food from less and less land. The problem gets aggravated because of the growing biotic and abiotic stresses and decline in the quality of environment and along with the menace of increasing global warming caused by the green house gases. There is an urgent need to focus attention on studying the impacts of climate change on growth, development, yield and quality of crops. The focus should also be on development of adaptation technologies and quantify the mitigation potential of agricultural crops. Hi-tech horticulture is to be adopted in an intensive way. Selection of plant species/cultivars is to be considered keeping in view the effects of climate change. Judicious water utilization in the form of drip, mist and sprinkler irrigation will be a key factor to deal with the water stress conditions. Development of new cultivars of crops tolerant to high temperature, resistant to pests and diseases, short duration and producing good yield under stress conditions, should be the main strategies to meet this challenge.

References

Abdalla AA, Verderk K (1968) Growth, flowering and fruit set of tomato at high temperature. The Neth J Agric Sci 16: 71-76.

Abdul-Baki AA, Stommel J (995) Pollen viability and fruit set of tomato genotypes under optimum and high-temperature regimes. HortScience 30: 115-117.

Ali M (2000) Dynamics of vegetables in Asia: A synthesis. In: Ali M (ed) Dynamics of vegetable production, distribution and consumption in Asia. AVRDC, Shanhua, Taiwan, pp 1-29.

AVRDC (1990) Vegetable Production Training Manual. Asian Vegetable Research and Training Center. Shanhua, Tainan, 447 pp.

AVRDC (1979) Annual Report. Asian Vegetable Research and Development Center. Shanhua, Taiwan. 173 pp.

AVRDC (1981) Annual Report. Asian Vegetable Research and Development Center. Shanhua, Taiwan. 84 pp.

AVRDC (2005) Annual Report. AVRDC – The World Vegetable Center. Shanhua, Taiwan.

AVRDC (2006) Vegetables Matter. AVRDC – The World Vegetable Center. Shanhua, Taiwan.

Bell GD, Halpert MS, Schnell RC, Higgins RW, Lowrimore J, Kousky VE, Tinker R, Thiaw W, Chelliah M, Artusa A (2000) Climate Assessment for 1999. Supplement June 2000 Bull Am Meteorol Soc Vol 81.

Boyer JS (1982) Plant productivity and environment. Science 218: 443-448.

Bray EA, Bailey-Serres J, Weretilnyk E (2000) Responses to abiotic stresses. In: Gruissem W, Buchannan B, Jones R (eds) Biochemistry and molecular biology of plants. ASPP, Rockville, MD pp 1158-1249.

Bray EA (2002) Abscisic acid regulation of gene expression during water-deficit stress in the era of the *Arabidopsis* genome. Plant Cell Environ 25: 153-161.

Capiati DA, País SM, Téllez-Iñón MT (2006) Wounding increases salt tolerance in tomato plants: evidence on the participation of calmodulin-like activities in cross-tolerance signaling. J Exp Bot 57: 2391-2400.

CGIAR (2003) Applications of molecular biology and genomics to genetic enhancement of crop tolerance to abiotic stresses – a discussion document. Interim Science Council Secretariat, FAO.

Chaves MM, Oliveira (2004) Mechanisms underlying plant resilience to water deficits: prospects for water-saving agriculture. J Exp Bot 55: 2365-2384.

Cheeseman JM (1988) Mechanisms of salinity tolerance in plants. Plant Physiol 87: 57-550.

Cuartero J, Fernandez-Munoz R (1999) Tomato and salinity. Sciencia Horticulturae 78: 83-125.

Cuartero J, Bolarin MC, Asins MJ, Moreno V (2006) Increasing salt tolerance in tomato. J Exp Bot 57: 1045–1058.

Déjardin A, Sokolov LN, Kleczkowski LA (1999) Sugar/osmoticum levels modulate differential ABA-independent expression of two stress responsive sucrose synthase genes in Arabidopsis. Biochem J 344: 503-509.

Drew MC (1979) Plant responses to anaerobic conditions in soil and solution culture. Curr Adv Plant Sci 36: 1-14.

Kawase M (1981) Anatomical and morphological adaptation of plants to waterlogging. HortScience 16: 30-34.

Kuo, DG, Tsay JS, Chen, BW, Lin PY (1982) Screening for flooding tolerance in the genus *Lycopersicon*. HortScience 17(1): 76-78.

Edelstein M (2004) Grafting vegetable-crop plants: Pros and Cons. Acta Horticulturae 65.

Erickson AN, Markhart AH (2002) Flower developmental stage and organ sensitivity of bell pepper (*Capsicum annuum* L) to elevated temperature. Plant Cell Environ 25: 123-130.

FAO (2001) Climate variability and change: A challenge for sustainable agricultural production. Committee on Agriculture, Sixteenth Session Report, 26-30 March, 2001. Rome, Italy.

FAO (2004) Impact of climate change on agriculture in Asia and the Pacific. Twenty-seventh FAO Regional Conference for Asia and the Pacific. Beijing, China, 17-21 May 2004.

FAO (2006) Agricultural data FAOSTAT. Food and Agriculture Organization of the United Nations. Rome, Italy.

Flowers TJ (2004) Improving crop salt tolerance. J Exp Bot 55: 307-319.

Foolad MR (2004) Recent advances in genetics of salt tolerance in tomato. Plant Cell Tissue Organ Culture 76: 101-119.

Foolad MR, Jones RA (1991) Genetic analysis of salt tolerance during germination in *Lycopersicon*. Theor. Appl. Genet. 81: 321–326.

Foolad MR, Zhang LP, Subbiah P (2003) Genetics of drought tolerance during seed germination in tomato: inheritance and QTL mapping. Genome 46: 536-545.

Brown P, Lumpkin T, Barber S, Hardie E, Kraft K, Luedeling E, Rosenstock T, Tabaj K, Clay D, Luther G, Marcotte P, Paul R, Weller S, Youssefi F, Demment M (2005) Global Horticulture Assessment. ISHS. Gent-Oostakker, Belgium. 135 pp.

Gur A, Zamir D (2004) Unused natural variation can lift yield barriers in plant breeding. PLoS Biol 2: 1610-1615.

Hanson PM, Chen JT, Kuo CG (2002) Gene action and heritability of high temperature fruit set in tomato line CL5915. HortScience 37: 172-175.

Hazra P, Samsul HA, Sikder D, Peter KV (2007) Breeding tomato (*Lycopersicon Esculentum* Mill) resistant to high temperature stress. Int J Plant Breed 1(1).

Hsieh TH, Lee JT, Charng YY, Chan MT (2002) Tomato plants ectopically expressing Arabidopsis CBF1 show enhanced resistance to water deficit stress. Plant Physiol 130: 618–626.

IDE (brochure) Affordable small-scale irrigation technologies. International Development Interprises, Lakewood, Colorado, USA.

Ingram J, Bartels D (1996) The molecular basis of dehydration tolerance in plants. Annual Review of Plant Physiol Plant Mol Biol 47: 377-403.

Jones RA (1986) The development of salt-tolerant tomatoes: breeding strategies. Acta Horticulturae 190: Symposium on Tomato Production on Arid Land.

Kuo CG, Chen BW, Chou MH, Tsai CL, Tsay TS (1979) Tomato fruit set at high temperatures. In: Proc 1st Intl Symp Trop Tomato p 95-109. AVRDC, Shanhua, Tainan.

Lee JT, Prasad V, Yang PT, Wu JF, Ho THD, Charng YY, Chan MT (2003) Expression of Arabidopsis CBF1 regulated by an ABA/stress promoter in transgenic tomato confers stress tolerance without affecting yield. Plant Cell Environ 26: 1181–1190.

Lin KH, Lo HF, Lee SP, Kuo CG, Chen JT (2006) RAPD markers for the identification of yield traits in tomatoes under heat stress via bulk segregant analysis. Hereditas 143: 142-154.

Liu Q, Kasuga M, Sakuma Y, Abe H, Miura S, Yamaguchi-Shinozaki K, Shinozaki K (1998) Two transcription factors, DREB1 and DREB2, with an EREBP/AP2 DNA binding domain separate two cellular signal transduction pathways in drought- and low-temperature-responsive gene expression, respectively, in Arabidopsis. Plant Cell 10: 1391-1406.

Maldonado C, Squeo FA, Ibacache E (2003) Phenotypic response of *Lycopersicon chilense* to water deficit. Revista Chilena Historia Natural 76: 129-137.

Martin B, Nienhuis J, King G, Schaefer A (1989) Restriction fragment length polymorphisms associated with water use efficiency in tomato. Science 243: 1725-1728.

Masle J, Gilmore SR, Farquhar, GD (2005) The ERECTA gene regulates plant transpiration efficiency in *Arabidopsis*. Nature 436: 866-870.

Matsubara S (1989) Studies on salt tolerance of vegetables-3. Salt tolerance of rootstocks. Agric Bull, Okayama Univ 73: 17-25. 19

Metwally E, El-Zawily A, Hassan N, Zanata O (1996) Inheritance of fruit set and yields of tomato under high temperature conditions in Egypt. First Egypt-Hung Hort Conf, Vol I.

Midmore DJ, Roan, YC, Wu, MH (1992) Management of moisture and heat stress for tomato and hot pepper production in the tropics. In: Kuo CG (ed) Adaptation of food crops to temperature and water stress. AVRDC, Shanhua, Taiwan pp 453-460.

Midmore, DJ, Roan YC, Wu DL (1997) Management practices to improve lowland subtropical summer tomato production: yields, economic returns and risk. Exptl Agric 33: 125-137.

McWilliam JR (1986) The national and international importance of drought and salinity effects on agricultural production. Austral J Plant Physiol 13: 1-13.

O'Connell MA, Medina AL, Sanchez Pena P, Trevino MB (2007) Molecular genetics of drought resistance response in tomato and related species. In: MK Razdan and AK Mattoo, (eds) Genetic Improvement of Solanaceous Crops, Vol. 2: Tomato, Science Publishers, Enfield USA pp. 261-283.

Okimura M, Matsou S, Arai K, Okitso S (1986) Influence of soil temperature on the growth of fruit vegetable grafted on different stocks. Bull Veg Ornam Crops Res Stn Japan. C9: 3-58.

Opena RT, Green SK, Talekar NS, Chen JT (1989) Genetic improvement of tomato adaptability to the tropics: Progress and future prospects. In: Green SK (ed) Tomato and pepper production in the tropics. AVRDC, Shanhua, Taiwan pp 70-85.

Opena RT, Green SK, Talekar NS, Chen JT (1990) Genetic improvement of tomato adaptability to the tropics. In: Green SK (ed) Integrated pest and management practices for tomato and pepper in the tropics. AVRDC, Shanhua, Taiwan pp 70-85.

Opena RT, Chen JT, Kuo CG, Chen HM (1992) genetic and physiological aspects of tropical adaptation in tomato. In: Kuo CG (ed) Adaptation of food crops to temperature and water stress. AVRDC, Shanhua, Taiwan pp 321-334.

Opena RT, LO SH (1981) Breeding for heat tolerance in heading Chinese cabbage. In: Talekar NS, Griggs TD (eds). Proceedings of the 1st International Symposium on Chinese cabbage. AVRDC, Shanhua, Taiwan.

Ort DR (2001) When there is too much light. Plant Physiol 125: 29-32.

Pandita ML, Singh N (1992) Vegetable production under water stress conditions in rainfed areas. In: Kuo CG (ed) Adaptation of food crops to temperature and water stress. AVRDC, Shanhua, Taiwan pp 467-472.

Park S, Li J, Pittman JK, Berkowitz GA, Yang H, Undurraga S, Morris, J, Hirschi KD, Gaxiola RA (2005) Up-regulation of a H+-pyrophosphatase (H+-PPase) as a strategy to engineer drought –resistant crop plants. PNAS 102: 18830-18835.

Pereira JS, Chaves MM (1995) Plant responses to drought under climate change in Mediterranean-type ecosystems. In: Moreno JM, Oechel WC (eds) Global change and mediterranean-type ecosystems. Springer-Verlag, Berlin, pp 140-160.

Phene CJ (1989) Water management of tomatoes in the tropics. In: Green SK (ed) Tomato and pepper production in the tropics. AVRDC, Shanhua, Taiwan pp 308-322.

IPCC (2001) Climate change 2001: Impacts, adaptation and vulnerability. Intergovermental Panel on Climate Change. New York, USA.

Reymond P, Weber H, Damond M, Farmer EE (2000) Differential gene expression in response to mechanical wounding and insect feeding in *Arabidopsis*. Plant Cell 12: 707-720.

Rick CM (1973) Potential genetic resources in tomato species: clues from observation in native habitats. In: Srb AM (ed) Genes, enzymes and populations, Plenum Press, New York, p. 255-269.

Romero L, Belakbir A, Ragala L, Ruiz MJ (1997) Response of plant yield and leaf pigments to saline conditions: effectiveness of different rootstocks in melon plant (*Cucumis melo* L) Soil Sci Plant Nutr 3: 855-862.

Sakuma Y, Liu Q, Dubouzet JG, Abe H, Shinozaki K, Yamaguchi-Shinozaki K (2002) DNA-binding specificity of the ERF/AP2 domain of *Arabidopsis* DREBs, transcription factors involved dehydration- and cold-inducible gene expression. Biochem Biophys Res Comm 290: 998-1009.

Sánchez Peña P (1999) Leaf water potentials in tomato (*L. esculentum* Mill.) *L. chilense* Dun. and their interspecific F1. MSc Thesis, New Mexico State University, Las Cruces, NM, USA.

Sato S, Peet MM, Thomas JF (2002) Determining critical pre- and post-anthesis periods and physiological process in *Lycopersicon esculentum* Mill. Exposed to moderately elevated temperatures. J Exp Bot 53, 1187-1195.

Seki M, Narusaka M, Abe H, Kasuga M, Yamaguchi-Shinozaki K, Carninci P, Hayashizaki Y, Shinozaki K (2001) Monitoring the expression pattern of 1300 Arabidopsis genes under drought and cold stresses by using a full-length cDNA microarray. Plant Cell 13: 61-72.

Stevens MA Rudich J (1978) Genetic potential for overcoming physiological limitations on adaptability, yield, and quality in tomato. HortScience 13: 673-678.

Sung DY, Kaplan F, Lee KJ, Guy CL (2003) Acquired tolerance to temperature extremes. Trends Plant Sci. 8: 179-187.

Villareal RL, Lai SH, Wong SH (1978) Screening for heat tolerance in genus Lycopersicon. HortScience 13: 479-481.

Von Westarp S, Chieng S, Scheier (2004) A comparison between low-cost drip irrigation, conventional drip irrigation, and hand watering in Nepal. Agric Water Mgt 64: 143-160.

Wang, WX, Vinocur B, Altman A (2003) Plant responses to drought, salinity and extreme temperatures: towards genetic engineering for stress tolerance. Planta 218: 1-14.

Weis E, Berry JA (1988) Plants and high temperature stress. Soc of Expt Biol, pp 329-346.

Yamaguchi T, Blumwald, E (2005) Developing salt-tolerant crop plants: challenges and opportunities. Trends Plant Sci 10(12): 616-619.

Yildirim E, Guvenc I (2006) Salt tolerance of pepper cultivars during germination and seedling growth. Turk J Agric Forestry 30: 347-353.

Zhang HX, Blumwald E (2001) Transgenic salt-tolerant tomato plants accumulate salt in foliage but not in fruit. Nat Biotechnol 19: 765-768.

Zhang JZ, Creelman RA, Zhu JK (2004) From laboratory to field. Using information from Arabidopsis to engineer salt, cold, and drought tolerance in crops. Plant Physiol 135: 615-621.

2014, Sustainable Rural Development through Agriculture *Pages* **62–80**
Editors: **Dr. Shobhana Gupta and Dr. S.S. Tomar**
Published by: **BIOTECH BOOKS, NEW DELHI**

Chapter 4
Dryland Farming in India

Narendra Kumawat, P.S. Shekhawat, Rakesh Kumar and U.N. Shukla

ABSTRACT

India's agriculture growth after independence has moved the country from the severe food crisis of the sixties to aggregate food surplus today and rainfed agriculture has played an important role in this. In India about two-third of total net sown area comes under rainfed lands. Rainfed crops account for 48 per cent area under food crops and 68 per cent under non-food crops. Realising the potential of drylands is by no means a simple task. Spread over nearly half of the country, the drylands cover cold arid regions, hot deserts, hilly and undulating uplands, forest areas, plateaus, ravines and coastal and non-coastal saline areas. They are the home to 43 per cent of our population. Water availability, soil conditions and the length of the growing season show wide variations here. Nine states (Rajasthan, Madhya Pradesh, Maharashtra, Gujarat, Chhatisgarh, Jharkhand, Andhra Pradesh, Karnataka and Tamil Nadu) account for over 80 per cent of the drylands. Annual rainfall in the drylands varies from less than 150 mm to 1600 mm. Indian agriculture is predominantly rainfed agriculture under which both dry farming and dryland agriculture is included. Dry faring was the earlier concept for which amount of rainfall (< 500 mm annually) remained the deciding factor for more than 50 years. In modern concept, dryland areas are those where the balance of moisture is always on the deficit side. In other words, annual evapotranspiration exceeds precipitation. Accordingly the production is either low or extremely uncertain and unstable which are the real problems of dryland in India. Out of 142 million ha of net sown area in the country, rainfed agriculture is practiced in 95 million ha (67 per cent). Nearly 67 m ha of rainfed area falls in the mean annual precipitation range of 500-1500 mm. Rainfed farming comprises about 91 per cent area of coarse cereals (sorghum, pearl millet, maize and finger millet), 91 per cent pulses (chickpea and pigeon pea), 80 per cent of oilseeds (groundnut, rape seed, mustard and soybean), and 65 per cent of cotton. Also, about 50 per cent area under rice and 19 per cent area under wheat is rainfed. During the past 25 years there occurred significant changes in the area and yield of imported crops of rainfed farming areas. The area under coarse cereals decreased by about 10.7 million ha and most of this was under sorghum. The area under oilseeds increased by 9.2 million ha and most of this increase was due to irrigated rapeseed and mustard and soybean. The total area under pulses and cotton remained constant but more of cotton became irrigated and shifts in the area under occurred from one agro-ecological region to others. Area under chickpea in the northern belt decreased but increase in the central belt. This change occurred due to increase in area under rice-wheat cropping system which displaced chickpea and also pearl millet to a great extent and maize to a small extent. We need not only viable agriculture packages but meticulously worked out land-use planning systems, which make careful use of available soil moisture through appropriate tree-crop mixes. Here, again, what works and what would fail will be known only after more detailed evaluation of such systems in the field. There are several dedicated institutes deputed to do this kind of work but their field presence is next to nil.

Introduction

By 2025 AD, India will have to produce around 300 million tonnes of food grains to feed his population. This target cannot be realized from irrigated areas alone as the irrigation potential is for 78 million hectares and is a function of rainfall received. Thus, appropriate technologies to cope with the uncertainties of rainfed agriculture have to be evolved. The second 'green revolution' in Indian agriculture should usurp rainfed/dryland agriculture since the first green revolution bypassed it. This is crucial not only for achieving self-sufficiency in food grains production, but also for equity and enhancing employment for betterment of standard of living of dryland farmers.

Indian agriculture is predominantly rainfed within which both dry farming and dryland agriculture are covered as 75 per cent of our arable land (143 m ha) is rainfed. More than 90 per cent of the area under coarse cereals and pulses, 80 per cent under oilseeds (groundnut, rapeseed-mustard and soybean), 65 per cent under cotton, about 50 per cent under rice and 19 per cent under wheat are rainfed. Rainfed agriculture accounts for more than 40 per cent of the total food grains produced, nearly 75 per cent of all oilseeds, 90 per cent of the pulses and 70 per cent of cotton.

By tradition, dryland farming refers to areas and situations wherein rainfall is below 500 mm per annum, a decisive factor for livelihood. In modern concept, there is no consideration of amount of rainfall as dryland areas are those where the balance of moisture is always on the "deficit" side, implying that annual evapo-transpiration exceeds precipitation. Thus, even areas receiving 1100 mm or more of rainfall can be categorised as dryland agriculture, if the 'evapo-transpiration exceeding precipitation' irrespective of the magnitude of rainfall received. Thus, the deficit in moisture affects crop production resulting in partial or total failure of crops. Accordingly, the production in rainfed areas is fraught with low, extremely uncertain and unstable production.

The factors contributing to worsening situation of dryland agriculture, other than traditional farming systems, are the stagnation of productivity of dryland crops, the decline of livestock, unsuitability of traditional technologies, and the mounting natural resource scarcity. The major achievement of past research was the generation of potential for high productivity arable farming, particularly in dry tropics with relatively dependable rainfall. Systematic research for different soil-rainfall zones forms an important part of future research and development. Other areas of concern include breaking yield barriers and competitiveness of dry crops, crucial role of institutions in new technologies, precision and management intensity, and the cost factor. The competitiveness of dryland crops is severely handicapped by the lack of HYVs in oilseeds and pulses, characterized as 'slow growth' crops. The imbalance due to focus on rice-wheat production has not strengthened the range of high productivity options and there is little in offing in terms of viable technologies for conservation and up gradation of depleted sub-marginal lands.

Though technologies are available, farmers seldom adopt them due to certain constraints. Therefore, improved dry farming is necessary for equity and prosperity. As such we cannot achieve stability in food production with unstabilized dryland agriculture. Therefore, we are required to adopt improved technology especially developed for dryland agriculture. Dryland research institutes have developed several dryland farming practices and tested for their sustainability. These technologies can be grouped into: (a) suitability of crops and varieties to regions and situations; (b) delineation of efficient cropping zones; (c) integrated farming systems; (d) optimum time of sowing; (e) better/best sowing techniques; (f) seed hardening; (g) efficient soil and moisture conservation techniques; (h) integrated nutrient and pest management practices; (i) contingency planning, and (j) watershed management.

Development of Dryland Farming in India

Dryland farming in India began centuries earlier than in North America. However, there are some striking similarities between the two regions with respect to the scientific study of dryland farming. Hegde (1995) reported that, in 1917, Aiyer had listed the important farmer practices and found them quite similar to those that Campbell (1907) had proposed for the Great Plains in the United States of America in the early 1900s. Field bunding, fall ploughing, frequent inter cultivations, drill sowing, and growing drought-resistant crops, such as finger millet, grain sorghum and pearl millet, were some of the practices listed. Scientific study of dryland farming was initiated by the Government of India in 1923. Early research focused on improving crop yields. Important practices included: (i) bunding to conserve soil and water; (ii) deep ploughing once in three years for better intake and storage of water; (iii) use of farmyard manure to supply plant nutrients; (iv) use of a low seeding rate; and (v) inter-cultivation for weed and evaporation control. These practices gave a 15–20 percent increase over the base yields (Hegde, 1995).

By the mid-1950s, the emphasis had shifted to soil management. Soil conservation research and training centres were established at eight locations, focusing on contour bunding. However, negative results were often obtained because of water accumulation and runoff problems, particularly on Vertisols. Even where yield increases were observed, they were again not more than 15–20 percent above the base yields. The importance of shorter-duration crops to match the soil-water availability period was recognized in the 1960s. It was also in the mid-1960s that high-yielding hybrids and cultivars became available that were responsive not only to fertilizers but also to management. An All-India Coordinated Research Project for Dryland Agriculture was established, and the research emphasis shifted to a multidisciplinary approach to tackle the problems. Similar efforts were initiated at the International Crops Research Institute for the Semi-Arid Tropicsat Hyderabad in 1972. Although many of the recommended practices for dryland farming in India are similar to those for North America, there are differences. A highly recommended practice for water conservation in India is the use of dust mulch (Hegde, 1995), similar to that recommended in the North America in the early 1900s, which is commonly considered to have contributed to the Dust Bowl. These contrasting recommendations illustrate the importance of recognizing and addressing the differences between semi-arid regions. Semi-arid conditions in North America are very different from those in India. For example, summer fallow has played an important role in North America because some precipitation occurs throughout the year, but at no time does monthly precipitation exceed PET. Many dryland regions in the Great Plains in the United States of America do not have any month with precipitation that even reaches half of PET. In contrast, most dryland areas in India have more than seven months with essentially zero precipitation. They have a monsoon season of varying length when the precipitation greatly exceeds the potential evapotranspiration for at least a portion of the growing season. Therefore, fallow for storing soil water is not a viable alternative because much of the water saved during the rainy season would be lost during the prolonged dry period. More importantly, there is usually more than enough precipitation during the monsoon season to fully wet the soil profile.

Rainfed Agriculture Research Network

CRIDA

All India co-ordinated research project for Dryland Agriculture was launched by ICAR in 1970 in collaboration with Government of Canada and later Central Research Institute for Dryland Agriculture (CRIDA) was established at Hyderabad.

Objectives

- ✰ Undertake basic and applied researches that will contribute to the development of strategies for sustainable farming systems in the rainfed areas,
- ✰ Act as a repository of information on rainfed agriculture in the country,
- ✰ Provide leadership and co-ordinate network research with state agricultural universities for generating location-specific technologies for rainfed areas,

- ✰ Act as a centre for training in research methodologies in the fields basic to management of rainfed-farming systems,
- ✰ Collaborate with relevant national and international agencies in achieving the above objectives, and
- ✰ Provide consultancy.

AICRPDA

All India Coordinated Research Project for Dryland Agriculture (AICRPDA) in 1970, based at Hyderabad with 23 cooperating centres spread across the country.

Objectives

- ✰ To evolve simple technologies to substantially increase crop productivity and viability.
- ✰ To optimise the use of eco-regional natural resources, *i.e.*, rainfall,land and water, and to minimise soil and water loss and degradation of environment
- ✰ To increase stability of regional crop production over years byproviding improvements in natural resources management, crop management systems and alternate crop production technologies matching weather aberrations
- ✰ To develop alternate and sustainable land use systems
- ✰ To evaluate and study transfer*abil*ity of improved dryland technology to farmers' fields

AICRPAM

All India Coordinated Research Project on Agro-meteorology (AICRPAM) was launched in 1983, also at Hyderabad, with 10 cooperating centres under different SAUs.

Objectives

- ✰ To study the agricultural climate in relation to regional crop planning and assessment of crop production potentials
- ✰ To establish crop-weather relationships for all the major rainfed and irrigated crops
- ✰ To evaluate different techniques of modification of crop microclimates for enhancing water-use efficiency and productivity
- ✰ To study the influence of weather on the incidence and spread of pests and diseases of field crops
- ✰ To provide agro-advisory support to farming communities

NPCC/NICRA

Recognising the vulnerability of rainfed agriculture to climate change, ICAR has launched the Network Programme on Climate Change (NPCC) during X Plan with 11 centres across the country and 12 more were added during XI Plan. The objectives

)f this programme are: quantify the vulnerability of Indian agriculture to increasing ·limatic variability and climate change, to develop adaptation strategies for ninimizing their negative impacts, and to identify mitigation strategies. Further ealizing the importance of climate variability and its impact on food security, ICAR aunched the National Initiative on Climate Resilient Agriculture (NICRA) during XI 'lan period with an outlay of Rs. 350 crores for two years (2010-12). The objectives of NICRA are to enhance the resilience of Indian agriculture covering crops, livestock nd fisheries to climatic variability and climate change through development and pplication of improved production and risk management technologies. Also to lemonstrate site-specific technology packages on farmers' fields to cope up with limate vulnerability, research infrastructure development and capacity building of cientists in new tools and techniques in climate change research.

;onstraints of Crop Production in Dryland

Dry farming crops are characterized by very low and highly variable and ncertain yields. Crop failures are quite common. These are mainly due to the following auses.

(i) **Inadequate and uneven distribution of rainfall:** In general, the rainfall is low and highly variable which results in uncertain crop yields. Besides its uncertainty, the distribution of rainfall during the crop period is uneven, receiving high amount of rain, when it is not needed and lack of it when crop needs it.

(ii) **Late onset and early cessation of rains:** Due to late onset of monsoon, the sowing of crop are delayed resulting in poor yields. Sometimes the rain may cease very early in the season exposing the crop to drought during flowering and maturity stages which reduces the crop yields considerably

(iii) **Prolonged dry spells during the crop period:** Long break in the rainy season is an important feature of Indian monsoon. These intervening dry spells when prolonged during crop period reduces crop growth and yield and when unduly prolonged crops fail.

(iv) **Low moisture retention capacity:** The crops raised on red soils, and coarse textured soil suffer due to lack of moisture whenever prolonged dry spells occur due to their low moisture holding capacity. Loss of rain occurs as runoff due to undulating and sloppy soils.

(v) **Low fertility of soils:** Drylands are not only thirsty, but also hungry too. Soil fertility has to be increased, but there is limited scope for extensive use of chemical fertilizers due to lack of adequate soil moisture.

rincipal Dry Farming Zones in India

Almost all the states have some area under rainfed culture depending upon pography and irrigation facilities, but only the major dry farming area are discussed re.

The Indo-Gangetic Plains of North India

This zone is the youngest in the geographical formation. This zone includec districts of Rajasthan, Punjab, Haryana, North-Western M.P. and U.P. This zone i: characterized by two soil types namely light loam, and heavy loam. The land i: nearly levelled with a modest slop of 2 ft/mile length. The soils are very deep anc situated at about 700 to 800 ft. above sea level. Because of heavy sand and silt fraction in the soil it has large pore space. The soils are rich in essential nutrients like nitroger phosphorus, potash, calcium etc. and, therefore, quite good for raising the cro excepting few with high water requirements. The cropping intensity, in this zon stands around 120 per cent and the major crops which are grown in this zone ar millets, cereals, oilseeds and pulses.As far as rainfall pattern in this zone is concernec it is observed that about 60 per cent or more of the total rainfall is observed betwee the end of July to the end of august, and the rainfall in remaining months is quit poor. Thus, due to very high intensity of rainfall, floods are of frequent occurrenc during the first week of September followed by a long spell of drought subsequentl

The Trapian Plateau of Peninsular India

This zone comprises the states of Maharashtra, Karnataka and Andhra Pradesl The soil of this zone has been derived from the Deccan trap. The tract is undulatin and consists of low ridge and valleys due to erosion which results in rapid run-of About 40 per cent of the land of this zone is not fit for cultivation. This tract is situate at an elevation of 1400 – 2 (xx) feet from sea level. The soil may be grouped into thre types based on its depth as deep, medium deep and shallow soils. Leaching of lim has resulted in the formation of lime nodules or kanker on the surface soil. The soil quite rich in total and available nitrogen, phosphorus and potash which favou production of crops if moisture is efficiently conserved.In this zone, two high peak of rain are observed because the area is affected by both south-west monsoon as we as northeast monsoon. About 40-55 per cent of total annual rainfall is obtained fro south-west monsoon and the rest from northeast monsoon. Mostly the millets an some oilseeds like groundnut are grown this zone.

Plateau of Granite Formation

The soils of this zone are grouped as red soils and black cotton soils. Red soi are shallow while black cotton soils are very deep like clayey soils. The topography of gentle undulations which favour run-off and soil erosion. The high pore space ar high swelling of soil obstruct the permeability of rain water in to the lower layers soil and its shrinkage results in hardening and clod formation on the true surfa which is unfavourable for plant growth. The red laterite and black cotton soils a deficient in nitrogen and phosphoric acids.

Improved Dryland Technology for Increasing the Productivit

To boost the crop production under dry farming, we will have to efficient manage our soil and water resources in the respective areas as dryland fanning ge more complex and intractable when drought occurs frequently. An efficient soil ar water conservation system will play a vital role in boosting the crop yield in d fanning. The different interdisciplinary approaches which are recommended for dryland fanning are categorized in to four major grouped namely engineering, physiological, genetic and agronomic approaches.

A) Engineering Approaches

These approaches are aimed at soil and moisture conservation through regulation of run-off, collection of surplus rain water checking evaporation and seepage losses of water and recycling of collected water as irrigation in times of critical need.

(i) **Contouring across the slop**: Contouring is practiced on the lands with 3-5 per cent slop. This system consists of constructing earthen bunds and the distance between the two bunds ranges from 30-50 m depending on the degree of slop. This is carried out with an object to provide a check to the flow of run-off water which then gets accumulated in the bunded area and is absorbed by the soil. Thus, contour bunding conserves moisture and prevents soil crosion.

(ii) **Smoothening of contour inter-bund areas**: This is practiced only in those areas which have a slope of less than 1 per cent. The smoothening may be achieved by running bullock drown harrows or cultivator but small undulation are levelled during the process so that impounding of water may take place and maximum water absorption by the soil may be achieved.

(iii) **Contour border strips methods:** This method is suitable for areas having a slop of 3-4 per cent in this case parallel strips across the slope ranging 10-15 m in width are laid downon contours and the soil surface is levelled by scrapping and placing the soil according to the need of the spot. It is done to reduce the run-off and to conserve soil and water from the field. It is however an expensive method as it requires culling and filling up the soil from higher spots to lower ones.

(iv) **Scooping of land:** In this practice, the land is generally scooped before the beginning of monsoon showers. By scooping, the soil is exposed for proper absorption and conservation of moisture. However, this is also a tedious as well as an expensive operation.

(v) **Opening of ridges and furrows:** In this practice, the entire land is laid out ridges and furrows across the slops. The ridges and furrows are opened before onset of monsoon so that the flow of water may be reduced and erosion may be controlled to the minimum. During rainy season, crops like maize, jowar, bajra, etc. may be grown in the furrows and legumes like soybean, arhar, urd, mung, cowpea etc. may be grown on the ridges. After the monsoon is over the land is again levelled. This way the furrows are used to accumulate maximum water which will supply moisture for winter season crops.

(vi) **Compartmental bunding:** Areas having a slop of 1 per cent or less are suitable for compartmental bunding. It helps in accumulation of more water and a uniform spread of water in the entire area. Levelling is also done with nominal or no additional expenditure.

(vii) **Bedding system:** In this system, small furrows are opened and the soil from the furrows is uniformly spread in space left between the furrows. Thus, inter furrows spaces from the raised beds of about 4-5 metres width. This method helps in the conservation of soil, moisture and checking the excess run-off of water. The raised beds, in this practice, are used for growing such crops which need less water like legumes and oilseeds crops, while the furrows are used for the crops which need more water.

(viii) **Broad based bunding**: This method especially suitable for heavy black soil. Water is allowed to spread over a vast area by constructing a broad based bund on a sloppy side. The water stays for a longer time because of high water holding capacity, lower leaching and seepage losses. The stored water may be used for fish culture and also for providing life saving irrigation grown in surrounding areas of catchment portion. These bunds are also called check dams and are give a regulated drain of outlet for protecting the bunds from breaking.

(ix) **Deep summer ploughing followed by surface tilling:** The field is ploughed deep by mould bold plough soon after harvesting *rabi* season crops with the objective of (a) exposing the soil for perfect drying, (b) killing the pathogens, (c) destroying eggs of insect pests, and (d) controlling weed sun drying. The surface tilling during other season forms natural mulch and thereby reduces evaporation loss of water from soil.

(x) **Water harvesting system:** Water harvesting is a technology of utilizing the collected and conserved water for purpose of crop production. It includes tillage practices for an efficient use of moisture between and within the crop rows. Frequent stirring of the land by surface tilling provides much and prevents the evaporation loss of water from the soil.Besides harvesting moisture from between and within the crop rows, run-off losses are considerably reduced. The store water is used for providing life saving irrigation to the crops grown in the surrounding areas. The water harvesting of this type can be done in areas situated near hill and on greatly undulated land. In these cases, check dam tanks, and other reservoir are constructed. The infiltration or percolation loss of water is prevented by spraying of asphalt compounds or by covering the bottom of the tank of ponds through thin plastic sheets. The seepage loss may also be checked by providing a plastic lining. The evaporation loos of water in controlled by pouring some burnt crude oil over water surface.

(B) Physiological Approaches

Hardly 1 per cent of the water absorbed by the plant roots is used for the growth and development of plants and remaining 99 per cent is wasted through transpiration back to atmosphere. Thus one of the greatest causes of soil water wastage is loss of soil water through evaporation.The extent of transpiration can be greatly influenced by using certain chemicals. These chemicals reduce transpiration, encourage root growth and protect the cytoplasmic proteins of the plants. These chemicals bring about more drought resistance in the plants. These compounds, according to their role are classified as given below :

(i) **Anti-transpirants:** Any chemical substance, which reduces rate of transpiration on its application to the plant surface is called anti-transpirant. Any substances which reduces the vapour pressure gradient in the stomatal cavity or increases stomatal resistance to water vapour diffusion, will act as anti-transpirant. These substances have been used for arresting water loos from plant body with various degree of success. These are Phenyl Mercuric Acetate (PMA), Hydroxy, Sulphonates (HS), Alkenyl Succinic Acid (ASA), Adol:-52 and S-600 (a plastic transpirating spray).

(ii) **Chemicals for improved cell membrane permeability of water**: Dry farming areas are characterized by scarce rainfall and usually the roots have lipid layers which lower the absorption of water from the roots. Some chemicals like Alkenyl Succinic Acid (ASA) and Decenyl Succinic Acid (DSA), when applied, penetrate into the root and increase its water absorption power 8 times. Therefore, these chemical are applied in the root zone for increased water absorption along with some chemicals to retards the transpiration from foliage.

(ii) **Use of plant hormones and growth retardants:** Some plant hormones like IAA and ABA may be used for reducing the frequency and period of stomatal opening thereby minimizing the water loss from the plant body. There are certain other chemicals known as growth retardants which either modify the plant structure or dwarf the plants by considerably reducing the total water requirement of the plants. The most important chemical of this group is cycocel or CCC (2-chloroethyl trichloromethylammonimum chloride). Cycocel is presently used in cotton to encourage production of more fruiting branches (sympodial rather than monopodial) and it thereby even under drought condition it results in higher yield.

(iii) **Use of chemicals**: There are certain chemicals which are used for seed treatment to bring about drought resistance in plants right from seeding. Soaking of seeds in calcium chloride solution (0.25 per cent) for 20 hours with frequent shaking results in better germination and drought resistance. Boron solution is also used for soaking seeds. Agrosan is a fungicide but also induces drought resistance in the plants when seeds are treated with this chemical.

(C) Genetic Approaches

Because of scanty and unreliable rains the farmers of dry farming areas are still practicing crop husbandry on the basis of traditional approaches like low intensity cropping little or no use of fertilizers or manures, raising low value crops. Crop varieties grown till the recent past were generally of long duration and slow growing, it they were poor yielders also.As nearly 70 per cent of our total agricultural land is rainfed and 45 per cent of rainfed area is dryland. There is no way out but to evolve suitable varieties as well as appropriate technology for getting the most from our rainfed or dryland areas. As such, the concerted efforts of our plant breeders have resulted in the cultivation of several new plant types which possess all the characters needed for rainfed areas or drylands. In terms of modern technology such plant materials are called "Ideotype". These are the suitable stains for dryland crops which are characterized by short growth duration, effective and extensive root system, drought tolerance, high yield potential having altered morphology of plants which hare conducive to drylands. According to breeder, an ideal "Ideotype" should have following to give desired results in dry farming:

(i) Early maturing and high early growth vigour.

(ii) Deeper root system with maximum branching at deeper root zone.

(iii) Dwarf plant type with lesser number of erect leave.

(iv) Moderate tillering, as profuse tillering cause competition.

(v) Good expression of ear heads even at higher planting density.

(vi) Resistance to disease and insect-pests.

(vii) Bolder grains with moderate dormancy.

(viii) Effective photosynthetic behaviour with greater sink capacity.

(D) Agronomic Approaches

The major objective of dry farming programme is to conserve the soil and moisture and achieve maximum production form the dry farming areas. In the past two decades, we have been able to solve many hurdles in the aforesaid areas but there had been no breakthrough as in case of irrigated crop production. Now we have the promising crop varieties and technology available with us about the maximum soil and water conservation. The agronomic approaches can be deals under the following four heads based on land types.

(a) **Agronomic approaches for high undulating lands:** These lands are confined to the hills of the locations which have suffered serious soil erosion problems and have been divided into gullies. The soils of these areas are more prone to further erosion If they are not properly managed. Therefore, in this category, the crop management practices are entirely different from other areas. The object of soil and crop management under such situation should be:

(i) To stabilize the soil by forestry and pasture management with a regulated grazing or no grazing at all.

(ii) To level the land gradually through contour bunding, terracing etc.

(iii) To practise strip cropping and pitcher farming.

(b) **Agronomical approaches for marginal lands:** Generally, marginal lands are very poor in fertility. The crop management in these lands is carried out in lines of crop management for levelled lands or flat lands which will be discussed later in this chapter.

(c) **Agronomical approaches for diara lands:** Diara lands are located on either side of rivers or between two rivers and are often flooded by these rivers.

These diara lands are formed due to flood and may have deposition of fine to coarse sands.These areas often lacking irrigation facilities and need a careful crop management. Since the land of this area is highly susceptible to floods, *kharif* cropping is practically impossible. But certain fodder crops can easily be grown soon after the onset of monsoon and harvested depending upon position of floods. The harvesting or cutting of crops is started from close to the river beds and as the water spreads the harvesting is also advanced. The life saving irrigation can be given by lifting water from the river or by drilling cavity wells or bamboo borings. However, in most of these cases there wells go out of order after a flood occurs. Therefore, the cavity well or bamboo borings are made at a distant location from river streams.

(d) Agronomical approaches for plain lands: Plain lands from the main dryland tracts of the country. There has been major emphasis on finding out ways and means through which the total soil productivity could be increased. Following recommendation should be followed on plain lands for improved crop productivity and efficient soil and water conservation.

(i) Tillage requirement of the crops: Tillage starts with the seed bed preparation and ends with mulching and control of weeds. Deep ploughing during summer helps in destroying weeds and suppressing insect pests and disease. It also helps in a efficient root penetration very deep into soil placement of seed at 5 cm and fertilizers at 7.5 cm in the same furrow followed by soil compaction have resulted in better germination, plant vigour, extensive root development and higher crop yields.

(ii) Selection or crops and varieties: There are a number of improved varieties of different crops which are drought tolerant or resistant to water stress. The most commonly grown crops in drylands are rice, maize, sorghum, bajra, finger millets, wheat, barley, pulses, oilseeds etc. The improved varieties of these crops have already been described area wise.

(iii) Sowing of crops: Sowing of crops deals with several associated factors namely sowing time, sowing methods, depth of sowing etc. It is important in the sense that once the ideal plant population is achieved the crop is bound to give yield. Sowing time can markedly influenced the production and productivity of dryland crops. Early sowing of *kharif* crops resulted in early crop maturity and thereby it facilitates early sowing of succeeding *rabi* crops. Early sowing of *rabi* crops helps in overcoming the moisture stress at later stages of plant growth, particularly at grain filling stage. Broadcasting of seeds should be avoided as it involves several losses and seed does not properly come in contact with moisture. Placing the seeds at about 5 cm depth through pora or seed drill is desirable.To get an ideal plant population it is necessary that about 25 per cent higher than required seed rate should be applied. Care must be taken to reduce plant competition for moisture by removing excess plant population about 2-3 weeks after the sowing depending upon the crops.

(iv) Fertilizer management: Use of fertilizers in dryland is limited as compared to irrigated areas. Today we use on an average only 20 kg/ha fertilizers in dry farming areas as against 60 kg/ha national average. Reasons for application of fertilizers in dry fanning are as follows; poor response because of faulty methods of application, poor financial conditions of farming to purchase fertilizers, wrong concept of the farmers that fertilizers will burn the seedling, and harm to the soil and application of organic manures only, which cannot meet the total nutrients requirement of the crop.

(E) Cropping Systems

Cropping system refers to an arrangement in which various crops are grown together in the same field. The cropping systems followed in drylands differ from those followed under normal conditions. Only those crops can be grown under dryland conditions which requires less water to complete their life cycle of which can stand or yield under drought conditions. This can include both drought resistant and drought tolerant plants. In addition, plant can be grown only where some water is available to sustain the growth of plants. Following are a few intercropping systems for different states under dryland areas:

State	*Suggested Intercropping*
Gujarat	Pearl millet + Green gram Pearl millet + Cow pea Pearl millet + Sesame
Haryana	Pearl millet + Cluster bean Pearl millet + Green gram Pearl millet + cow pea
Uttar Pradesh	Pearl millet + Green gram Pearl millet + Cow pea Pearl millet + Sesame
Madhya Pradesh	Pearl millet + Pigeon pea Pearl millet + Cow pea Pearl millet + Soybean
Karnataka	Pearl millet + Pigeonpea/soybean Pearl millet + sun flower Pearl millet + Green gram
Andhra Pradesh	Pearl millet + Pigeonpea/soybean Pearl millet + Green gram
Tamil Nadu	Pearl millet + Pigeonpea/soybean Pearl millet + Green gram Pearl millet + Cow pea Pearl millet + Sun flower

Contd...

State	Suggested Intercropping
East Rajasthan	Pearl millet + Cluster bean
	Pearl millet + Cow pea
	Pearl millet + Green gram
Western Rajasthan	Pearl millet + Moth bean
	Pearl millet + Cluster bean
Manarashtra	Pearl millet + Pigeon pea/soybean
	Pearl millet + Black gram
	Pearl millet + Green gram
	Pearl millet + Cow pea
	Pearl millet + Moth bean
	Pearl millet + Sun flower

Mixed cropping may be defined as sowing of two or more crops simultaneously on the same piece of land in certain time. Mixed cropping is also followed to minimize the effect of unpredictable of rain. Mixed cropping may have low yield potential but it works as a buffer against failure under possible unfavourable conditions.

Example: Guar + Arhar + Mungbean, Bajra + Arhar + Mungbean and Maize + Urd etc.

(F) Cropping Patterns

Cropping pattern is defined as sequence of growing crops in a particular field at a particular period. To most common cropping pattern for dryland farming are discussed below :

For North Indian Conditions

- ☆ Sorghum – Safflower / mustard
- ☆ Sorghum – mung / urd / cowpea- Gram / Wheat – Gram
- ☆ Rice – Gram (for low laying areas)
- ☆ Bajra – Gram + Linseed
- ☆ Bajra + Urd / Mung / Soybean – Wheat / Barley + Gram / mustard
- ☆ Maize – Gram / Safflower

For Central Indian Condition

- ☆ Mungbean – Horse gram
- ☆ Mungbean – Safflower

(G) Weed Control

Presence of weeds in the crop field, especially in case of dryland, causes a severe crop weed competition for water nutrient and light. The reduction in yield due to weed varies from 30-75 per cent depending upon the crop and nature and extent of weed infestation. Weeds may be controlled by hand weeding, intercultural operations and herbicides application or by adopting an integrated approach.

(H) Plant Protection Measures

In light textured soils of arid and semi-arid regions termites and white grubs cause intensive damage to emerging seedlings and also to grown up plants. Use of Imidacloprid 30.5 per cent or Endosulphan in the soil and incorporating it well into the soil at the time of the last ploughing will control termite. The white grubs may be controlled by drilling of Thimet 20 G @ 15 kg/ha along with seeds. Aphids in mustard are very destructive; therefore, they are controlled by spraying 0.2 per cent Metasystox or Dimecron. Similarly, the borers in pulses are controlled by spraying 0.05 per cent Endosulfon.

(I) Alternate Land Use

All drylands are not suitable for crop production. Same lands may be suitable for range/pasture management and for tree farming and ley farming, dryland horticulture, agro-forestry systems including alley cropping. All these systems which are alternative to crop production are called as alternate land use systems. This system helps to generate off-season employment in mono cropped dryland and also, minimizes risk, utilizes off-season rains, prevents degradation of soils and restores balance in the ecosystem. The different alternate land use systems are alley cropping, agri-horticultural systems and silvi-pastoral systems, which utilizes the resources in better way for increased and stabilized production from drylands.

(i) Alley Cropping

For imparting stability and providing sustainability to the farming system, a tree-cum-crop system will be one most appropriate for such situations. One such system called 'alley cropping' - a version of agro-forestry system, could meet the multiple requirements of food, fodder, fuel, fertilizer etc. Alley cropping is a system in which food crops are grown in alleys formed by hedge rows of trees or shrubs. The essential feature of the system is that hedge rows are cut back at planting and kept pruned during cropping to prevent shading and to reduce competition with food crops.

For example, fast growing leguminous trees such as Leucaena leucocephala or liliricidia spp. are planted in rows. During the cropping season, trees are lopped at about 0.5 metre height. These loppings are used as much to reduce moisture loss and improve the nutrient status of soil. Arable crops like maize, rice, pearl millet, legumes, oilseeds etc. are planted in the alleys formed by the two rows of threes. This is also known as agri-silvi culture. Alley cropping is also a form of conservation farming which enhances soil fertility and prevents erosion. One very strong argument in favour of alley cropping is its ability to produce usable material even in years of severe drought. At Rajkot in 1985, rainfall received during the season was only 30 per cent of the normal. There was total failure of grain production of the three legume crops tried in the system. In sole crop plots production was limited to 5.0 q/ha to 17.0 q/ha of green fodder. However, in alley cropped plots, Leucaena hedge-rows produced over 50.0 q/ha of green fodder.

(ii) Agri-horticultural System

Agri-horticultural system palys an important role in dryland areas, especially in semi-arid regions where production of annual crops is not only low but also highly unstable. Fruit trees if suitably integrated in dryland farming system could add significantly to overall agricultural production including food, fuel and fodder, conservation of soil and water and stability in production and income. Dryland fruit trees being deep rooted and hardy, can better tolerated monsoonal aberrations than short duration seasonal crops. Hence, in drought year when annual crops usually fail or their production is highly depressed; fruit trees species yield considerable food, fodder and fuel.

A suitable example of agri-horti-system is growing of cow pea/green gram/horse gram in inter space of ber (*Zizyphus mauritiaria*) at 6 x 6 m spacing at Hyderabad. Phalsa (*Grewia asiatica*) may be planted in between two ber plants in a row with a view to intensify the system. A well-managed dryland orchard of ber should give 50 kg fruits per tree/year. There should be 250 plants/ha for optimized production. The grow income would touch around Rs. 50,000/ha (250 x 50 x 4), assuming that one kg ber fetches Rs. 4. One could get an additional income of Rs. 800-Rs. 1000 from green gram/cow pea (2.5-3.0 q/ha).

(iii) Silvi-pastoral System

This system is suited to marginal drylands and is most preferable where the fodder shortages are experienced frequently. The system essentially consists of a top feed tree species carrying grasses on legumes (preferable perennial) as understory crops. Dryland farmers having larger holdings and keeping a land follow for a longer period for one reason on the other, should go in for this system which could provide both fodder and fuel. In a survey carried out in Andhra Pradesh, Karnataka and Maharashtra by CRIDA scientists, it was revealed that after food it is the fodder which is of paramount importance for sustaining animal wealth in rural areas. In years to come, fuel will assume greater importance.

Recommendations for Dry Farming Areas

The research programmes of All India Coordinated Research Project for Dryland Agriculture have concluded into certain recommendations to the farmers of dryland areas which are described below:

1. Bunding across the slope and levelling the land should be done before onset of monsoon.
2. Deep summer ploughing should be followed by surface tillage during monsoon months and also rest of the year.
3. Application of organic manures like FYM, compost etc. @ 15-20 tonnes/ha or green manuring should be done. These manures should be applied about 20-25 days before sowing and should be well mixed in the soil.
4. Fertilizers should be basal placed at a depth of 7.5 to 10.0 cm in the soil and the seed should be sown in the same furrow about 3 cm above the fertilizers. This is important especially during winter season. The nitrogen (20-50 per cent of total) should be top dressed by side or band placement method at about 10-15 cm apart. The crop rows should be done soon after the rains but if there is not sufficient moisture in the soil, the nitrogen should betrayed

over the foliage with urea solution containing 3-5 per cent nitrogen. Zinc and sulphur should be applied as basal if needed.

5. Soil application of Endosulphan/methyal parathion (5-10 per cent) for termite and Thimet 20 G @ 15 kg/ha for white grub should be done. These chemicals must be mixed with soil properly while ploughing or at the time of sowing.
6. Selection of suitable crops and their varieties should be done according to their suitability to a particular region.
7. Seeds must be treated with suitable fungicides and that of legume with *Rhizobium*/PSM culture before sowing. Soaking seeds in plain water for *rabi* sowing helps in getting higher germination, better seedling vigour and an early maturity.
8. Proper crop rotation should be followed which should preferably have at least one legume every year.
9. For better seed soil moisture contact through soil compaction should be done by running a rubber wheel or roller especially for *rabi* crops.
10. At the event of total crop failure during *kharif* season a suitable catch crop like urd/mungbean or toria etc. should be sown.
11. Intercropping of oilseeds and pulses should be done with jowar, bajra, pigeonpea and maize crops for the purpose of making best use of soil and inter row moisture harvesting.
12. Line sowing by drilling the seed at a depth of 5-7.5 cm or even more depending upon the situation should be practiced because it helps in better seed germination. This also helps in stabilizing the require plant population and thereby in getting higher yield.
13. Proper weed management practices should be followed by adopting integrated weed control measures.
14. Mulching should be done by providing frequent interculture and pulverizing the soil. If intercultural operation are not possible then use of artificial mulches like covering the surface with tree leaves, uprooting weeds, sugarcane leave, saw dust or polythene sheets are used to check the evaporation from soil.
15. Water harvesting between the rows should be done by growing some pulse crops and run-off water should be collected in some nearby location ponds and used as life saving irrigation.
16. An efficient plant protection measures should be adopted to protect the crops from insect pest and diseases damage.
17. The crop should be harvested at proper physiological maturity so that the following or succeeding crop may be sown slightly earlier than the scheduled time and best use of rain water or residual may be made for crop production.

18. Crops like cotton, chillies, etc. should be sprayed with CCC or cycocel and groundnut should be sprayed with planofix for modified growth, higher drought resistance and better yield.

Future Strategies and Need to Work

- ✰ Address critical problems of rainfed agriculture through basic and strategic research using frontier science tools.
- ✰ Generate location-specific technologies on rainwater management, soil management, farm mechanization, cropping systems and alternate land use systems through network research.
- ✰ Promote cost-effective water harvesting and recycling technologies for supplemental irrigation and drought proofing of rainfed crops.
- ✰ Enhance the resilience of rainfed agriculture to climate change by developing adaptation and mitigation strategies through network research.
- ✰ Undertake action research to evolve innovations in technology dissemination and up-scaling for enhancing livelihood security.
- ✰ Carry out impact assessment of rainfed agriculture technologies and suggest policy reforms that lead to better technology adoption in rainfed agriculture.
- ✰ Develop linkages and collaborate with national and international agencies in advancing rainfed agriculture.

References

Anonymous 2006. Samaj Pragati Sahayog, Food Security in 2020: Drylands are the Answer, Bagli.

Anonymous 2011. Dryland Farming and Dryland Farming in India http: // www.world-agriculture.com

Campbell H 1907. Campbell's 1907 soil culture manual; a complete guide to scientific agriculture as adapted to the semi-arid regions. Lincoln, Nebr. The Campbell soil culture co. (inc.) Publisher.

Chandrasekharan B, Pandian BJ 2009. Rainwater harvesting and water saving technologies. Indian J Agron 54 (1) 90-97.

Creswell R, Martin FW 1998. Dryland farming: Crops and Techniques for Arid Regions. ECHO, 17391 Durrance Rd., North Ft. Myers FL 33917, USA.

Hegde NG 1995. Sustainable Livelihood through Watershed Development. Yojana (Marathi) Apr-Jun: pp 30-31.

Kanwar JS 1982. Rainwater and Dryland Agriculture-An Overview. Symposium on Rainwater and Dryland Agriculture. Indian National Science Academy, New Delhi, pp 1-9.

Reddy SR 2011. Principles of Agronomy. Kalyani Publishers, New Delhi, pp 236-283.

Reddy TY, Reddy GH 2012. Principle of Agronomy. Kalyani Publishers, New Delhi, pp 337-383.

Singh S, Rathore MS 2010. Rainfed Agriculture in India: Perspectives and Challenges. Rawat Publications, Jaipur pp 125-156.

Venkateswarlu B 2011. CRIDA Vision 2030, Central Research Institute for Dryland Agriculture, Santoshnagar, Hyderabad, India pp 31.

2014, Sustainable Rural Development through Agriculture *Pages 81–99*
Editors: **Dr. Shobhana Gupta and Dr. S.S. Tomar**
Published by: **BIOTECH BOOKS, NEW DELHI**

Chapter 5

Shifting Cultivation: A Perspectives in North Eastern Hilly States

Rakesh Kumar, U.N. Shukla and Narendra Kumawat

ABSTRACT

The existing practice of shifting cultivation in north eastern states of India is an extravagant and unscientific form of land use. The effects of shifting cultivation are devastating and far-reaching in degrading the environment and ecology of these regions. The earlier 15-20 year cycle of shifting cultivation on particular land has reduced to 2-3 years now. This has resulted in large-scale deforestation, soil and nutrient loss, and invasion by weeds and other species. The indigenous bio-diversity has been affected to a large extent. Thus, reduction in the cycles of jhuming, adversely affects recovery of soil fertility and nutrient conservation by the ecosystem and also repeated short-cycle jhuming has created forest-canopy gaps which are evident from the barren hills.

1.1 Introduction

Jhum/shifting cultivation is considered as one of the oldest practice in north eastern hill region of India. Almost 85 per cent of the total cultivation in northeast India is occupied by shifting cultivation (Singh and Singh, 1992). It is the form of agriculture in which a piece of forest land is slashed, burnt and cropped without tilling soil, and cropped land is subsequently fallowed to attain pre-slashed forest status through natural succession (Ramakrishnan, 1993). In this practice, first a forest area is chosen mostly at the top of the hills and all big trees and other vegetations are cut down in the month of January. After that, leftover dried twigs of the plants are burned in the month of April. Then with the onset of monsoon when the first rainfall received seeds of cereals, vegetables, oil seeds *etc* are sown on *jhum* land. This cultivation is continued for one-two cropping season(s) only. Then the site is left and the same site is chosen after approximately 7-8 years. However, the existing shifting cultivation practice in north eastern India is an injudicious form of land use. This has resulted in large-scale deforestation, soil and nutrient loss and losing of soil and forest bio-diversity which lead to the environment and ecology degradation. Arunachalam (2002) reported that soil organic carbon, available-P, total nitrogen, ammonium-N and nitrate-N decreased as duration of cultivation increased under *jhum*. However, microbial biomass C, N, and P were high in the forest stand. Microbial biomass carbon increased gradually as cultivation progressed, while microbial biomass N and P showed a post-burn decreasing trend. Bacterial and fungal populations drastically reduced following slash burning. Over past two decades, due to increasing human population, *jhuming* cycle in the same land, which extended to 20-30 yr in older days has now been reduced to 3-6 yr (Arunachalam, 2002). The earlier duration of shifting cultivation cycle on a particular land has reduced due to huge population pressure. Now, a stage has come that it has already affected 2.7 million ha of land, and each year 0.45 ha of land fall under shifting cultivation in northeast India (Ranjan and Updhyay, 1999). Presently, it is estimated that number of people practicing shifting cultivation are around 367,000 tribal families and the area affected by this practice is 385,400 ha annually (Patiram and Verma, 2001). Loss of 100-250 metric tonnes of top soil per ha per year are depleted due to *jhum* cultivation in Bangladesh hills (Karim and Mansor, 2011). On an average, an area of 3,869 km^2 is put under shifting cultivation every year. Excessive deforestation (net loss of 1,577 km^2 forest cover between 1999 and 2001 (Government of India, 2001) coupled with shifting cultivation practices have resulted in tremendous soil loss of 200 t ha^{-1} yr^{-1} (Prasad, 1987), poor soil physical health, and an uncountable loss of bio-diversity. Studies about soil fertility status under different *jhum* cycles at various elevations suggest that plots with short *jhum* cycles (5 yr) generally have lower fertility than those with longer cycles of 10 or more years (Mishra and Ramakrishnan 1983). Saha *et al.* (2011) studied soil erodibility characteristics under 6 land-use systems *i.e.*, agriculture, agrihorti-silvi-pastoral, natural forest, livestock-based land use, natural fallow, and shifting cultivation and observed that shifting cultivation showed the highest erosion ratio (12.46), followed by agriculture (10.42) indicating the need to adopt tree-based land-use systems for resource conservation. They also reported that soil loss was significantly higher in shifting cultivation (30.2–170.2 t ha^{-1} yr^{-1}), agriculture (5.10–68.20 t ha^{-1} yr^{-1}) and the livestock-based land-use systems (0.88–14.28 t ha^{-1} yr^{-1}) as compared to other modified land-use systems.

Since a huge number of tribal farmers are involved in this cultivation. Therefore, complete eradication of this method is practically impossible. Thus only two ways are left to check the damage of the environment either by increasing the *jhum* cycle and or take some immediate reclamation process to areas already affected by *jhum* through wasteland development plan. Arunachalam and Arunachalam (2002) suggested utilization of bamboos in eco-restoration of *'jhum'* fallows in Arunachal Pradesh. Soil pH increased after burning and decreased as cultivation progressed in the *jhum* field. In different parts of north-eastern India, land is oft abandoned after first year of *jhum* cropping and second year cropping is sometimes practiced with plantations of banana and pineapple (Kushwaha and Ramakrishnan 1987). Rathore and Bhatt (2008) observed that rice-vegetable pea-beans cropping system was most suitable under *jhum* land of Nagaland and integration of fish, pig, dairy cattle, duck, and the crops such as rice, vegetable pea and beans showed maximum system productivity, *i.e.*, 126.5 t ha^{-1} (rice-equivalent yield), followed by cultivation of rice, vegetable pea and beans along with dairy cattle (free grazing), having system productivity of 105 t ha^{-1} of rice-equivalent yield. Tawnenga and Tripathi (1996) hypothesized that if second year cropping on *jhum* fields is essentially introduced, the dependence of the shifting cultivators on the forest could be reduced to almost one-half. They suggested that a combination of inorganic and organic manuring is more suitable to improve economic yield during second year cropping. According to the findings of the Central Forestry Commission of India in 1984, 6.7 m ha land of cultivable neighborhood was affected by *jhum* in the country. Population explosion and emergence of new generation of youth cultivators encouraged increasing demand for cultivable land which resulted reduction of the cycle of cultivation from 25-30 years to 2-3 years due to abandoning and re-occupying of fallow land frequently. Fallow cycle of 20-30 year prevalent during earlier period helps the land to return to its natural condition after the anthropogenic disturbances.

1.2 Shifting Cultivation and its Crop Scenario

Almost 80 per cent of the states in the northeast *viz.* Manipur, Tripura, Sikkim, Arunachal Pradesh, Nagaland, Meghalaya except Assam have a large chunk of areas under hilly tracts or terrain. These regions are mainly cultivated with only few cereal crops like rice, maize etc. and oilseed crops like rapeseed and mustard, soybean etc. cash crops like potato, sugarcane, etc. and a number of vegetables. There is imbalance proportion of area under crops in *rabi* and *kharif* seasons. Lack of irrigation facilities, cattle menace and sparsely distributed rainfall during winter season are the main reasons why all the available land used during *kharif*/summer seasons are not fully utilized during *rabi*. Upland rice in *jhums* through shifting cultivation systems inter-cropped/mix-cropped with maize, pulses etc. is the main crops in hills. There are restrictions of modern or scientific cultivation of every crop due to natural topography or geographical situations in the hill slopes. It was noted that loss in forest cover in the northeastern states was mainly due to the shifting cultivation. In the State of Assam, other factors have also been shown to contribute to reduction in the forest cover. From 1993 to 1995 and 1995 to 1997, loss in forest cover was, respectively, 783 sq km and 316 sq km. According to the 1995 and 1997 reports, although 1078 sq km and 1700 sq km areas were gained from the shifting cultivation, they constituted only scurby vegetation. Nevertheless, these growths can also help in checking soil erosion from the hilly slopes which are catchment of number of streams and rivers of the region. But due to reduction of cycle to 2-3 years, the resilience of ecosystem is interrupted and the quality of the land is get worsening day by day.

A recent study based on satellite data carried out by ICAR, area under shifting cultivation is 4.37 million hectare which constitutes nearly 5.02 million families (Table 5.1). Odisha state (2.0 mha and 1.60 million families) occupied at first position on acreage and family engaged in shifting cultivation. Among 7 eastern states, Nagaland (23 per cent) top in shifting cultivation followed by Mizoram (22 per cent) and Manipur (21 per cent) on percent basis. The practice of shifting cultivation was not harmful or considered rather useful during the time when it was started. Least disturbance to soil, natural fertility build up of soil, mixed cropping on slopes and totally depends on rainy season and local resources, were some of its merits. When system emerged there was no population pressure and cycle of rotation was 10 to 20 years or above, thus leaving enough time for the soil to revive. But, now restorer period has about 5-10 year due to heavily demographic pressure on land and resources utilization.

Table 5.1: Shifting Cultivation in different States of India

State	*Tribal Families (Million)*	*Total Area (Million hectare)*
Andhra Pradesh	0.11	0.15
Arunachal Pradesh	0.43	0.21
Assam	0.31	0.31
Bihar	0.23	0.19
Madhya Pradesh	0.19	0.38
Manipur	0.36	0.26
Meghalaya	0.61	0.47
Mizoram	0.40	0.19
Nagaland	0.19	0.12
Odisha	2.00	1.60
Tripura	0.19	0.49
Total	**5.02**	**4.37**

Source: Shifting cultivation in India, ICAR.

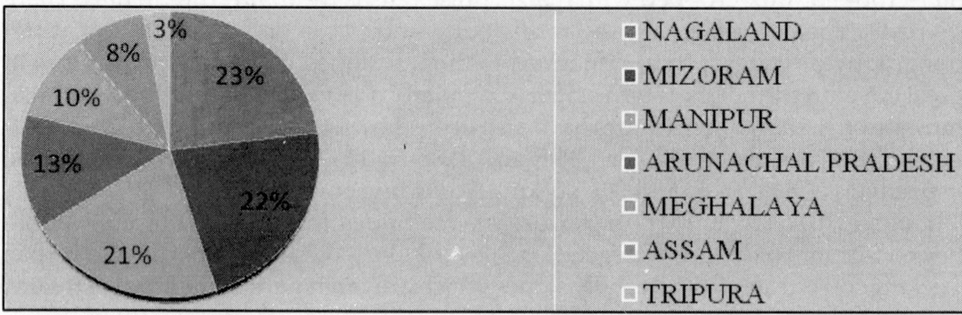

Chart 5.1: Total Area Under Shifting Cultivation in NE India.

Figure 5.1: Slash and Burn Practice.

Figure 5.2: Cultivation Practices.

Figure 5.3: Harvesting Process.

Figure 5.4: Restorer Period.

Figures 5.1–5.4: Step in Shifting Cultivation.

1.3 Soil Composition Under Shifting Cultivation

Shifting cultivation or *Jhum*/slash and burn method of crop cultivation is the system where farmers shift their fields after taking one crop in a single season. The field after harvest is abandoned for at least 5-10 years during which field regains its fertility due to various factors. The main factors for fertility restoration during this period are diverse. The leaves, twigs and other plant parts falling to soil get decomposed which later acts as good feed to numbers of beneficial micro-organism thriving in soil under a congenial soil environment. The earthworms regarded as "**friends of the farmers**" are the major role players in regaining natural soil fertility by enriching the soil with their castings. Bird litters, animal droppings, biotic activities of flora and fauna, decomposed plant debris acting as natural mulch contribute to soil health building. Hence, soil composition with fair amounts of major nutrients like nitrogen, phosphorus and potash besides, creating a good soil structure makes the soil virgin in a cycle of 10 years period.

1.4 Classification of Shifting Agriculture

Tiwari *et al.* (1998) classifies shifting agriculture into four types: traditional, distorted, innovated and modified. This classification aids in understanding how *jhum* has changed over the years and what is going wrong with newer practices of *jhum*.

1.4.1 Traditional Shifting Agriculture

These types of shifting cultivation are mostly found in villages which have not experienced much pressure of population increase. It is sustainable but may not fulfill all the needs of a modern society. It helps conserve forests as the pattern observes a long fallow period followed by a short cropping phase.

1.4.2 Distorted Shifting Agriculture

In land scarce villages for reasons such as population increase, fallow periods are reduced and shifting agriculture has got distorted. Elsewhere *jhum* has spread to lands with steep slopes due to non-availability of lands with lesser degrees of slopes. In Mizoram at places less than 1000 m altitude, distorted shifting agriculture has converted much of subtropical evergreen and semi evergreen forests.

1.4.3 Innovated Shifting Agriculture

The farmers have switched to newer methods of cultivation which are more adapted to the current availability of resources (for example green biomass, technology and labour) and present day societal requirements (for example cash). In some cases they have also introduced new crops (for example kollar beans in Nagaland, pisum in Meghalaya) which help maintain fertility and also have good market.

1.4.4 Modified Shifting Agriculture

This was introduced during past decade with implementation of two development projects. These are Nagaland Environmental Protection and Economic Development (NEPED) in Nagaland, and North Eastern Region Community Resource Management Project (NERCORMP) in Meghalaya, Manipur, and hill districts of

Assam. These projects have demonstrated that through multi-pronged external intervention, the productivity of shifting agriculture can be enhanced.

From the above classification that emphasis should be on controlling distortions or retrogressive developments rather than on controlling shifting agriculture itself.

1.5 Harmful Effects of Shifting Cultivation

Slashing or felling down of trees, herbs and shrubs in *Jhum* cultivation reduce the oxygen generation, burning down of the sun-dried vegetation pollute air with carbon-monoxide, nitrous oxides and many other harmful gases. Various emissions from burning the fossil fuels including dry vegetation contribute to depletion of ozone layer (a layer in the troposphere which protects ultraviolet rays from the sun. Not only these, destruction of forest trees, flora and fauna, animals, birds and all other living beings in the forest are against the law of the Forest, Environment and Wildlife Department/Ministry of Govt. Destruction of natural habitats of these living organisms that brings ecological imbalance in the ecosystem is also forbidden by the forest laws and acts. Soil erosion and top soil degradation is the direct outcome of the clearing the natural vegetation. However, system of field operations after sowing of crops involves no use of synthetic chemicals. Cultivation under *Jhum* is done with "no tillage" or "zero tillage" where heavy machineries are not employed causing air pollution. Synthetic fertilizer, weedicides, insecticides, fungicides and other plant protection chemicals are not at all used. The *Jhum* cultivation is hence eco-friendly with less cost of cultivation but comparatively low in yield or productivity.

1.6 Abolishing Shifting Cultivation

1.6.1 Is it Feasible?

The destruction of natural vegetation in *Jhum* cultivation is alarmingly on the rise. The natural topography compels people in the hills of the Northeast India to adopt the age-old traditional cultivation which is the only means to sustain living. The risks before the *Jhumias* are to be considered before taking a step to ban it. The Govt. of India's policy to abolish the *jhum* cultivation and replacing it with growing of horticultural crops is in one sense, a good gesture towards economic development of people. However, change of one's occupation over night is something which is not feasible on the part of hill farmers. Provisions of funds and food to feed farmers and their families till reaping fruits of the newly adopted horticultural crops cultivation in lieu of *jhum* cultivation would only motivate farmers to switch over to the new methods. This is the reason why many farmers are not willing to change their profession of cultivation in the hills of Northeast. Hence, swift change over from shifting cultivation to any other forms of cultivation under the present conditions would be something impracticable for the poor farmers.

1.6.2 Role of the Government

The Government should be trust-worthy to the farmers. The scientific systems of cultivation should be imparted to the farmers and fulfill the promises which were laid before the farmers. Special incentives should be provided to the farmers of *Jhums* to change their cultivation to a scientific method. Crop insurance should be implemented in toto. Food stock and essential items should be available to the farmers who adopt Government's policy of new Agricultural technologies. The farmers should be compensated properly against any natural calamity. Replacing shifting cultivation and adopting an alternative cultivation system would then be a success. The current practice of shifting cultivation in eastern and north eastern regions of India is an extravagant and unscientific form of land use. The evil effects of shifting cultivation are devastating and far-reaching in degrading the environment and ecology of these regions. The earlier 15-20 year cycle of shifting cultivation on a particular land has reduced to 2-3 years now. This has resulted in large-scale deforestation, soil and nutrient loss, and invasion by weeds and other species. The indigenous biodiversity has been affected to a large extent. The increase in human population, particularly in developing countries, has put tremendous pressures on land. The extension of crop lands for increasing food production has been directly responsible for the reduction in areas under forests and grass lands. According to one estimate, about 40 per cent of the land surface of the earth was converted into crop lands and permanent pastures by early 1990s. More than 6 per cent area under tropical forests was converted to shifting cultivation between 1980 and 1990 across all tropical countries. On abandoned land, natural regeneration starts from available root stocks and seed bank. Bamboo comes up naturally; and *kendu*, *mahua*, *Terminalia* along with certain other climbers also regenerate. Generally, this land is not cultivated for the next 10 years. During this period, tribals use this land to collect root suckers which are used as eatables; *Mahua* and *Eleocarpus* (Salpa) trees are used for the preparation of liquor for their consumption. With reduction in *jhum* cycle from 20-30 years to 2–3 years, the land under shifting cultivation looses its nutrients and the top soil.

1.7 Impact of Shifting Cultivation

Frequent shifting from one land to the other has affected the ecology of these regions. The area under natural forest has declined fragmentation of habitat, local disappearance of native species and invasion by exotic weeds and other plants are some of other ecological consequences of shifting agriculture.

The shifting cultivation is generally practiced in the following sequence:

- ✰ Selecting a forest patch and clear fell the vegetation normally in December and January
- ✰ Burning of the vegetation and small, cut-trunks portion and roots are normally not removed.
- ✰ Herbs, shrubs and twigs and branches (slashed vegetation) are burnt in February and March
- ✰ Sowing of seeds by dibbling generally of cereals, vegetables and oil seeds in April-May
- ✰ Continuing cultivation for a few years
- ✰ Abandoning cultivated site and shifting to other forest sites
- ✰ Returning to the former site, and once again practice shifting cultivation on it.

- Area having *jhum* cycle of 5 and 10 years is more vulnerable to weed invasion as compared to *jhum* cycle of 15 years.
- The area with fifteen-year *jhum* cycle has more soil nutrients, larger number of species, and higher agronomic yield with ratio of energy output to input as 25.6 compared to *jhum* cycle of 10 and 5 years (4.6–9.8).
- Similarly, while studying *jhum* ecology in Meghalaya, it was reported that water and nutrient losses in shifting-cultivation areas were far greater than in the virgin areas, and areas left for 50 years after *jhuming*.
- Thus, reduction in the cycles of *jhuming*, adversely affects the recovery of soil fertility, and nutrient conservation by the ecosystem. Repeated short-cycle *jhuming* has created forest-canopy gaps which are evident from the barren hills.

1.8 Strategies for Minimizing the Shifting Cultivation

1.8.1 Initiatives

Although shifting cultivation is a non-viable resource-utilization practice, tribals are still clinging to this primitive practice to sustain themselves and their families mainly due to non-availability of timely employment avenues. Various attempts have been made by the Government to settle the tribal's involved in shifting cultivation.

- Arable land is provided to the tribals for carrying out agriculture and also to settle in the area; a few schemes are being implemented under integrated tribal development programme in the districts of Koraput, Keonjhar and Phulbani in Orissa. These schemes have however, not yielded the desired results perhaps because of the ignorance of the authorities about socio-economic and agro-ecological conditions of shifting cultivation and also due to minimal involvement of Forest Department officials, who are more informed about the above factors, in implementation of the scheme.
- An Agro-forestry project known as Nagaland Environment Protection for Economic Development (NEPED) funded by Canadian International Development Agency (CIDA) through India-Canada Environment Facility (ICEF) was initiated in 1995 to make Nagaland self-sufficient in agro-forestry. Hence project does not aim at eliminating *jhum* cultivation, but making it more stable and profitable. This novel project may give us a more scientific way to tackle tribal-forest conflict.

1.8.2 Strategies

- Providing employment opportunities and income generation on a regular basis through proper utilization of land resources, *i.e.* by equitable distribution of waste land among the tribals. But, various schemes of Government, under the tribal plan, will have to pump in sufficient resources for proper reclamation and development of the wasteland through agro-forestry and silvi-pasture.

- By encouraging cooperative efforts for carrying out forest-based activities *i.e.* basket making, rope making, cane furniture processing of minor forest produce, honey collection, etc. have to be made commercially viable by providing proper marketing facilities. This will not only discourage tribals from practising shifting cultivation but will also help them monetarily.
- By forming Village Forest Committees for protection and development of degraded forests. These committees by providing suitable incentives to tribals, after the time of harvest can divert some of the tribals away from the shifting cultivation. Employing the tribals for collection of *kendu* leaves and *sal* seeds and also involving the tribals in the various rural employment schemes is also the need of the hour.
- By ensuring implementation of total literacy campaign; which due to remoteness and un-supportive attitude of tribals, has not been successful? For educating tribal women and children, services of various non-Governmental organizations and voluntary agencies, besides the regular Government machinery, are on required sustainable basis rather than with a targeted approach.

1.8.3 Eco-development Plan for Areas under Shifting Cultivation

Ecologically, these regions are far worse than realized. Apart from losing vegetation and bio-mass due to practice of shifting cultivation, many other ecological factors too have been affected. Due to shifting cultivation practice on slopes, down-stream siltation of water bodies is apparent in many districts. Protection and repair of drainage basins for conservation of ecological resources including water need large amounts of financial input. The shifting cultivation areas normally receive moderate to high rainfall. Due to splash forces generated from the rain drops, erosion of precious top soil occurs. Thus, major factors which influence the rate of soil erosion are the rainfall, the topography of the terrain, and the kind of vegetation and soil conditions. The mountain eco-systems of these regions with shifting cultivation practice; have to be made ecologically sustainable. Formulating an eco-development plan for the region for environmental sustainability, could consider completely replacing agricultural practice with farm forestry. Agricultural practices are at the cost of loss of biodiversity resources; estimates indicate that one unit of energy in agronomic production costs loss of greater energy from the forests. However, in Central Himalayan eco-systems, where agriculture practice is more scientific compared to shifting cultivation, one unit of energy in agronomic production entails an expenditure of about 10-12 units of energy from the surrounding forests as firewood, fodder and leaf manure. Loss of energy from forests per unit of agricultural production may be far greater in shifting-cultivation areas.

- The protective values of trees are far greater than those of annual crops
- Unlike annual crops which require frequent ploughing, tree plantations cause minimal soil disturbance
- Net above-ground primary productivity of forests is notably greater than that of agricultural crops and grassland

- ☆ By providing fodder and firewood, farm forestry would create a favourable situation for revival of the natural forests.
- ☆ A revived forest cover could offer several other advantages.

Determining the population-supporting capacity (PSC) may be one of the major aspects for checking the degradation of environment and depletion of resources (Tawnenga and Tripathi, 1996). The PSC should be based on:

- ☆ Land and resource management, in order to restore and improve sustainable yields
- ☆ Prevailing standards of living of the population, in relation to sustainable yields of subsistence materials and changes in technology, innovation and Government policy
- ☆ Flow of trade for food items, rather than for timber, fuel wood and fodder; with emphasis on information on transportation system, administration, natural disasters, regional conflicts in relation to import and export
- ☆ The state of ecological balance and level of exploitation of natural resources.

The forests surrounding a hill village is considered as 'support area' which provide firewood, fodder, timber, water and animal bedding to the people. The tribal population mainly depends on renewable resources and is not willing to move out of its natural habitat. The PSC study will provide a direction for wise allocation of resources, and change in existing practice of cultivation to achieve higher yields. While examining six models for a hill district of Central Himalayas, it was observed that collection of revenue from forest resources and milk products for purchasing food grains is more economical than cultivating same land for food-grains. Per unit produce from forest is much higher than agriculture. In the hills of eastern and northeastern region the above findings can be applied, as production from these areas under shifting cultivation cannot be increased even by applying modern agricultural practices. Inputs like irrigation facilities, good seeds, and fertilizers may prove to be counter-productive due to slope, topography, elevation and other environmental constraints. In case this eco-development plant of farm forestry is taken up, and tribals discontinue practice of shifting cultivation, daily livelihood and income generation avenues for tribals could be met using earlier-discussed four strategies.

1.9 Main Reasons Behind Continuing Jhum Cultivation

1.9.1 Adjustment Problem with Non-tribes in the Settled Area

Since tribal are very much fond of God hence they faced difficulty for building worship place of a particular religion when there is mixed population. The tribal of Nagaland are Christians and there is also other religious group where they live.

1.9.2 Lack of Sufficient Attraction towards their Colonial Home

1. Colonies are not set according to religion and culture of tribes.
2. Rehabilitation of the tribal family is far away from their original habitat

3. Lack of social environment and freedom in rehabilitation area.
4. Lack of special training for plan land cultivation.

1.9.3 Financial Problem

Government schemes are not enough to make them financially reliable to stay in the new colony for initial years. According to rehabilitation schemes in the year 1953-54, each family was allotted 5 acres of arable land over and above a grant of Rs. 500/- for purchasing essential requirements to support cultivation. Out of this cultivation later on rubber plantation was one of significant cultivation which was provided to them. But it took at least 7-8 years to turn into a mature productive plant to provide earning. But they were neither economically strong nor skilled enough to go for an alternative source of earning in that period.

1.9.4 Lack of Proper Education

Available data on dropouts of students in the state revealed existence of educational wastage both at State and national level. Dropout rate at the primary stage was estimated to be significantly higher in schedule tribes than general category of pupils and scheduled caste. Same trends were also observed in the middle and secondary stages. This proves the unsteady as well as pathetic conditions of the tribal communities in India.

1.10 Constraints

1.10.1 Agricultural Development under *Jhumias*

1.10.1.1 Uneconomic and Fragmented Holdings

The marginalization of farmers is a dominant factor adversely affecting household income. Over 60.2 per cent of the operational holdings are below 1.0 ha and 22.1 per cent of the holdings are in the farm size group 1-2 ha. Except Nagaland, in other N.E.states, the size of operational holding is very small. Such small holdings are uneconomic and results in under-investment in agriculture leading to low input use and low production.

1.10.1.2 Low Adoption of Improved Technology

The adoption of high-yielding varieties of rice varies between 42 to 50 per cent in Meghalaya, Manipur and Assam and 23 to 34 per cent in Arunachal Pradesh, Nagaland and Mizoram. Highest adoption of HYV is observed in Tripura at 96.7 per cent. Although more than 85 per cent area of wheat is under HYV but covers only around 2 per cent of gross cropped area. The main causes of slow growth of HYV area are non-availability of suitable seeds, predominance of traditional seeds in hill areas under Jhum cultivation, short supply of recommended seeds and defective distribution system.

1.10.1.3 Fertilizer Consumption is Extremely Low and Variable

Fertilizer use is the lowest at 2.29 kg/ha in Arunachal Pradesh and the highest in Manipur at 72.46 kg/ha. In Meghalaya, the fertilizer use is declining from 15.55 kg per ha in 1986-87 to 13.39 kg in 1997-98, the same is stagnant at around 3 kg in

Nagaland. There are number of factors limiting the expansion of fertilizer use such as defective distribution system, poor transport and communication system and inadequate institutional credit.

1.10.1.4 Irrigation is Another Crucial Constraint to Agricultural Production

Which, at present covers less than 10 per cent of the gross cropped area. It is ironical that the region with annual average rainfall as high as 3,400 mm, also faces water problems. It implies that development of water resource is extremely poor and the state support is weak. For example, on account of inefficient management system, about 109 state-owned watershed projects in Assam are currently unused, making the investment infractuous. Under the circumstance, people's initiative is required in proper rainwater harvesting and redistribution in the dry season crops.

1.10.2 Technological Constraints

- ✰ Lack of suitable high-yielding rice varieties for diverse upland situations, flood affected areas, moisture stress conditions, and hill areas
- ✰ Alternative crops for escaping pre-monsoon showers to avoid the problem of pre-harvest sprouting of crop in flood, free period
- ✰ Develop improved crop management practices for shifting cultivation
- ✰ Improvement and standardization of production techniques of fruits and vegetable crops
- ✰ Use of improved post harvest management including pest and disease management and processing techniques for the major cash crops and horticultural crops
- ✰ Land and water management technique specifically for acid soils
- ✰ Economic packages for integrated farming systems combining crop cultivation with livestock, fishery, etc.
- ✰ Integrated livestock management system for increased livestock products as well as draught power. The facilities of storage, processing and marketing are particularly deficient for perishable commodities.

1.10.3 Rural Transport and Communication Network

Transport and communication facilities also need marked improvement in the difficult terrains of hills, dense forest, rivers, etc. At present, availability of road per 100 sq. km is low at 45 km against 62.8 km at all-India average. Most of rural areas remain inaccessible in rainy season for non-availability of all-weather roads.

1.10.4 Attitude towards Agriculture

Agriculture is the most preferred activity of a large section of the population in the region, yet the method of cultivation is indigenous. In view of the rapid technology turn over, appropriate strategy for HRD is required to maximize farm income through adoption of cutting-edge modern technology. The farmers usually stick to old practices and younger generation distracts, which makes agriculture an occupation of elderly people living in rural areas. To attract enterprising youths to take up farming as profitable occupation and to reverse out-migration, innovative strategy such as commercialization of agriculture and adoption of improved methods must be promoted.

1.10.5 Socio-economic Constraints

The NE India has diverse ethnic groups and social systems bound with customs and traditions. These factors clearly differentiate type of economic activities and economic status of the population, which inhibit the adoption of modern methods. Carefully prioritized strategies for agricultural development of region may become instrumental to break deadlock.

1.11 Prioritization and Policy Perspectives

1. Development of irrigation facilities and promotion of water-harvesting methods for assured water supply particularly in *rabi* season.
2. Creation of "single window" input delivery system in the rural areas to ensure timely supply
3. Surveillance of major pests and diseases and adoption of timely control measures
4. User-friendly information system through improved method such as on-farm trials, demonstration, training, farmer-participatory interaction programme along with programmes for updating knowledge and skill of field level officers
5. To identify need-based programmes for overcoming technology gaps through extensive field surveys under diverse agro ecological and socio-economic situations
6. Programmes to promote development of cash crops like jute, sugarcane, maize, horticultural crops, etc.
7. Creating storage facilities particularly cold storage for perishable commodities
8. Introduction of value-addition to agricultural produce through research and development activities.
9. Agricultural development programmes must take care of cropping pattern in pre and post-flood situations particularly in flood-affected areas including development of allied agricultural activities
10. Improved crop cultivation practices for *Jhumming* in hill areas
11. Financial support for creation of agricultural infrastructure such as strengthening irrigation facilities, farm machineries, processing and storage facilities, rural roads and communication
12. Marketing infrastructure to be created at the primary markets in rural areas and regulated markets in district level. The dominance of traders and middlemen to be reduced so that farmers is a powerful force for distress sale.

1.12 The Strategies

- ☆ Providing employment opportunities and income generation on a regular basis through proper utilization of the land resources, *i.e.* by equitable distribution of waste land among the tribals. But, various schemes of Government, under the tribal plan, will have to pump in sufficient resources for proper reclamation and development of the wasteland through agro-forestry and silvi-pasture practices.
- ☆ By encouraging cooperative efforts for carrying out forest-based activities, *i.e.* basket making, rope making, cane furniture processing of minor forest produce, honey collection, etc. have to be made commercially viable by providing proper marketing facilities. This will not only discourage tribals from practising shifting cultivation but will also help them monetarily.
- ☆ By forming Village Forest Committees for the protection and development of the degraded forests. These committies by providing suitable incentives to the tribals, after the time of harvest can divert some of the tribals away from shifting cultivation. Generating employment opportunities in lean season of forestry operations will also prevent tribals from shifting to other areas. Employing the tribals for collection of *kendu* leaves and *sal* seeds and also involving tribals in various rural employment schemes is also the need of the hour.
- ☆ By ensuring implementation of total literacy campaign; which due to remoteness and un-supportive attitude of tribals has not been successful? For educating tribal women and children, services of various non-Governmental organizations and voluntary agencies, besides the regular Government machinery are on required sustainable basis rather than with a targeted approach.

1.13 Tips to *Jhumias*

- ☆ Farmers in the hill region should be aware of new technologies emerging from time to time.
- ☆ Uneconomical practices of hill agriculture should be avoided while selecting the crops.
- ☆ Adequate cropping systems should be adopted to boost their income and maintain a sustainable self sufficiency in the household.
- ☆ Use of certain synthetic plant food to replenish vanishing food reserve in the soil should be adopted but with an adequate quantity or dose.
- ☆ Modern systems of plant protection and care should be given priority to harvest a bumper yield.
- ☆ Continuous cultivation of the once selected spot can be cultivated/tilled year after year with such amelioration of the soil.
- ☆ Soil amendments should be done where nutrient reserve is limited due to continuous cultivation through organic matters, mulching etc.

- ✩ The expertise of concerned authorities/personnel should be gained to tap potential of region.

1.14 Conclusion

It is apparent that *jhum* cultivation has its adverse affect on the species diversity of a region as this unscientific form of agricultural practice continuously degrading quantity and quality of the natural habitats of various floras and faunas. Destruction of the natural habitats of living organisms that brings ecological imbalance in the ecosystem is also forbidden by the forest laws and acts. But still good numbers of new class of shifting cultivators are practicing the shifting cultivation throughout the northeastern states unaware of those facts. This is the time to bring our hand together to help people to be educationally sound enough to understand existing ecological hazards as well as to create a strong awareness about the deleterious effects of shifting cultivation among all tribes and non-tribes of the North East region to restore the ecosystem of this potential hot spot region.

Shifting agriculture sums up the issues as follows:

- ✩ A meaningful solution to the problem of *jhum* has become critical, not only from the point of biodiversity conservation but also for productive agriculture in the region. The solution to problem of the shortening fallow cycle lies in strengthening this already weakened agro-forestry system, through fallow management using appropriate tree species, rather than imposing an alien technology from outside the region. Nepalese Alder and bamboos are examples amongst many others to be selected through intense community participation.
- ✩ Scientifically analysed traditional ecological knowledge available with local communities is a powerful tool to redevelop *jhum* as part of a short-term strategy and as a component of a more comprehensive landscape management plan.
- ✩ A variety of other opportunities exist to redevelop land use systems other than *jhum* such as valley wet rice cultivation, through appropriate designed traditional water harvesting and soil fertility management technologies. The strengthening of this land use could be also seen as a means to take away the pressure from *jhum* itself to the extent possible

References

Arunachalam A 2002. Dynamics of soil nutrients and microbial biomass during first year cropping in an 8-year *jhum* cycle. *Nutrient Cycling in Agroecosystems,* 64: 283–291.

Arunachalam A and Arunachalam K 2002. Evaluation of bamboos in eco-restoration of '*jhum*' fallows in Arunachal Pradesh: ground vegetation, soil and microbial biomass. Forest Ecology and Management, 159: 231–239.

Borthakur DN 1992. Agriculture of the north eastern region with special reference to hill agriculture. Guwahati, India: Beecee Prakashan.

Karim SMR and Mansor M 2011. Impact of *jhum* cultivation on agro ecology of mountains and socio economy of tribal peoples. *Asian Journal of Agricultural research.* 5: 109-114.

Kushwaha SPS and Ramakrishnan ÑS 1987. An analysis of some agro-ecosystem types of north-eastern India. *Proceedings of the Indian National Science Academy*, B53, pp. 161-168.

Mishra BK, and Ramakrishnan PS 1983. Slash and burn agriculture at higher elevations in north-eastern India. II. Soil fertility changes. *Agriculture, Ecosystem and Environment*, 9: 83–96.

Patiram and Verma ND 2001. Reformed system to increase the productivity of shifting cultivation in north-eastern hills. *Journal of North Eastern Council*, 21: 24–32.

Prasad RN 1987. Degradation of soil and water resources and crop production in NEH Region. *Indian Journal of Hill Farming*, 1: 1–8.

Ramakrishnan PS and Toky OP 1981. Soil nutrient status of hill agro-ecosystems and recovery pattern after slash and burn agriculture (*jhum*) in north-eastern India. *Plant and Soil*, 60: 41–64.

Ramakrishnan ÑS 1993. Shifting agriculture and sustainable development: an interdisciplinary study from north-eastern India, Unesco and Oxford University Press, pp. 424.

Ranjan R and Upadhyay VP 1999. Ecological problems due to shifting cultivation.*Current Science*, 77(10): 1246-1250.

Rathore SS and Bhatt BP 2008. Productivity improvement in *jhum* fields through integrated farming system. *Indian Journal of Agronomy*, 53: 167-171.

Saha R, Mishra VK and Khan SK 2011. Soil erodibility characteristics under modified land-use systems as against shifting cultivation in hilly ecosystems of Meghalaya, India. *Journal of Sustainable Forestry*, 30: 301–312.

Singh JS and Singh SP 1992. Forests of Himalaya, Gyanodaya Prakashan, Nainital, p. 294.

Tawnenga US and Tripathi RS 1996. Evaluating second year cropping on *jhum* fallows in Mizoram, north-eastern India-Phytomass dynamics and primary productivity. *Journal of Biosciences*, 21: 563-575.

Uhl C, Jordan CF and Montagnini Å 1983. Traditional and innovative approaches to agriculture on Amazon basin terra firma sites. *In:* Lowrance, R. R., Todd, R. L. Asmussen L. Å. and Leonard, R.A. (eds) *Nutrient cycling in agricultural ecosystems*, Athens: University of Georgia, College of Agriculture Experimental Station, special publication 23, pp 73-95.

Borthakur DN, Singh A, Awasthi RP and Rai RN 1980. Proceedings of National Seminar on Resources, Development and Environment in Himalayan Region, D. S. J., New Delhi, pp. 330–342.

Singh J and Ramakrishnan PS 1982. *Proc. Indian Acad. Sci.*, 1982, B91, 269–280.

Ramakrishnan PS, Shankar UT and RS Tripathi 1996. Evaluating second year cropping on jhum fallows in Mizoram, North-eastern India – phytomass dynamics and primary productivity. Journal of Biosciences 21: 563-575.

Ministry of Environment and Forests, 1987, The State of the Forest Report – 1987. Forest Survey of India, Government of India, Dehradun.

Ministry of Environment and Forests, 1993, The State of the Forest Report - 1993. Forest Survey of India, Government of India, Dehradun.

Ministry of Environment and Forests, 1995, The State of the Forest Report – 1995. Forest Survey of India, Government of India, Dehradun.

T.R.S. Raman, 1996, 'Impact of shifting cultivation on diurnal squirrels and primates in Mizoram, North East India: a preliminary study', Current Science 70: 747-750;

Raman TRS, Rawat GS and Johnsingh AJT 1998. Recovery of tropical rainforest avifauna in relation to vegetation succession following shifting cultivation in Mizoram, North East India. Journal of Applied Ecology 35: 214-231.

Tiwari BK, Barik SK and Tripathi RS 1998. Biodiversity value, status, and strategies for conservation of sacred groves of Meghalaya, India. Ecosystem Health 4: 20-31.

— Theme II –

Responding to Challenges Through Agriculture

2014, Sustainable Rural Development through Agriculture *Pages 103–112*
Editors: **Dr. Shobhana Gupta and Dr. S.S. Tomar**
Published by: **BIOTECH BOOKS, NEW DELHI**

Chapter 6
Exploring Indigenous Technical Knowledge

Jitendra Chauhan and Shobhana Gupta

What is ITK ?

ITK have been understood that it is eminently practical and utilitarian. ITK differed from science as it only encompassed areas of direct practical value. Those who have looked at the world from the viewpoint of organised science or of the culture of which it is a part, have conventionally regarded the knowledge of other cultures as 'pre-logical' or 'irrational' and played down its validity.

Levi-Strauss (1966) argued forcefully against such a distinction on the grounds that human societies could not, for example, possibly have acquired the skills to make water-tight pots without a genuinely scientific attitude and a desire for knowledge for its own sake. ITK, like scientific knowledge should, therefore, be regarded in the first instance as something which became possible as a result of a more general intellectual process of creating order out of disorder, and not simply as a response to 'practical' human needs such as sustenance and health. Some of the knowledge arising in this way would of course have direct practical applications, and equally new knowledge about the way in which the world worked might arise as the result of a process of inquiry triggered initially by the wish to solve a problem of a 'practical' kind.

Difference between Science and ITK

First and important difference between science and ITK lies in the way in which phenomena are observed and ordered.

- The scientific mode of thought is characterised by a greater ability to break down data presented to the senses and to reassemble it in different ways.
- The mode of ITK, on the other hand, is 'concrete' and relies almost exclusively on intuition and evidence directly available to the senses.

A second distinction derives from the way practitioners of the two modes of thought represent to themselves the nature of the enterprise in which they are engaged.

- Science is an open system whose adherents are always aware of the possibility of alternative perspectives to those adopted at any particular point of time.
- ITK, on the other hand, as a closed system, is characterised by a lack of awareness that there may be other ways of regarding the world. This is not to say that ITK does not change, but rather that those changes which occur are in nearly all instances comparable to the achievements of what Kihn (1962) termed 'normal science', or to the detailed working out of relatively minor 'puzzles' within an established 'paradigm' of thought. Science, in contrast, constantly carries with it the possibility of 'revolutionary change' in which one paradigm would be destroyed and replaced by another.

Put slightly differently, science and ITK can be contrasted and evaluated according to three criteria:

- As systems of classification;
- Ss systems of explanation and prediction;
- In terms of speed of accumulation.

While ITK and science are comparable on the first criterion, science is generally superior on the second and markedly superior on the third.

ITK can itself be classified in various ways, including:

- In terms of the idioms and conceptual tools through which ITK becomes possible. This can be separated into two clusters - the propensity to classify and the propensity to quantify;
- In terms of the objects towards which thought is directed. Possible subdivisions here include:

  physical/inanimate (*e.g.* soils, water, climate); biological (*e.g.* crops, weeds, pests, domestic and other animals, insects); medical; and energy related;
- In terms of knowledge about fabrication and use of artefacts;
- In terms of knowledge of the operation of the social and economic structures within which production is embedded.

This final category is arguably only admissible under a broad definition of ITK. It includes readily articulate knowledge about such things as markets and cooperatives. It may also include mechanisms of ecological adaptation bound up in rituals such as the intermittent slaughtering of pigs in parts of New Guinea. This raises the question whether people themselves conceive of production activities as separable from social and economic relations.

Regarding the concept of ITK, there are reservations on two grounds.

First, it can imply an old / new distinction which is not helpful, since at any time the knowledge available to people is the outcome of processes of transmission and generation which have occurred both within and beyond the local environment. Assimilation of 'outside' knowledge, and synthesis and hybridisation with existing knowledge, are continuing processes.

Second, it may overemphasise the static notion of a stock of knowledge available to be tapped to the neglect of knowledge-generation as a dynamic process.

Changes in ITK

The idea of knowledge as a process is useful in showing that ITK cannot be understood independently of the ways in which it changes. Apart from assimilation and synthesis or hybridisation, the basic process of accumulation is, as with scientific knowledge, through experiment. In addition to the examples given in Howes' paper, two further instances of indigenous experimentation can be cited. In one case, in Nigeria, people experimented with cassava when it was first introduced. As cassava can be poisonous, it was important to establish the conditions in which it could safely be eaten. The procedure adopted was to feed it first to goats and dogs. In another case, also in Nigeria, a scientist believed he had made a breakthrough when he found a way of breeding yams from seed, propagation normally being vegetative. A farmer was casually encountered, however, who had not only himself succeeded in doing this, but had also discovered that whereas the first generation of tubers were abnormally small, the second and subsequent generations were of normal size. The scientist reportedly exclaimed 'Thank god these farmers don't write scientific papers'. It was also noted, in support of the prevalence of experimentation by farmers, that there is a Yoruba word for 'experiment'.

The rate at which new knowledge can be acquired through such forms of experiment is, however, slow compared with science. Stress can trigger innovation; and the development of the bamboo tubewell in India is a recent example of this. But this process can work in reverse, as in the case of the Dogon who abandoned their elaborate system of water use when moving from densely populated upland areas on to the plains. It should also be noted that in general ITK lacks means for systematic and rapid R&D.

The most significant changes in ITK come with the assimilation of small-scale societies to national and international systems. Some of these changes involve uncontroversial adoption of new knowledge.

In Botswana, for example, farmers are said to have abandoned traditional categories for classifying cattle in favour of those used in marketing meat. Elsewhere, especially in medicine, there have been cases of synthesis between ITK and science-based knowledge.

But generally, it seems that when ITK and scientific stocks of knowledge come together, synthesis does not occur. One of two things tends to happen: either the two sets of knowledge are isolated from each other (as with the head of an agricultural research station who tried to persuade farmers to adopt monocropping while still intercropping on his own land); or ITK is ignored and squeezed out as inferior. This squeezing out is more common and can lead to loss of confidence among the possessors of ITK as well as to irreversible loss of knowledge.

At the root of the problem lies the fact that officials - agricultural extension staff, planners, research workers, 'experts' and others - depend on scientific knowledge to legitimise their superior status. They thus have a vested interest in devaluing ITK and in imposing a sense of dependence on the part of their rural clients. This suggests that change may only be brought about through an assault at the level of ideology, and through a reorientation of reward systems.

The problem, however, is not just one of stocks of ITK, but of undermining the foundations for indigenous participation in the process of generating new technical knowledge. Thus Mali pastoralists are said to have accepted the dependent status which has been thrust upon them, and now believe that their major hope for salvation lies with the World Bank; and more generally, rural people tend to lack the confidence or inclination to engage in self-help activities in spheres where they have past experience of external assistance. In principle, there is no reason why this process should not be made to operate in reverse - with people gaining confidence and acquiring knowledge as a result of being drawn into the processes of generating technology - but in practice, there is little evidence that this happens.

How to Elicit ITK

Some conventional approaches to research have serious limitations for eliciting ITK and finding out how it is organised. Questionnaires impose the compiler's categories upon the respondent and do violence to the latter's meaning system. This may not always be immediately apparent since respondents often adapt to the logical framework implied by their questioner. Difficulties arise where, for example, an extension agent asks for information on yields per acre from a farmer who is more concerned with yields per unit of labour. Problems are compounded when the questioner has a different native tongue from the respondent. The boundaries delineating colours, for example, vary between languages, but these variations may not be recognised; and culturally specific concepts are often hard to translate. Full-scale anthropological methods of observer-participation can overcome these difficulties but they are time-consuming and probably rarely cost-effective. Methods of investigation are needed which are open-ended, quick, and reliable.

One such approach is to take part with informants in their work. While this may not enable the observer fully to see the world through the informants' eyes, a high degree of empathy can be achieved by working together, and information and insights may be provided which informants would not otherwise have thought to mention. Another approach is to observe and learn the games people play since these are often how important skills are acquired and practised. It s also often particularly useful to find out about indigenous systems of quantification and to calibrate these against formal scientific measures.

Other ways of eliciting ITK can simultaneously stimulate the creativity of informants. These approaches include the use and adaptation of games as described by Barker and Richards.

Uses of the Stock of ITK

Can the stock of ITK be used either to economise on the use of scarce trained scientific manpower or to extend the range of observations upon which science can draw?

Instances where this has happened are few, but suggest a considerable potential. Pastoralists, for example, have detailed genealogical knowledge of their animals which can quickly be translated to give a picture of fertility and age-specific mortality. Similarly, work on the variegated grasshopper (Zonocerus Variegatus) in Nigeria, which drew on indigenous perceptions, provides a useful basis for determining the seriousness of the problems which they generated, and hence the priority to be attached to remedial action (Barker *et al.*, 1977).

Other ways can be suggested in which indigenous observers might - in theory at least - act as 'the eyes and ears of science'. Knowledge of micro-environmental conditions could be used in the preparation of soil maps; local people could be consulted to determine the milk yields of animals under 'real' conditions where scientific testing had not been carried out; indigenous observers might be encouraged to report back on changes in the species composition of pasture as an early warning system for environmental deterioration; farmers could be used in crop reporting systems instead of extension personnel; and so on.

Many such possibilities might be opened up with little technical difficulty: often all that is required is standardization of systems of measurement. However, one should not simply think in terms of how ITK can be used in isolation, but rather consider ways in which it can be brought into creative synthesis with science. In the environmental sphere, for example, the ideal form of monitoring might well involve a combination of sophisticated satellite technology with observers operating at the local level.

In attempting to mount such an exercise it is also important to recognize that ITK is not distributed evenly among the members of a society. It is likely to be controlled and manipulated by certain groups and classes in the pursuit of their own interests. Sometimes particular types of knowledge are the preserve of 'caste-like' groups such as Tuareg smiths; in other cases religious groups like the Marabuts in West Africa are paid and respected as repositories of knowledge. Such interest groups may provide a basis for collaboration, but equally they may stand in the way of change. Elsewhere, variable access to knowledge can arise out of the differentiation of a society into economic classes. In all societies systematic variations in knowledge are likely to be associated with sex and age. In addition, individuals always differ in ability and aptitude.

There are further important practical questions about the way in which knowledge is transmitted between individuals and generations. An understanding of established learning processes might provide a useful starting point for seeing how people could 'draw-down' on scientific knowledge more effectively.

Implications for R&D

How can ITK contribute to the generation and exploitation of technology to benefit rural populations? This can be seen as a question of finding an optimum mix and balance between indigenous participation and scientific participation in R&D processes rather than a choice of either one or the other. What mix is optimal will vary.

It can be argued that formal R&D systems are efficient for generating new knowledge quickly. Whatever the merits of ITK and of R&D activities which involve rural people themselves, the means and methods of scientific research can, in many fields, achieve far more far faster than would ever be possible through reliance on indigenous experimentation. In this view, the urgency of rural development is such that rapid advance to major breakthroughs is essential, and some at least of these have to come primarily through the formal R&D system.

On the other hand, rural people already take the final and crucial decision whether to adopt a new technique. In addition, they often adapt the standard packages with which they are presented to fit their particular needs and conditions. However, it may be only certain people, notably the relatively powerful and wealthy, who normally take part in such decisions.

Certain aspects of knowledge-generation will always have to be centralised and formally organised. Opinions differ, however, about the extent to which this is desirable. Much formal R&D has three phases: problems; a period of development and testing removed from that environment - on a research station or in a laboratory; and a period of re-entry and testing, during which the innovation is brought into the rural environment. For any technology, the question is what balance is optimal between these three. For mechanical and engineering technology, the case appears strong for much more work in the rural environment and with rural people. With seed-breeding programmes, in contrast, a phase in the controlled conditions of a research station is desirable for efficiency. Similarly, in developing a vaccine for cattle, some work in a well-equipped laboratory may be essential. Although opinions differ, it may be generally more efficient, in terms of ultimate benefits to rural people, for much more R&D to be conducted in rural environments and with rural people than is current practice.

Substantial efforts have been made in this direction. Before any radical proposals are put forward, attention should be paid to the experience gained by the International Agricultural Research Centres and by national research institutions. At the same time, there is scope for making these formal systems more responsive to the views and needs of those whom they are supposed to serve. Formal R&D is still struggling to get to grips with the variability of tropical environments, and with the accordant need to decentralise research to involve local people more actively in it. A further general failing is the tendency to see he end product of a research programme as a report or an article rather than a proper evaluation of adoption, benefits and lessons. Also, research activities still tend to carve up reality in a manner which hinders a holistic view of local-level conditions.

To overcome or reduce these problems, six proposals seem worth considering:

(1) Rural Exposure for Extension and Research Staff

Extension and research staff could be confronted more directly than is usual with the realities to which their work relates. This could be done both during initial training and at intervals thereafter. The repertory grid method (see Richards, this volume) might serve as a starting point for enabling professional personnel to appreciate the difference between their way of looking at the world and that of the people who were supposed to benefit from their work.

(2) Checklists

Checklists could be used to draw attention to factors which might otherwise not be considered in determining research priorities or extension advice. Some examples of factors that may be overlooked with an innovation are implications for women, profitability, effectiveness and efficiency, availability and access to inputs and complementary items, whether a farmer can afford an innovation, risk, socialsignificance and acceptability, lightness for carrying and 'mendability', labour requirements, and effects on diet and on the variety and timeliness of food supply. Checklists have their uses but can be criticised for the implicit assumption that decisions will be made by a small group of people who will determine what is good for others.

(3) Local-Level Influence on Research Priorities

To improve the criteria chosen in research and then to see they are acted on, producers could sit on the boards of agricultural research stations, following the model of the Kenyan commodity boards. Further, priorities could be set by national research committees which consulted at the local level, although there would be a danger that this would merely reinforce elite preconceptions.

(4) A Cafeteria System

Farmers could be offered different packages and left to decide for themselves which they would adopt. In Sri Lanka, for example, farmers were provided with 'mini-kits' of different seed varieties, with which they could experiment on their own farms.

(5) Starting with Indigenous Practice

A more radical proposal is that research should take existing indigenous practice as its starting point, seeking to refine this in various ways and then to feed results back into the system. This would go hand in hand with the actual and metaphorical removal of the 'fences' surrounding research institutions so that no aspect of the process of knowledge-generation fell beyond the purview of those whose livelihoods would ultimately be affected. An objection here, however, is that indigenous practice, as with intercropping, growing two or more types of crop together, may be so complex as to be laborious and difficult to test under controlled research conditions.

(6) Experimental Work in Rural Conditions

The process might be taken a stage further, perhaps through full-blown experimental work on farmers' fields and with farmers' collaboration. In general, people are more likely to operate and exploit a new technology successfully if they have themselves taken part in its creation.

The validity of this sixth proposal is supported by the extent to which important technical change has taken place and can take place outside formal R&D systems. It turns part of the earlier discussion on its head; instead of asking how experts and scientists can better understand the potential of ITK, the question now is how rural people themselves can assess and utilise the potential of science. To pursue this approach, more has to be known about the way in which knowledge is generated and hybridised and about the potential for different modes of participation. A further need is to see whether ITK can in some way help to stimulate demand which will make R&D respond to the needs of neglected groups and classes.

One objection to this sixth proposal is the earlier arguments in favour of formal science with its implied centralization. Another is that people can and often do use and benefit from techniques without understanding the technology underlying them. Opinions differ on these points, suggesting a need for research to identify optimal and feasible degrees of decentralization and modes of participation according to type of technology and social conditions.

Values and Rewards

Proposals for using the stock of ITK and for local involvement in R&D can only be adopted easily when lack of awareness is the only constraint. In practice this is rarely the case. In situations where change seems desirable, deep-rooted structural impediments will frequently be encountered. Junior field extension staff, for example, being low in the government service, have a vested interest in exaggerating differences between themselves and local people; and the distinction between 'superior' scientific and 'inferior' indigenous knowledge protects and legitimates their status. In addition most of the proposals presuppose flexibility and initiative at the lower levels in the bureaucracy, but this conflict with bureaucratic norms. There are also likely to be problems among more senior staff engaged in R&D. Established professional values dictate that rewards should be given to those who make original contributions to knowledge, achieve breakthroughs at the level of theory, and publish their findings in internationally reputable journals; but offer relatively little incentive to individuals to go out on a limb with approaches involving ITK. Changes in values and reward systems are necessary preconditions of progress.

Such changes can be sought directly and indirectly. Possible direct approaches include the award of Nobel prizes and of other international and national medals and distinctions for outstanding work with ITK and for exceptional local-level breakthroughs. For their part, academics can encourage research related to ITK and publish the results in international and national journals. A system of rewards for villages, perhaps along the lines of the former 'village of the year' competition in Uganda, might promote self-confidence and creativity and be linked with ITK. Finally, R&D staff might be rewarded according to the practical result of their work, possible through an assessment by local people themselves; but in the case of agricultural research, at least, this would prove difficult in practice.

Less direct approaches might involve an attack on prevailing ideology. Initiatives through education can be suggested. Primary school teachers with extensive ITK could be accorded high status and encouraged to communicate their knowledge through the formal educational process. Knowledgeable local people could also teach in schools. Third world universities could be encouraged to extend fieldwork for students, on the lines of the useful studies already carried out by Makerere University, the University of Dar es Salaam, and the University of Nairobi. Such exercises need only small research budgets.

Research workers in the richer countries also have an important role to play. By studying and recording ITK and making it academically respectable, they can counteract the ideologies in the name of which it is being destroyed. By encouraging students - particularly those from third world countries - also to adopt this perspective, the effect can be multiplied.

Some Outstanding Questions

Questions which remain unresolved and questions which may deserve further research include the following:

ITK

1. Do rural people conceive production systems separately from the social and economic structures in which they are embedded? In other words, to what extent, or in what senses, are they aware of their technical knowledge as technical knowledge?
2. How is established knowledge transmitted between generations and individuals? What implications, if any, do such processes have for the appraisal and acquisition of scientific and other knowledge?
3. What are the strengths and weaknesses of different categories of the stock of ITK and what are their potential contributions to rural development?
4. Why does the meeting of ITK and science sometimes lead to constructive synthesis (as sometimes in medicine) but more frequently to the subjugation of ITK by science? How are ITK and scientific knowledge synthesized, and how might that synthesis be improved?

R&D and the Generation of Knowledge

1. How is ITK generated?
2. In developing scientific R&D programmes how useful is it to start with ITK and with current rural practices?
3. How useful are checklists?
4. What degree of decentralization and of work with rural people in rural environments is optimal, by type of technology, by phase of R&D, and by social conditions? In particular, how important and feasible is active participation in R&D by the ultimate users of the technology?

5. What demands are exerted or might be exerted by rural people upon formal knowledge-creation systems, and through what modes of participation?
6. To what extent and how successfully have the International Agricultural Research Centers and national research organizations adapted their programmes to take account of ITK, of local environmental conditions, and of particular social groups, and what can be learnt from their experiences?

Professional Training and Values

In modifying professional values and behaviour, what is the potential of:

1. New reward systems?
2. Games played with farmers and others as part of the training of staff?
3. Research on ITK required to be carried out by extension and research workers, and by their trainers?

References

Barker, D., Oguntoyinbo, J. and Richards, P. (1977) The Utility of the Nigerian Peasant Farmer's Knowledge in the Monitoring of Agricultural Resources: A General Report, Monitoring and Assessment Research Centre of the Scientific Committee on Problems of the Environment, International Council of Scientific Unions, MARC Report 4.

Kuhn, Thomas S. (1962) The Structure of Scientific Revolutions, Chicago and London: University of Chicago Press.

Levi-Strauss, C. (1966) The Savage Mind, London: Weidenfeld and Nicholson.

2014, Sustainable Rural Development through Agriculture *Pages* ***113–126***
Editors: **Dr. Shobhana Gupta and Dr. S.S. Tomar**
Published by: **BIOTECH BOOKS, NEW DELHI**

Chapter 7

Mitigation and Adaption Strategies in Agriculture against the Changing Climate

Sarju Narain, O.P. Maurya, Vikas Kumar and R.R. Kushwaha

We are all witnessing the negative role of climate on agriculture. There are number of factors responsible for this. Still the human can play a great role in minimizing the negative impact of climate on agriculture. There are some symptoms of climate change as increasing global temperature, irregular pattern of rainfall melting of glaciers, increasing frequency of Cyclone and Tsunami, desertification and increasing dry area, depletion or hole in Ozone layer, global cooling especially in European and American countries, irregular seasonal change that affect leaves shedding, flowering and fruiting time in mango, neem, etc., increasing incidence of insect-pests and diseases on crops especially in paddy, cotton, pulses, etc.

What is Climate Change?

The Earth's climate is mainly influenced by solar energy, but also by the amount of green house gases and aerosols in the atmosphere. The concentration of 'green house gases' such as carbon dioxide (CO_2), methane (CH_4) and nitrous.

Abundance and Life Time of Greenhouse Gases in the Atmosphere

Parameters	CO_2	CH_4	N_2O	*Chlorofluoro Carbons*
Average concentration 100 years ago (ppbV)	290,000	900	270	0
Current concentration (ppbV)	380,000	1,774	319	3-5
Projected concentration in the year 2030 (ppbV)	400,000-500,000	2,800-3,000	400-500	3-6
Atmospheric life time (year)	5-200	9-15	114	75
Global warming potential (100 years relative to CO_2)	1	25	298	4750-10900

Source: NSFSC, 2010, Key note lecture of H. Pathak, Division of Environmental Sciences. IARI, New Delhi – 110 012.

Nitogen oxide (N_2O) in the atmosphere significantly increased since the beginning of the Industrial Revolution, mostly as a result of human activities, such as fossil fuels burning, biomass utilization and deforestation. The burning of fossil fuels has also resulted in emission of aerosols that absorb and emit heat and reflect light. These additions of greenhouse gases and aerosols have changed the composition of atmosphere and lead to climate changes that influence temperature, precipitation, winds and sea level on global and regional scale. It seems that climate change processes are already happening, being one of the greatest environmental, social and economic threats facing the planet (IPCC, 2007). In the seasonally dry regions even a slight warming (1-2°C) may have significant negative effects on livelihoods of people (**Easterling *et al.*, 2007**).

The inter–governmental Panel on climate change (IPCC, 2007) projected a temperature increase between 1.1.and 6.4°C by the end of 21st Century. Fishery sector may be adversely affected due to rise in sea level and production of meat and milk may be affected due to decreased water and fodder availability. Socio-economic determinants of food supply are thus expected to be serious distressed. (**G.C. Tewari, 2010**).

Except above impacts shortage of irrigation water, high cost of maintaining climate controlled environment for bio-life, rural urban migration as a result of drought (**impact seen in some Indian pockets like Bundelkhand region of U.P. as reported by Sarju, 2011**). Thus, climate change is a product of the unsustainable consumption/ life style of non-renewable forms of energy, products, etc. by industrial countries, the harmfull impact of climate change will be felt more by poor nations and the poor in all nations due to their limited coping capacity (**A draft National Policy for Farmers, National Commission on Farmers, 2006, India**).

Source of Green House Gases (GHGs)

Atmosphere is a mixture of gases in which major GHG are Carbon dioxide (72 per cent share in GHG), methane (18 per cent share in GHG), nitrous oxide (9 per cent share in GHG) and chloroflouro carbon (1 per cent share in GHG) etc. popularly known as Green House Gases (GHGs). They cause Global Warming (GW) and GW is responsible for climate change. The share of Agriculture in GHG production is about 20 percent per annum.

Carbon dioxide (CO_2)

A sample of pure air consisting 0.03 percent CO_2 by volume. The main source of CO_2 is decay organic matter, forest fire, volcanoes, burning of fossil fuels, deforestation, land use changing, etc. Among all sources, fossil fuels burning and deforestation played important role in increasing concentration of CO_2. Agriculture is not considered to be main source of CO_2.

Methane (CH_4)

In atmosphere the share of methane is very minor in terms of volume. The main source of methane is wet land, organic decay, termites, natural gas and oil extraction, biomass burning, rice cultivation, ruminant animals, etc. In which increasing area of rice, increasing population of ruminant animals considered to be major source. Methane is about 25 times more effective than CO_2 as a heat – trapping gas.

Nitrous Oxide (N_2O)

It is also found in very minor amount in atmosphere by volume. Forests, grassland, oceans, soils, nitrogenous fertilizers, burning of biomass and fossil fuels are the source of N_2O. Soil contributes about 65 per cent of the total N_2O emission. The major sources are soil cultivation, chemical fertilizer application and burning of organic materials especially paddy residues/straw in paddy growing areas and fossil fuels. N_2O is 298 times more effective than CO_2. Appropriate crop management practices can reduces the N_2O emission from soil and can decreases the loss of nitrogen from soil.

Chloro Flourocarbon (CFC)

Refrigerants- cooling industry, cleaning fluids, aerosol propellants, etc. are is the main producer of CFC. When these chemicals reach in stratosphere of atmosphere, ultraviolet rays of sun, break them apart and release chlorine which destroy ozone. For every one molecule of chlorine from chloroflouro carbon, about 1,00,000 molecules of ozone are removed from ozone layer.

Share of Agricultural Emission in GHG

According to IPCC, the global average emission from agriculture is only 13.5 per cent while Indian agriculture share in total Indian emission is 28 per cent. The emission from agriculture are primarily due to methane emission from rice fields (23 per cent), enteric fermentation in ruminant animals (59 per cent) and nitrous oxide from application of chemical fertilizers and manures to agriculture soils. Anaerobic condition of rice field and in ruminants produces methane while nitrogenous fertilizers and their mall application practices as well as bacterial activities *i.e.*, Nitrification and de-nitrification are responsible for the production of nitrous oxide.

Influence of Climate Change on Agriculture

Agriculture is the riskiest profession in the world and now it is affected by global climate change. Global climate change is human induced condition which can affect Biological production system through their direct and indirect effects on ecosystem in gradual way.

Reduction in Yield

Rise in the mean temperature above a threshold level will cause a reduction in yield. However, change in minimum temperature is more crucial than that of the maximum temperature. Grain yield of rice, for example declined by 10 per cent for each 1°C increase in growing season when minimum temperature above 32°C (Pathak *et al.*, 2003). The climate change impact on productivity of rice in Punjab, India showed that with all other climate variabilities remaining constant, temperature increases of, 2 and 3°C would reduce the grain yield of rice by 5.4,7.4 and 25.1 per cent, respectively (Aggarwal *et al.*, 2005).

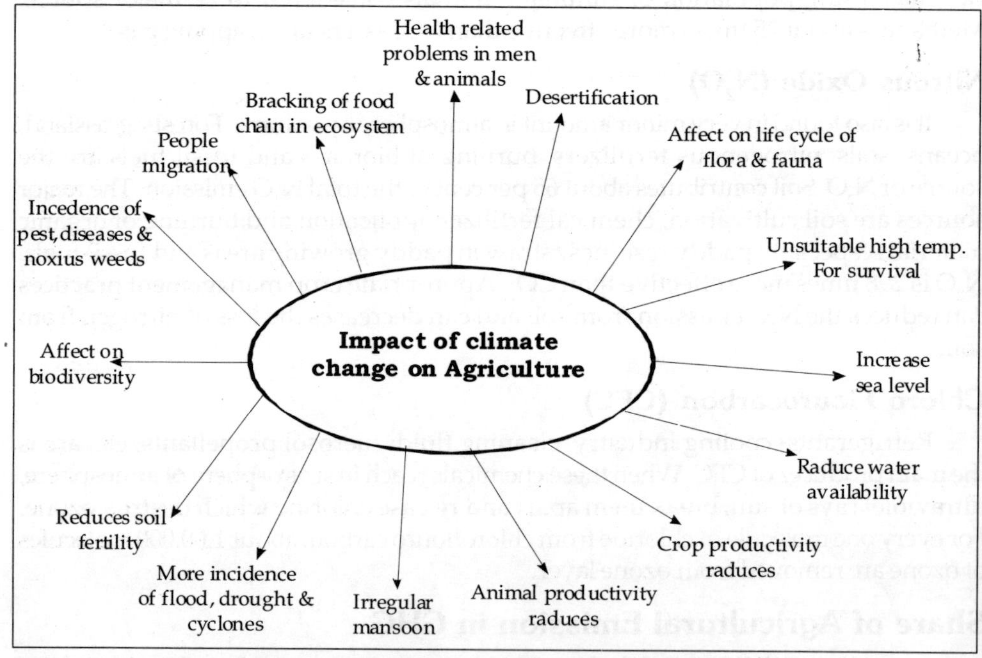

Impact of Climate Change on Agriculture.

Shortage of Water

Irrigation demand for water is projected to increase in a warmer climate **(Gitay *et al.*, 2001)**. Demand of irrigation water as well as safe drinking water would increase. Increase in temperature directly related to increase in the evapo-transpiration as well as demand of drinking water. Shortage of irrigation water decline the many crops areas and production especially rice cultivation. Overexploitation of ground water reduces the availability and green area/zone converted into gray/dark zone. In many states like Punjab, Haryana, Uttar Pradesh, etc. having many dark zone/block. For the solution of these problem watershed management and water conservation in fields is very important issue.

Irregularities in Onset of Monsoon, Flood, Drought and Cyclone

It is well established fact that agriculture is the gamble of monsoon because monsoon precipitation particularly in rainfed areas play important role in agricultural production and making continue of biological life. Climate modelers and IPCC document have projected possibilities of increasing variability in Asian monsoon circulation in a warmer world. In many part of the world drought caused inadequate and uneven distribution of rainfall which is badly effected to agriculture production. The warmer pockets of world will be more affected from warming, flooding and cyclone frequency may be increased in some packets of world may cause heavy losses. In Indian sub continental flooding would be major problem.

Rise in Sea Level

Rice production regions near to sea areas of World especially south, southeast and East Asian regions would be suffer badly to increase 1-m sea level. If no adaptation measures are taken then nearly 4 per cent of rice yield may be reduces which can increases food security problem. Spreading of salt water of sea on cultivated and uncultivated areas soil salinity increases which reduces agricultural productivity as well as serious concern to ecology of these region.

Soil Health

High temperature increases the evapo-transpiration, soil erosion through mineralization and decomposition rate of organic matter. At high temperature, though nutrient availability will increases in short term but in long term loss of inorganic content as well as release of nutrients from soil may be increases. Thus, micro-climate of crop plants will be badly affected. Therefore, soil health is badly affected which may be increases/accelerated rural migration.

Loss of Biodiversity

Loss in biodiversity is serious concern and our future hope is depend upon it. Due to changes in climate several flora and fauna habitat may be badly affected and they couldn't maintain their life cycle. Therefore, more chances may be possible for disappear of many species of flora and fauna, who can't adopt/acclimatize in changing climate.

Resurgence of Pests, Diseases and Noxious Weeds

Insect pest are very diverse and dynamic organism in the earth. If temperature increases then more possibilities will spreading the insect-pests population. Their development growth and migration habit may be change. Many insect-pests and their natural enemies as well as their host life cycle will be disturbed. So climate change may alter the balance between insect pests and their enemies.

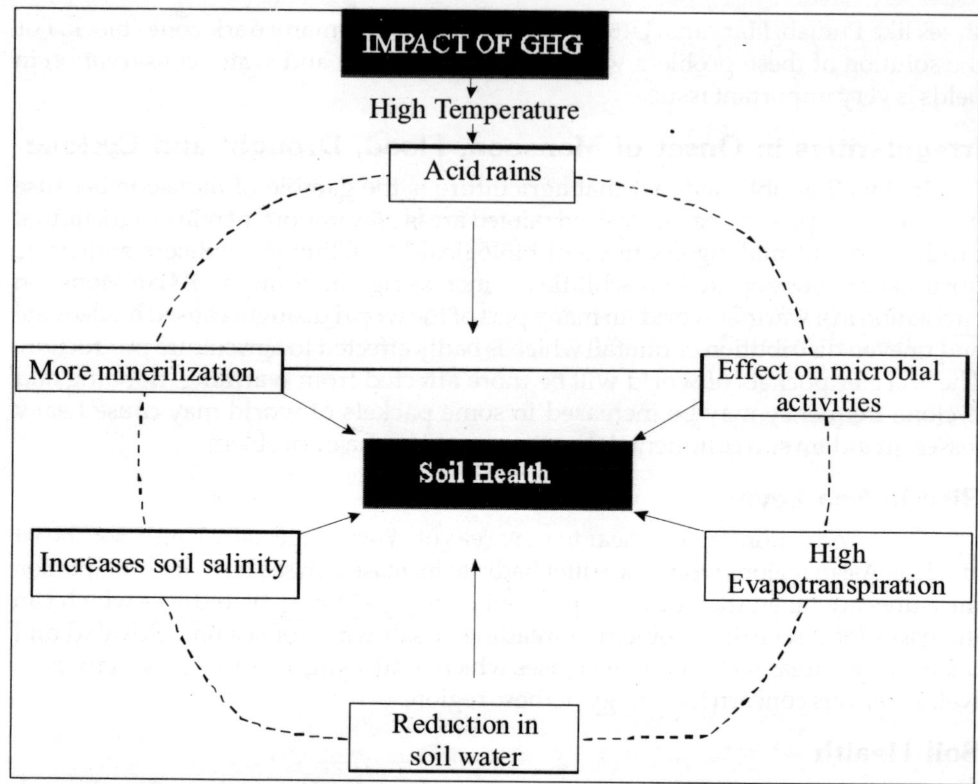

Impact of Climate Change on Soil Health.

Many noxious weeds phenology, morphology, etc. will be change and they may be more vigour and noxious in habit. Therefore, all managerial measures may be redefined in changing scenario.

Socio-Economic Aspects Related to Agriculture

Socio-economic consequences of climate change will be influenced by adaptation made by farmers, consumers, government agencies and other related institutions. Issue-wise aspects are as.

Food and Fodder Demand may be Imbalanced

As per projection based on 9 percent GDP growth of India, the demand in 2026 will be 102.1 mt rice, 65.9 mt wheat, 57.7 mt pulses and 40.9 mt edible oil (Surabhi Mittal, 2008). This increasing demand of food items could not achieve in changing climate scenario without any especial attempts. As same a big question would be arise related to fodder security for milch animals.

Cost Benefit Ratio may be Change

Climate change also affect inputs use efficiency and especial managerial efforts for crop/plants. This condition may be promoted for more use of inputs to the crops/

plants. Therefore, it may be possible that cost of cultivation will be increases. Ultimately cost – benefit ratio may be changed.

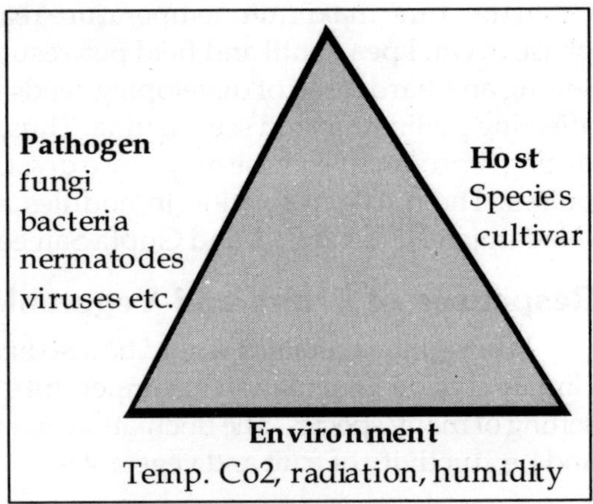

Relationship of Environment, Pathogen and Host.

Trade Related Policy may be Change

The impact of climate change would be directly effect to production system and trade related policy of nations and world.

Farmers Related Issues

Rural poverty and rural migration may be too increases. In changing scenario small and marginal farmers of India may be out from agriculture production system due to specialized production technology and heavy/costly infrastructure.

Climate Change and Indian Agriculture

Indian agriculture and their production scenario also affected by global climate change. High temperature affected the life cycle of winter crops. It is possible that temperature inhibit the pollination of winter crops, so our production system may be badly affected. Today India produces 241 million tonnes of food grain which would increases about 294 million tonnes upto 2020 for full filling our food security needs. The evidence from last years as showed by different research centers of wheat and rice indicate negative growth rate in production. This negative growth rate of food production presents a grim scenario of future food security policy. According to experts opinion 10-40 percent crop production losses may be occurs in coming years in India.

Responses of Wheat to Climate Change

If increases 1°C temperature in wheat growing pockets of India during wheat growing period, then it may be possible that about 4-5 million tons wheat yield losses occurs up to 2020-30 as predicted by experts. A 0.5°C increase in winter temperature would reduce wheat yield by 0.45 ha^{-1} in India (Kalra, *et al.*, 2003) In north Indian region December month is best suited for wheat tillering in field, if temperature increases then tillering badly affected. In the month of February if temperature increases more than 30°C, then pollination affected and yield is reduces.

Responses of Pulses to Climate Change

All India coordinated trials on pulses results indicated that reproductive parts and grain filling process are extremely sensitive to temperature extremities. Pigeon pea is highly sensitive to temperature fluctuation, causing massive flower drop and dropping forced drying and bending of apical leaves subjected to cold stress (<5°C). In green gram high temperature (> 42°C) causes hardening of seed during summer.

Often day time maximum temperature reaches beyond 40°C during reproductive phase of chickpea, lentil and field pea resulting in complete failure of anthesis, pod setting and hardening of developing seeds while day temperature above 35°C start affecting pollination and seed setting. Thus, high temperature under field condition during reproductive phase may cause drastic reduction in the productivity of winter pulses. The nitrogen fixation in nodules also retards at high night temperature. (Nadarajan N., Basu P. S. and Gupta Sanjeev, 2010).

Responses of Fruits and Vegetables to Climate Change

Fruits and vegetables would be also badly affected by high temperature due to climate change. Fluctuation in temperature directly affected to pollination and fruit setting of many species. The fluctuation in temperature will also affect the production and productivity of fruit and vegetables.

Response of Livestock to Climate Change

High temperature also reduced milk yield, affect reproductive performance, inferior milk quality, increased health care costs and reduce heifer growth (Robert J. Collier, *et al.*, 2008). Many scientific studies indicate that heat stress reduced the length and intensity of estrus, making it more difficult to detect cows in estrus for artificial insemination. Except these activities it may be need of cow comfort and cooling as well as nutritional modifications.

Mitigation and Adoption Strategies to Climate Change

Mitigation Strategies

Mitigation refers to strategies to reduce the probability of climate change through sustainable practices that mitigate the increased occurrence, severity and unpredictability of weather patterns resulting from climate change.

Management of Methane Emission

Rice cultivation in submerged condition is the main source of methane production, which can be minimize by altering water management and growing of rice cultivar/species/varieties those well grown in aerobic condition, *i.e.* non submerged condition. Ruminants are more useful animals to us. Methane emission from ruminants can be reduced by altering the feed composition.

Management of Nitrous Oxide

In agriculture production system, site specific nutrient management can minimize the nitrous oxide emission. The emission could also be reduces by nitrification inhibitors such as nitrapyrin and dicyandiamide (DCD), etc. Some organic and nature available products like *neem oil, neem cake and karanja seed extract* inhibit nitrification process.

Management of Carbon Dioxide

Carbon dioxide emission from agriculture may be manage up to some extant by soil management practices such as reduced tillage/zero tillage technology, mannuring, residue incorporation, improving soil biodiversity, micro-aggregation and mulching can play important role in sequestering carbon in soil. Carbon sequestration through agro forestry is a potential viable option to add trees on the farm lands alone or in association with crops to mitigate climate change.

Management of Chloroflouro Carbon

Sustainable consumption and sustainable resource use is a good option for managing this problem. Green technology like use of Floro-Flouro Carbon (FFC) in place of CFC is a big example for solution of this problem.

Adaptation Strategies

Intelligence is the ability to adopt to change.

(Stephen Hawking, Physicist)

If intelligence is ability to adopt to change, then-wisdom is to know how and why to adopt for more than personal or economic gain.

(Ivan Urlaub, executive Director, N- Carolina, Sustainable Energy Association)

Thus, intelligence and wisdom form the foundation of individual and collective capacity to adopt and cope with adversity and change. Increasing climate risks in future, associated with global climate change are likely to further compounded the problem. Therefore, it is an urgent need of adaptation strategies.

Developing New Genotype Suitable for Changing Climate

The first and most important adoptive option includes identification and development of heat, drought, insect- pests resistance and other desirable genes from the biodiversity of flora and fauna, which fit for changing climatic condition. ICAR institutes/centres, SAUs are busy in this line. Wheat varieties like Halna, Golden Halna, Shatabdi, etc. are heat tolerent in nature developed by Chandra Shekhar Azad University of Agriculture and Technology (CSAUA and T, Kanpur) and several other crop varieties of Indian Agricultural Research Institute (IARI), New Delhi are the suitable example for solving the problems. Another wheat variety PBW-17 is also suitable in high temperature areas compare to general wheat varieties. Development of photo-in-sensitive and temperature insensitive wheat and photo in sensitive rice varieties has permitted their cultivation in new areas/high temperature condition. Change in maturity duration may permit new changing environment.

Conservation of Local Cultivars/Varieties – 'Save Desi Seeds'

There is an urgent need to conserve desi varieties/cultivars of crops grown in specific pocket/regions, for example paddy variety *'kala namak'* popularly grown in Chhattisgarh pocket. This type of diversity spread in whole India which need attention because desi varieties/cultivars having more adjustability with changing climate. In this direction participation of all stalk holder is very necessary for 'saving desi seeds' and establishing 'community seed bank'. Underutilized small millets like *' Sanwa' 'Kondo', 'Mandua', 'cheena', 'kakun', ragi' ramdana,* etc. provide nutritional security.

Therefore they are called as 'future food'. Due to spreading of irrigation facility farmers don't grown or poorly grown these minor millets, therefore the availability of these seeds are badly affected. These minor millets are having good survival rate under harse climatic condition. Orissa cyclone and Tsunami extent salt in productive soil. These salt affected saline soil best suited for salt tolerant deshi paddy varieties like, *'kalambank'*, *'kartikapatini'* *'chakakhi'*, *'Doodheshwar'*, *'Leelawati'*, *'Luna'*, etc. which provide food security to local people. Drought susceptible and suitable for rainfed deshi paddy varieties like *'Bhatkala'*, *'Loha'*, *'Gora'*, *'Nata'*, *'Raja'*, *'Manik'*, *'Dhumka'*, *'Kaya'*, etc. need to conserving seeds for facing any adverse condition in future.

Management of Diversification

Agriculture diversification demands more resources and more inputs while crop diversification demands only technical knowledge. Both systems having the potential against drought and heat tolerance. In this line mixed cropping, mixed farming, multi-stories cropping, agro-horti, agro-silvi, etc. systems proved mile stone against adverse condition.

Changes in Land Use Management Practices

New adjustment of cropping sequence changing the timing of sowing planting, spacing arrangement, mulching, time and method of irrigation, etc. will be based son occurrence of small changes in climatic parameters.

Relocating Crops into Alternate Areas

Climate changes direct effecting to production system but the impact will be different for different corps/cultivars/varieties and regions. There is need to identify the crops and regions that are more sensitive to climate change/variability and relocate them in more suitable areas.

Development of Cloud Seeding Technology

To managing rainfall this technology may be more helpful to production system. Israel and China is a present example of managing cloud seeding technology for their Agriculture

Adjusting Cropping Season

Changing the cropping calendar to take advantage of wet period and to avoid extreme weather events during the growing season. New season crop varieties like maize for zaid and rabi are also example of adjusting cropping season as demand of hour. Farmer will have to adopt to changing hydrological regimes by changing crops, such as farmers of central U.P. replace rice cultivation (in kharif season when rainfall late) by Bajra cultivation.

Weather and Climate Forecasting Services

To identify the changing climate short, medium and long term weather and climate changing forecasting is necessary; so that our farmers adopt new strategies. In this line ICAR, New Delhi taking initial steps to strengthening the weather forecast services at local level by Krishi Vigyan Kendra (KVK).

Adaptation of Resource Conservation Technology (RCT)

Resource conserving technologies involving zero or minimum tillage with direct seeding, lesser leveling, mulching with residue, crop rotation with sod crops, sowing of some crops on bund or raised bed planting, line sowing, micro/drip-irrigation, fertigation, etc. have the potential to be adopted over area. These technologies lead to a saving of irrigation water, fertilizers, seeds, etc. besides significantly improving soil health. In paddy and sugarcane growing pockets of India where paddy straw and sugarcane leaves are good mulching materials which can be used for moisture conservation in crop field.

Conservation of Biodiversity and/Ecology

Due to pressure of human kind activities the ecology and biodiversity system is badly affected which increases natural imbalance. Therefore, our food chain system is disturbed. The impact of imbalance situation come in the form of resurgence of pest. Defforestation of social/religious plants especially *'Bargad'*, *'Goolar'* *'Pakakr'*, *'Peepal*, etc. which loss the natural habitate of various small birds who feed many type of insects-pests; result the population of lepidopteran insects are increasing and now many of them are converted as polyphagous pests.

Disaster Preparedness/Risk Management and Crop Insurance

Agriculture is the riskiest profession in the world. Due to influence of climate change the severity of risk and urgency of preparedness is increasing. Therefore, need of establishing, climate risk management center (CRMC) in different micro level agro climatic zone. This center will help farmers to adopt contingency plans and action.

Harnessing the Indigenous Technical Knowledge (ITK) among Farmers

Farmers experience and local technologies development is the result of many generations or centuries exercise. Collection of these need based ITKs and deshi cultivars are the another option which protect from climate change. There is a traditional wealth of knowledge which found among the farmers community that can help in developing new technologies to overcome climate vulnerabilities.

Developing Advance Strategies of Pests Management

Pests are not a problem but a symptom. Disturbance in the ecological balance among different components of crop ecosystem (biotic and abiotic) makes certain insects reach pest status. Rainfall, humidity and temperature would affect pests and diseases incidence and virulence of major crops. Climate change will potentially affect the relationship of host and pests effectiveness. Several new strains of insects, diseases, weeds, etc. may be more survive in changing climate, therefore, advance strategies of pests' management need to be strengthened by research. For this purpose the direction of movement are (i) development of pests' resistance cultivars/varieties; use of (ii) integration of agronomic and biological means of control; (iii) use of simulation models, ICT and other pest forecasting strategies.

Heat Effect Management in Milch Animals

Climate change also affected to milch animals. Nutritional management strategies may provide opportunities to improve lactation performance of dairy cattle during period of thermal stress.

Infrastructure Development and Policy Setting

Weather watch groups monitor the wealth impacts and enabling appropriate policy responses need to be established. Food and forage banks in community partnerships to manage scarcity during drought and floods are called for developing mechanisms for integrated management of water. Green research fund for strengthening research on adaptation, mitigation and impact assessment needs to be established. In this line ICAR launched a new programme/project named as 'National Initiatives on Climate Resilience Awareness (NICRA). Thus. Infrastructure development and policy decision are the backbone of future development in context of climate change.

Capacity Building Through Extension-cum-Developmental Program for Farmers

Due to climate change some regions appears winners and some appear as looser. Extension can helps farmers prepare for greater climate variability and uncertainty. Create contingency measures to deal with exponentially increasing risk and alleviate the consequence of climate change by providing advice on how to deal with droughts, floods and so forth. Extension can also help with mitigation of climate change.

In conclusion, climate change will bring unreliable and less predictable conditions which affect our agricultural production system. It is likely to reduce yields of most crops in long term; in short term effects may be small. Costs of adaptation are less understood but likely to be high; cost of inaction, however, could be even higher. It should be noted that adaptation practices take time to become effective. Therefore, we need to act now.

References

Aggarwal P.K., Pathak and Bhatia A. (2005). Global warming : Indian estimates of green house gas emissions from agriculture fields. *Success Story Series No. 7*, Indian Agricultural Research Institute, New Delhi, pp. 16.

CCSP, (2008). The effects of climate change on agriculture, land resources, water resources and biodiversity in the Untied States. A report by the US climate Change Science Program and the Subcommittee on Global Change Research (Ed. P. Backlund, A. Janetos, D. chimael, J. Hatfield, K. Boote P. Fay, L. Hahn C. Isaurralde, B.A. Kimball, T. Mader, J. Morgan, D. Ort, W. Polley, A. Thomso, D. Wolfe, M.G. Ryan, S.R. Archer, R. Birdsey, C. Dahm, L. health, J. Hicke, D. Hollinger, T. Huxman, G. Okin, R. Oren J. Randerson, WI Schlesinger, D. Lettenmaieir, D. Major, L. Proff. S. Running L. Hansen, D. Inouye, B.P. Kelly, L. Meyerson, B. Petrson, R. Shaw), US Department of Agriculture, Washington DC, USA.

Easterling, W.E. Aggarwal P.K., Batima P., Brander K.M., Erda L., Howden S.M. Kirrilenko A., Morton J. Sousana J.F., Schmidhuber J. and Tubiello F.N. (2007). Food fibre and forest product In *Climate Change 2007; Impacts Adaptation and Vulnerability. Contribution of working group II to the fourth assessment report of the intergovernmental panel on Climate Change* (Eds. M.L. Parry, O.F. Canziani, J.P. Palautiko, P.J. Vander Linden and C.E. Hanson) pp 274-313. Cambridge University Press. Cambridge U.K.

Gitay H., Brown S., Easterling W.E., Jallow B. Antle J.O., Apps M., Beamish R. and Cerri. C. (2001). Ecosystem and their goods and services. In *Climate Change 2001. Impacts, Adaptation and Vulnerability to Climate Change* (Eds. J.J. Mc Carthy, O. F. Canziani, N.A. Leary, Contribution of Working Group II to the Third Assessment Report of the Intergovernmental Panel on Climate Change. Cambridge University Press. Cambridge.

IPCC. (2007). Climate change, 2007 : Synthesis report, *summary of policy makers*. IPCC, Geneva, Switzerland, pp 22.

Kalra. N., Aggarwal P.K., Chander S., Pathak H., Choudhary R., Chaudhary A.S, Sehgall M., Soni, U.A., Sharma A., Jolly M., Singh U.K., Ahmed O. and Hussain M.Z. (2003). Impacts of climate change on agriculture. In *climate changed and India : Vulnerability Assessment and Adaptation* (Eds. P.R. Shukla, S.K. Sharma, N.H. Ravindranath A. Garg and S. Bhattacharya), pp. 193-226.University Press. India

Nadarjan N., Basu P.S. and Gupta Sanjeev (2010). Effect of climate adversities in pulse Productivity. Crop adaptation to climate change, souvenir of National Symposium on Food Security in Context of changing climate, 30 Oct – 01 Nov. 2010, Kanpur India; pp 43-15.

Narain, Sarju. (2011). Impact of drought on rural migration in Bundelkhand region. of U.P. (India) unpublished work.

National Commission on farmers, India (2006). A draft- National policy for farmers.

Pathak H. Ladha J.K. Aggarwal P.K., Peng S., Das S., Yadvinder Singh, Bijay, Singh, Kamra S.K., Mishra B., SastriA.R.A.S., Aggawal H.P., Das D.K. ad Gupta R.K. (2003). Climatic potential and on farm yield trends of rice and wheat in the Indo-gangetic plains. *Field Crops Res.* 80 (3); pp 223-234.

Pathak, H. (2010). Climate change and Agriculture. Impact, mitigation and adaptation, souvenir of National Symposium on Food Security in Context of changing climate, 30 Oct - 01 Nov2010, Kanpur, India; pp 01-07.

Reilly J. Hohman N. (1993). Climate change and Agriculture : The role of International trade, *An economic Association pop. Proceeding* 83; pp 306-312.

Robert J. Collier *et al.* (2008). Effect of climate change on dairy cattle production, *Annals of Arid Zone* 47 (3 and 4); pp : 393-421, 2008.

Simon J. Lloyd and R. Sari Kovats. (2008). Health and climate change in Arid zones, *Annals of Arid Zone* 47 (3 and 4); pp 443-455.

Sing, A.K. and Burman R.R. (2010). Adaptation and mitigation of climate change : role of Agricultural Extension. Souvenir of National Symposium on Food Security in Context of changing climate, 30 Oct – 01 Nov. 2010, Kanpur, India; pp 159-161

Surabhi Mittal. (2008). Yojna, October, 2010. Vol. 54 pp. 38-39.

Tewari G.C. (2010). Sustainability of Indian Agriculture in context of climate change, Sonvenier of National Symposium on Food Security in Context of changing climate, 30 Oct – 01 Nov. 2010, Kanpur India, Preamble.

World Bank, (2008). *The world development Indicators*. The world Bank Publications.

2014, Sustainable Rural Development through Agriculture *Pages* ***127–134***
Editors: **Dr. Shobhana Gupta and Dr. S.S. Tomar**
Published by: **BIOTECH BOOKS, NEW DELHI**

Chapter 8

Strategies for Drought Mitigation in Western Rajasthan

Lokesh Kumar Jain

Drought is a common phenomenon in the Rajasthan particularly in western Rajasthan of the country. The average annual rainfall in western Rajasthan is less than 300 mm but is markedly variable ranges from 200 mm in the west to about 370 mm in the East. Rainfall is the main climatic constraints to dryland agriculture. The lack of rainfall coupled with other adverse climatic components resulted in droughts and disturbed the normal life of resource poor farming communities in terms of loss of crops and livestock. The environment degrades and livelihoods largely affected as their lifeline destroyed. Arid western Plan Zone 1A of Rajasthan state has cultivated area of 27 m ha contributes 53 per cent of the geographical area. Sand dunes and desert soils occupy major area in the zone. There are aeoline soils and loamy fine to coarse and calcareous at places. The zone covers four tehsils of Jodhpur (Phalodi, Shergarh, Osian and Jodhpur) and all tehsils of district Barmer. Rainfed agriculture is complex, diverse and risk prone and is characterized by low levels of productivity and low input usage. Vagaries of the monsoon result in wide variation and instability in yields. Pearl millet is the predominant crop of the zone followed by cluster bean, moth bean, sesame and green gram in kharif. The crops like castor during Kharif and cumin, rapeseed and mustard, wheat, Isabgol etc. during Rabi are grown on only 7 per cent irrigated area. The ecosystem of western Rajasthan is very fragile because rainfall is being the major source to sustain entire production system. Since 1971 western Rajasthan has suffered recurrent droughts. According to literature, the worst years with below average rainfall are 1972, 1974, 1979, 1982, 1986, 1987, 2002,2004, 2007,2008 and 2009. In a survey Barmer district of Rajasthan, farmers recalled 64 drought years from 1901 to 2010 (63 in Western Rajasthan). Irrespective of the irrigation facilities created, the successful agriculture is still governed by the rains. The district has low food and fodder productivity and peoples are characterized by low literacy, poor socio economic status and low risk bearing capacity. Land degradation due to soil erosion is a more acute problem faced by the farmers in dryland regions and has social impact like shortage of water, loss of cultivable area, low productivity etc. The major agent of soil erosion is runoff. According to the farmers, droughts seriously affected them both in the normally wet and dry seasons also as high rains causes flood situation due to impermeable layer beneath soil hinders infiltration.

Impacts of Drought

The 2002 drought had the most crippling effect in Indian economy. The year 2002-03 witnessed a setback in agriculture and GDP growth declined to 3.2 per cent as compared to 5.7 per cent growth in 2001-02 due to severe drought conditions in several states. In the state more than 30 thousand villages was affected by drought. The seasonal deficit in Barmer district was up to 74 per cent of normal rainfall (277mm). There were unprecedented crop failures due to no rains in July. The regional (western Rajasthan) grain production fall by 50 per cent short of expected levels. The production of major food grain *e.g.* pearl millet and kharif pulses was reduced by 98 percent in Barmer district while in western Rajasthan it was 86 and 89 per cent, respectively. Rainfed areas therefore need to contribute substantially to incremental output by producing marketable surpluses more reliably. The droughts led to widespread suffering with loss of cattle (10 per cent) also. Farmers in the zone summarized the major effects of these droughts are as follows:

- ☆ Partial or complete crop failure (because of low soil moisture content and disease outbreaks).
- ☆ Livestock deaths.
- ☆ Productivity of pastures and grazing land reduced significantly.
- ☆ Wild life damaged the crops.
- ☆ Trees drying and dying.
- ☆ Boreholes, rivers, springs and other water points drying and caused scarcity of drinking water for both livestock and human.
- ☆ Shortage of basic commodities in the local markets.
- ☆ Price hikes across all sectors.
- ☆ High government expenditure because of food imports.
- ☆ Malnutrition, especially in children.
- ☆ Unemployment, coupled with rampant crime and robbery.
- ☆ Migration to urban areas for casual labour and herds to other state or districts for fodder.
- ☆ Loss of natural resources.
- ☆ Land degradation in terms of desertification and salinity.

- ✫ Increased asthmatic problem due to increased suspension of sand in air.
- ✫ Mentally disturbance because of unemployment and loss of earnings.
- ✫ Conflicts increased for water and inability to repayment of credits.

Strategies for Coping Droughts on Individual Basis

Farmers' responses to the effects of drought have been varied. Below are some of the actions being undertaken to mitigate drought in western Rajasthan especially in Barmer districts.

Water Harvesting

In western zone farmers harvest water individually nearby their houses in villages and from rooftops in urban areas and divert natural springs into in locally structured tanks known as *Tanka* but water available hardly for a year. After facing this drought they constructed big tanks as per family size and drought in mind and permanent it. The size of tanks also constructed keeping herd's size as water is also used for drinking purpose of small ruminants. This ensures that they have a substantial amount of water stored up to mitigate drought. This step also reduced drudgery for farm women because they take water for drinking from miles.

Livestock

As in western zone the livelihood of 70 percent farmers depend on livestock particularly on small ruminants. Farmers who have few ruminants graze them on common pastures or feed them leafs of trees like *P.cineraria*. Farmers store fodder of pearlmillet and pulses which failed to produce grain in indigenous storage techniques like Karai and Pachawa. The capacity of storage in these techniques up to 1000 q. Farmers feed their animal with urea treated fodder and chaff cutting to avoid wastage during droughts also.

Granaries

A majority of the farmers store food to be used in case of a drought. They have a specific granary (kothi, Kanana etc) stocked with grain (pearlmillet for a shorter period of time) with protection against post harvest pests/stored grain pest indigenously treated with ash and dry neem leafs. This granary is kept untouched and out of bounds for children. Only the head of the household is allowed into it.

Seed

Farmers generally stored some seed of current season for next year sowing so they have not need to pay higher prices for purchase the new seeds. They treat the grain that will be used for seed with Kerosene or Methyl Parathion 2 per cent dust.

Savings

Farmers save some money in order to purchase food in times of drought. These are the affluent farmers with flourishing business enterprises who can set aside enough money to buy food for a whole year if the rains fail. Farmers also earned some money from labour and farm women and children's collect materials from perennial trees like gum and seed from *Acacia Senegal*, fruit from *P.cineraria*, fruit from *Salvadora oleoides* (Pilu), *Capparis decidus* (Ker) and *Ziziphus spp.* (Ber) etc. and sold in local market to earn income. Farmers also collect and stored shankhpusphi a medicinal crop and collect guggal from tress to sold in market for purchase food grain also. These are principal source of cash income on western Rajasthan in rainfed conditions. The fruiting time do not match with Kharif season so family members collect these without affecting agriculture.

Drought Tolerant Crops

Farmers are slowly discarding the idea of growing pearlmillet as the main crop because sowing of this crop cannot be recommended after 15 July. They are shifting towards growing pulses and oilseed. These crops are drought resistant and therefore gave a sizable yield even with little rain. Farmers are also adopting recently released high yielding, drought tolerant and early maturing cultivars resistant to postharvest pests instead of deshi cultivars susceptible for biotic and abiotic stress likewise *RMO 40 and RMO 435 of* moth (*Vigna acontifolia*), *HHB 67 Improved of pearlmillet and RGC 936 of clusterbean. Watermelon,* a cucurbit, is popular ripens fast and saves people from possible starvation. It is not only drought resistant but can survive in poor soils.

Composting

As the soils of western Rajasthan is sandy with poor or no organic matter. The infestation of crop with polyphagous insects like termite and white grub is common. The major reason of termite infestation is directly use of row dung instead of composting. Now-a-days farmer's composting it and incorporate at right time so income was increased because of increased yield and down the cost of cultivation. Some farmer's also using chemicals for managing these pests.

Strategies for Coping Droughts on Community Basis

Seed Banks

Seed is the primer in improving crop yields. It is meant for sustainable, ecological and economic development of the farmer. The availability of seeds, which is a critical and basic input in attaining the higher productivity level, has increased substantially in the country over the years but low risk bearing capacity avoids farmers of rainfed areas to use improved seed. To strengthen the seed sector, particularly the production and distribution of quality seeds to farmers, government/state Ag. Department/ gram panchyat through cooperatives should implement the seed bank scheme so the requirement of pure and reliable seed to the grower may be supplied on time. Seed production of recommended variety if produced by the village through identified farmers will be no cheating and timely supply with good quality will ensured is most important for rainfed agriculture. Promotion of cultivation of nutrient rich varieties of staple or traditional food like ICTP 9802 of pearlmillet that is rich in iron and the deficiency of Iron was observed among children and pregnant ladies in western Rajasthan on wide scale. Also ensure proper distribution through co-operatives on community basis.

Fodder Banks

The population of livestock is more in western Rajasthan in comparison to human population because the livelihood of marginal and small farmers is dependent on small ruminants. Similar to seed banks, fodder production and storage progamme may be initiated on local level and marketed through cooperatives so every farmer may be benefitted. Also panchayats should regenerate and develop common pasture land in centre and state government funded schemes and control the grazing pattern ultimately improve the overall productivity of land and livestock will reflects in better livelihood. A large volume of community land known as gochhar and orans in western Rajasthan provides additional income and compensate the cost of maintenance after development. Some efforts from state agriculture departments have been made by large scale distribution of seed and other input of grasses and fodder to promote the farmers in irrigated belts of western Rajasthan improved the fodder productivity. Through MNERGA panchyats may develop new pastures with adequate water harvesting structures also with control grazing.

Agroforestry and Plantation

Several grass species like Sewan Grass (*Lesiurus sindicus*), Anjan/Buffelgrass (*Cenchrus ciliaris*), Dhaman/Birdwood grass (*Cenchrus setigarus*), Bharut (*Carchrus biflorus*), Lampra (*Aristida funiculata*), Bekaria (*Indigophora cordifolia*) etc may be grown because some of them have mechanism to tolerate high salt concentration in the root zone soil. Also some of these highly tolerant grasses either exclude the absorption of salts from the soil and/or deposit the absorbed/translocated salts at points within the plant system which do not allow them to interfere in metabolic processes. The performance of these grasses in association with salt tolerant trees like *Prosopis juliflora, Acacia nilotica, Prosopis cineraria* in a unified agroforestry system has vast potential to yield of green biomass per annum in saline soil.

Plantaion of multi-purpose trees *e.g.* Jaal/Peelu (*Salvadora persica*), Mithi jaal (*Salvadora oleoides*), Kummat (*Acacia Senegal*), Khejri (*Prosopis cineraria*), Rohida (*Tecomella undulata*), Kair (*Capparis decidua*), Jharberi (*Zizyphus numularia*), Babool(*Acacia tortilis*), Vilayati babool (*Prosopis juliflora*), Fog (*Calligonum polygonoides*) etc on waste land improve environment and pay sustainably for a long periods.

Water Harvesting Structures

A community based approach for harvesting rain water for drinking to human beings and livestock for sufficient time according to catchment and population will be a good step and advantageous for livelihood. In western Rajasthan the ground water level is 200 feet and salinity of ground water causes health hazards for rural poor is serious concern. During water scarcity period, farm women consume three to four hours for a pitcher after travelling a mile. It may be reclaim when water harvesting structure with sufficient storage like nadis, tanks etc. may be constructed. It will improve health, reduce the drudgery and the time spent for watering that may be utilized in either agriculture or other income generation activities.

Custom Hiring Centre

Farm mechanization is another area which can make a significant impact on labour productivity and cost reduction. Rainfed farmers cannot afford high cost agricultural implements developed for cultivation in arid sandy soils and hiring during peak period face unfair rates taken by implements holders. Deficit of human labour during peak periods and fast declining draught animal population has changed the farm power scenario of rainfed agriculture. Thus there is a growing demand for powered mechanization in this sector. We are likely to witness more and more custom hiring centres and outsourcing of major farm operations to these centres by farmers of all categories. The mechanization is today's ultimate need and it become many fold in reference to western Rajasthan where soils are sandy, large holding size, undulated topography, high temperature, high evaporation rates etc. Custom hiring centre must be equipped with locally required powered driven machineries for timely completion of agriculture operation. The water holding capacity of soils of western Rajasthan is very low due to soil texture and poor organic matter content. The available time for agriculture operation like sowing and intercultural operations is very short. In this situation farmers follow broadcasting resulted in low yield coupled with uneven plant growth. Broadcasting also hinder use of farm implements in standing crop like hoeing, spraying of plant protection chemicals, harvesting etc. increases the cost of cultivation by reducing labour efficiency. Undulated topography once leveled permit the power driven machinery for field preparation to harvesting. Efforts to make available energy efficient and time saving farm machines, implements and tools to the farmers, so a custom hiring centre on community basis may be established, maintained and distributed fairly as per priority. It will reduce the drudgery as well as cost of production. The energy and time wasted in such operation may be diverted to other productive and income generating works. The majo equipment and machinery required for pool will be tractor, disc harrow, disc plough leveler, seed drill, power sprayer, power weeder and chaff cutter etc. The chaff cutte will solve the problem of most scare resource *i.e.* fodder for animals. Thus, the comin decades will be crucial for improving productivity and profitability in rainfed area by reducing costs besides reducing drudgery and increasing employment.

Development of Common Pool Resources (CPR)

Intensive development of CPR of the village will be carried out for the benefit c farming community as whole which constitutes the key for success. Self help group (SGH's) will be constituted to implement and monitor the development of commo pool resources financial assistance for community works would be mobilized in th form of revolving funds by the government support. Development and effectiv management of CPR's will help to sustain the productivity of each farming unit i the village on long term basis. To update the knowledge systems IT need be use wherever feasible through Resource Centres (RCs) funds for outsourcing exper (either from GO or NGO) by the village communities, and facilities for encouragir INM, NPM, PHM and in-village seed production.

Steps to Mitigate Droughts from Governments

Extension Services

Efforts should be made to strengthen the infrastructure for agricultural extension and to develop human resource for extension activities in the States It was calculate that an extension personnel employed by the government hardly pay ½ hour per farmer available per year for their attention. The programmes taken up are agricultural extension through voluntary organizations; strengthening of research extension; farmer's linkages, agricultural extension through farmer's organizations, human resource development in extension; and training of women in agriculture. Farm women's Self Help Groups were mobilized and agricultural support services were provided to farm women through these groups. Training and extension services were provided to farmers and farm women through different training programmes. The on-going schemes on agricultural extension have been integrated and restructured to bring in reforms in agricultural extension at State level. The major thrust areas for reforms in policy framework of agricultural extension to encourage private sector partnership in agricultural and augment media support to extension for capacity building and skill up gradation of farmers/farm women and extension functionaries.

Agri-Clinics and Agri-business Centres

Establishment of Agri Clinics and Agri-business Centres at every block to involve private sector in agricultural extension information and support services, for supplementing efforts of Govt./Public sector agencies. The scheme aims at utilizing unemployed agriculture graduates or those willing to establish their own service/ business centers to provide extension services to farmers on payment basis. Training programmes for the identified graduates to set up their Agri-clinics/Agri-business centers in different areas of activities also.

Cooperation

The cooperative sector has been playing a significant role in the area of disbursing agriculture credit, providing market support to farmers and distribution of agricultural inputs etc. An intensive trainings to the cooperative personnel for enhancing their efficiency in the management of cooperative societies specially to procure perishable commodities to protect the growers from making distress sale in the event of bumper crops, particularly during the peak arrival period. The small ruminants (the domain of the shepherds) had been considered primarily by the MoRD under DDP and DPAP programmes. Fortunately the individuals through cooperatives can earn daily income via generating employment or income by sale of milk and milk products, manure (dung and stall-refuse as compost) etc.

Credit

The total quantum of ground level credit flow for agriculture and allied activities provided by cooperative banks, commercial banks, Regional Rural Banks (RRBs) and other agencies. The Kisan Credit Card (KCC) scheme which aims to facilitate adequate and timely credit support from banking system to the farmers for cultivation needs including purchase of inputs in a flexible and cost-effective manner has been gaining popularity over the years Government should provide assistance to cooperative credit institutions for professionalization, training and capacity building, improving their infrastructure facility, computerization, improving mobility of the field staff, etc.

Agricultural Insurance

The National Agricultural Insurance Scheme (NAIS), introduced from Rabi 1999-2000 provides greater coverage in terms of farmers (non-loanee farmers brought under coverage), crops (annual, commercial/horticultural crops included) and risks (*i.e.* upto the value of threshold yield). Now-a-days there is urgent need to include damage caused by other climatic factors like frost etc. Further research need at to understand monsoon behavior and to develop effective and dependable strategies for short term and long term forecasting.

Medicinal and Aromatic Crops

Government should promote or provide funds or project for growing medicinal and aromatic plants to farmers individually or through non-profit organizations so peoples can get additional income during lean period. A number of medicinal and aromatic crops have been screened for salinity and sodicity tolerance in India. Crops like Isabgol (*Plantago ovata*) can be successfully cultivated in soils having pH of 9.5 and EC between 8-10 dS/m. Similarly, Salvadora, a non-edible oil tree can be grown in salt affected soils very successfully. Industrial species like *Euphorbia* and *mulethi* (*Glycyrrhiza glabra*) also have good scope for cultivation in salty environments. Other medicinal and aromatic crops showed potential yields are Guggal (*Commiphora whight*), *Aloe vera*, Sonamukhi (*Cassia angustifolia*), Tumba (*Citrullus colocynthis*), Shankhphusphi etc.

Other Policy Issues

A compressive national policy on "Drought Management" may be prepared including both short term and long term measures for drought mitigation should be in place specific to arid and semi arid areas. Also develop and strengthened linkage between research, disciplines and institutions involved in drought prediction and drought management for efficient short term, medium term and long term planning. The indigenous methods for weather forecasting may be promoted and standardized. Forage production and research should be considered as a national agenda. Shift in investment from irrigated to rainfed areas and emphasis must be focus on aforestation and sand dune stabilization.

The main challenge for researchers, development practitioners and policy makers is to facilitate this process and modify as per farmer's needs by involving his active participation. The main focus and goals for all drought mitigation projects have been to strengthen the community livelihoods. Governments need to work together towards fulfilling the concept of sustainable livelihoods systems, as elaborated above. This way, food security and community resilience to drought can be achieved.

2014, Sustainable Rural Development through Agriculture *Pages 135–140*
Editors: **Dr. Shobhana Gupta and Dr. S.S. Tomar**
Published by: **BIOTECH BOOKS, NEW DELHI**

Chapter 9

Sustainable Crop Production Techniques in Western Rajasthan

Lokesh Kumar Jain and P.D. Kumawat

ABSTRACT

The India is an agrarian country and more than 60 per cent of total cultivated area is still depends on rainfall from SE monsoon. There is always an uncertainty of weather in the country especially in dryland areas or hot arid western zone of India mostly comprises of Thar desert of western Rajasthan. The hot arid western zone comprises 208751 km^2 with 57 per cent of total geographical area of Rajasthan. Improved agronomical and engineering practices leads to sustainable income from these drought prone areas.

The hot arid zone of western Rajasthan has highest animal and human population density among the deserts of world. The climate of western Rajasthan is very uneven. Due to the uncertain and erratic weather conditions including very high temperature difference (-2° to 48°C), late onset of monsoon, frequent droughts, early cessation of monsoon, low rainfall (<300mm annually), less no. of rainy days (15-20) or hot winds (10-35 km/hour) during summer results in failure or very poor crop production is a common phenomenon. In western Rajasthan where annual mean rainfall ranged 150-300 mm annually and drought appear in every three years. The soils are sandy to sandy loam with poor organic matter and low water holding capacity. The water and soil quality also hinder the crop production. The Barmer districts have highest area under pearlmillet 1.01 mha but the productivity is very low (103 kg/ha) in comparison to state (825 kg/ha) during 2008-2009. In order to bring stability and increase productivity, a certain techniques for such situations is very essential. The following steps may be of great help in solving the problems to a great extent.

1. Soil and Moisture Conservation Practices

i. *In situ* Water Harvesting: In the western Rajasthan sorghum is grown as rainfed crop primarily for fodder because the animal population is higher than human population and livelihood of most of farmers depends on livestock only. In area where sorghum is grown farmers are advised to go for ridge and furrow system of planting. This will be helpful in moisture conservation and in some cases avoid water logging conditions in the cropped area. For water harvesting in situ board bed and furrow system is suggested. The furrow may be spaced 12-15 meter along the contours or across the stop. The spacing is selected according to the cropping pattern. This will increase the moisture status of soil profile so that crop can with stand during dry spell as well as early recession of rainfall, which ultimately increase the yield potentials of the crops.

ii. To get better and uniform germination of pearlmillet in desert areas, an recommendation for better contact of seed with soil moisture, the seeded row to be compacted with 4 kg heavy rubber wheel inbuilt on tynes of seed drill.

iii. Water harvesting and recycling-run-off is inevitable in the tropical and sub-tropical climates. It varies from 10 to 40 per cent of the total rainfall. Of this, at least 30 per cent could be harvested and harvested water could be re-cycled into the donar area during dry-spell period. The concept of run-off re-cycling can be immediately transmitted only in the deep black soil where the seepage losses are very much less.

iv. Re-charging of ground water- Exploitation of ground water for various purposes results in decline water table. The only alternative is to replenish the ground water by artificial re-charging, therefore the construction of percolation tanks, check dams and anicuts across the water crosses are considered quite suitable for artificial re-charging of ground water. Harvest the water in Khadins also.

v. Contour bunding: In western Rajasthan, sloppy sand dunes are abundant and naked, sowing of crops across the slope or planting grasses on contour helps in reduction of soil erosion and muddy rain water which block the pores on cultivated areas and will reduce crust formation. As pearlmillet is common cereal crop in this area, the seed size is small than other cereals and crust creates problem in germination ultimately reflects poor yield. An idiom is common "No rainfall after sowing of pearlmillet in next nine days".

vi. Land leveling: In western Rajasthan, leveling is very tedious and costly because of big sand dunes however on cultivated area leveling may be practiced for better yield by high infiltration and uniform growth.

2. Suitable Crops and their Varieties

Crops which do well under normal years may not do so under abnormal weather situations. The selection of crop and their varieties must be suited to dry condition and short duration. The varieties must be tolerance to biotic and abiotic stress also. For example, under south Rajasthan condition sorghum perform better than maize (traditional crop) during varied climate in *kharif* and barley found better than wheat (traditional crop) in normal years while taramira, mustard and safflower during drought years in rabi. Growing large areas under improved varieties of cereals, pulses and oilseeds in kharif and rabi. These varieties have been found most suitable for dryland areas:

Bajra	HHB 67 Improved, HHB 67, MH 169, RAJ 171, CZP 9802
Sorghum	CSH 6, CSH 9, SPV 245, SPV 346
Groundnut	JL 24, AK-12-24
Greengram	RMG-62, RMG 268
Castor	GCH 7, DCS 9
Sesame	RT 125, RT 346, RT 127
Clusterbean	RGC 936, RGC 1003, Maru Guar
Moth	RMO 40, CAZRI MOTH 3, RMO 435, RMO 257
Wheat	RAJ 1482, RAJ 3077
Barley	RD 2055, RD 2052
Gram	Dahod yellow, BG 209, C 235, ICCV 10
Mustard	Varuna, Durgamani, Bio 902
Taramira	RTM 314

3. Crop Rotation and Cropping Systems

As in sandy soils the infection of polyphagous pests and soil borne diseases are common due to low fertility and undecomposed material. Also the choice of crop is limited tremendously increase the infection of insect pest and diseases. In such situation farmers must follow crop rotation with legumes followed by cereals. In dryland conditions intercropping systems showed superiority over sole cropping systems. Intercropping systems increases production and income. It also gave stability and assured successive crop production in addition to other indirect benefits in the soil health. Some important inter cropping systems have been found suitable is as under

i. Paired planting of bajra at 30/75 cm and raising of greengram or cluster bean in the inter space was found most suitable intercropping system.

ii. Growing of two rows of greengram in the inter row space of castor grown at 120 cm row spacing was better than growing sole crops.

iii. Gram + mustard in 4:1 row ratio at 30 cm spacing was found remunerative.

iv. Inter cropping combination like grass+greengram and sunflower+ greengram found better for western Rajasthan.

4. Integrated Nutrient Management

In rainfed areas the crop production is a risky occupation and farming community don't interested for investment. The soil status of western Rajasthan particularly Barmer and Jaiselmer are very low in nitrogen, medium in phosphorus and medium in potash. Also hike in prices of inorganic fertilizers farmers don't afford results in imbalance nutrition. In hot arid zone particularly in Barmer and Jaiselmer district of Rajasthan the fertilizer consumption rate per hectare is only 1.69 and 10.41 kg during Kharif 2009-10 (State average 32.27 kg/ha). To maintain the productivity of soils farmers must follow the integrated approach by using organic, inorganic and bio-fertilizers in balanced ratio for long term sustainability. In organic manures farmers may apply FYM, compost, green manuring, cakes of neem and tumba while *Rhizobium*, PSB, *Azotobactor* etc. as bio-fertilizer.

5. Alternate Land Use

The western part of Rajasthan have lot of marginal land not suitable for agricultural production but may be put for cultivation with trees, shrubs, grasses, medicinal plants etc so farmers harvest additional income besides arable farming. The medicinal plants especially shunkhphusphi, aloevira, sonamukhi, mulathi etc found best and gave profitable remunerative. The discriminate use of land lead to severe soil erosion problems, silting-up of reservoirs and consequently the occurrence of flash floods. To achieve greater efficiency in utilizing the resources in dryland and also to combat the "Energy crisis" and deficiency of animal feed, major approaches in this direction should be

i. Putting the land (class IV and above) under suitable grasses, legumes and fodder crops and integrating with animal productive systems.

ii. Using the same land or a portion of it simultaneously or sequentially for food, fodder and fuel, due attention being given to interaction among various uses aiming to obtaining greater sustained production and securing both immediate benefits and long term environmental concerns.

   The various alternate land use systems which may be adopted are as follows

   a) Agro-forestry system
   b) Silvi-pastoral system
   c) Farm forestry system
   d) Agro-horticulture system
   e) Silvi-horticulture system
   f) Alley cropping system

For these systems the probable suitable crops, grasses, legumes and trees are given below. For hot arid areas (<300 mm annual rainfall)

Crops	Grasses and Leguems	Trees
Sorghum	*Cenchrus ciliaris*	*Zizyphus numularia*
Pearlmillet	*Cenchrus setigerus*	*Acacia tortilis*
Greengram	*Sehima nervosum*	*Calligonum poly gonoides*
Clusterbean	*Panicum antidotale*	
Moth (Kidney bean)	*Lasiurus sindicus*	*Prosopis juliflora*
Kulthi (Horsegram)	*Heteropogon controtus*	*Capparis Decidua*
Groundnut	*Dichanthium annulatum*	*Prosopis cineraria*
Mustard	*Atylosia scarabaeoides*	*Leucaena leucocephala*
Grain ameranth	*Stylosanthes humilies*	*Tecomella undulata*
Barley	*Clitoria ternatea*	*Salvadora persica*
Safflower	*Carchrus biflorus*	*Salvadora Oleoides*
Gram	*Aristida funiculata*	*Acacia Senegal*
Taramira	*Indigophora Cordifolia*	

In agri-hotriculture system, crops with erect nature and early maturing habit like greengram, moth and cluster bean may be intercropped in orchard of ber, datepalm, pomegranate or during early stage of establishment of orchards of pomegranate, castor may be intercropped to utilize the inter row space and incorporate organic matter and irrigated with drip for water saving.

6. Other Steps

i) Crop Life Saving Practices

a) Life saving irrigation from harvested water
b) Frequent shallow inter culture operation to keep the field free from weeds and create soil mulch or use of surface mulch for prevent water losses
c) Removing lower leaves of the plants
d) Reduced plant population by the way of thinning
e) Use of anti-transpirants etc.

ii) Mid-Season Corrections

In drought years, complete crop failure can be avoided by adopting mid-term corrections *i.e.*

a) Removing the susceptible component and allowing the resistant component in intercropping system.
b) Removing main component and sowing of rabi crop, if rain received late.
c) Ratooning in case of sorghum and bajra can be done successfully.

iii) Community Nurseries

Transplanting of sorghum and bajra seeding at 20-25 days stage can be done successfully for gap filling and for main crop when rains occur after long drought. This is possible by growing community nurseries.

iv) Other improved agronomical practices

a) Timely sowing at proper spacing

b) Use of balance fertilizers

c) Proper and timely plant protection measures

d) Timely hoeing and weeding operations

e) Sowing improved and healthy seed with proper seed treatment

f) Seed soaking before sowing etc.

Keeping above point in mind a sustainability in dry region may obtained after adopting or manipulating the different agronomical practices as per climate situation.

2014, Sustainable Rural Development through Agriculture *Pages* ***141–148***
Editors: **Dr. Shobhana Gupta and Dr. S.S. Tomar**
Published by: **BIOTECH BOOKS, NEW DELHI**

Chapter 10

Sustainable Use of Agricultural Biodiversity in Western Rajasthan

Lokesh Kumar Jain, R.K. Rathore and P.C. Meena

Agricultural biodiversity refers to the variety and variability of animals, plants, and micro-organisms on earth that are important to food and agriculture which result from the interaction between the environment, genetic resources and the management systems and practices used by people. It has spatial, temporal and scale dimensions. It comprises the diversity of genetic resources (varieties, breeds, etc.) and species used directly or indirectly for food and agriculture for the production of food, fodder, fibre, fuel and pharmaceuticals, the diversity of species that support production (soil biota, pollinators, predators, etc.) and those in the wider environment that support agro-ecosystems (agricultural, pastoral, forest and aquatic) as well as the diversity of the agro-ecosystems themselves. Agricultural biodiversity encompass the variety and variability of animals, plants and micro-organisms which are necessary to sustain key functions of the agro-ecosystem, its structure and processes for, and in support of, food production and food security (FAO). When the environment provides congenial conditions for the living beings, a variety of plants and animals flourish in the region. This is called Biodiversity. A major reason for a great diversity of life found on the earth is the tremendous variety of environment available for life. A rich biodiversity is observed in all the five ecosystems and has come to stay with us. The term biodiversity was first defined in the earth summit held at Rio-de- Janeiro, Brazil in 1992. Biodiversity basically helps to maximize productivity and preserves ecosystems (B. S. Siddaramaiah, 2013).

Agricultural biodiversity is essential for sustainable food production along with livelihood security and sustainable agriculture. Drought is an unexceptionally common phenomenon in the particularly western Rajasthan of the country. The average annual rainfall in western Rajasthan is less than 300 mm but is markedly variable from 200 mm in the west to about 370 mm in the East. The lack of rainfall coupled with other adverse climatic components has been the cause of many droughts causing widespread disruption to many farming communities with the significant loss of crops and livestock. The rural poor are also most affected by environmental degradation and their livelihoods largely depend on the available natural resources. In a survey Barmer district of Rajasthan, farmers recalled 64 drought years from 1901 to 2010 (63 in Western Rajasthan). Farmers of these areas adopted mixed cropping as well as mixed farming because the crop production is totally rainfed and the risk bearing capacity is very low due to unsecure production during Kharif. They grow more than 7 crops in a field by mixing seeds of pearlmillet, greengram, moth, sesame, clusterbean, cowpea as field crops and water melon and cucumber for vegetables. The basic factor behind this is to get dietary requirement for foodgrain, pulse, oilseeds and vegetables for family. The livelihood for more than 70 per cent farm families is directly depended on livestock also. The rear small ruminants like sheep and goats in herds. The alternate source of income is products of forest trees like *Prosopis cineraria (khejri)*, Jaal/Peelu (*Salvadora persica*), Mithi jaal (*Salvadora oleoides*), Kummat (*Acacia senegal*), Rohida (*Tecomella undulata*), Kair (*Capparis decidua*), Jharberi (*Zizyphus numularia*), Babool (*Acacia tortilis*), Vilayati babool (*Prosopis juliflora*), Fog (*Calligonum polygonoides*), etc. They generally use animal dung as manure instead of chemical fertilizers; limited uses of plant protection chemicals resulted in improvement and maintain the soil health, increase predators and keep the parasites to a limited extent. In other words they practice organic farming under the shade of diversification and contribute beneficially to the ecosystem.

Normally, it is assumed that if the number of crops grown in an enumeration unit is large, say about ten, each occupying only 10 per cent of the cropped area, it would mean that the cropped area is uniformly distributed among all the ten crops. In such a case, crop diversification is of a very high degree and may be taken as an extreme case of crop diversification. Diversification of agricultural systems supports strong populations of predators and parasites that keep pest populations at manageable levels in western Rajasthan particularly. Crop diversification minimizes nutrient losses and to reduce pressure from insect-pests, weeds and diseases. Mixing of crops minimizes the incidence of leaf roller/capsule borer and bud fly to greater extents. A variety of decomposing organisms starting from microbes to macro fauna like earth worms etc. help in decomposition of organic matter, maintain soil fertility, soil structure and act as growth promoters. A rich biodiversity at all levels is a boon to agricultural growth and production. Crop diversification is intended to give a wider choice in the production of a variety of crops in given area so as to expand production related activities on various crops and also to lessen risk. It is generally viewed as a shift from traditionally grown less remunerative crops to more remunerative crops. The crop shift also takes place due to governmental policies and thrust on some crops over a given time *e.g.* Green revolution, technology mission on oilseed and pulses, market (growing cash crops and high yielding crops) became major threatened top bio diversity. Also agricultural biodiversity plays a huge role in maintaining resilient local economies, balanced diets and balanced ecosystems. The rapid disappearance of agricultural biodiversity and the lack of measures to protect it are therefore great causes of concern. Mainstream agricultural policies, which generally promote monoculture agriculture, Genetically Modified Organisms (GMOs) and Intellectual Property Rights threaten such agricultural biodiversity, having an impact on agricultural landscapes, species, varieties, breeds, the wild relatives of crops and livestock, pollinators, micro-organisms and genes. These policies and practices lead to the disappearance of plant and animal species, and the knowledge embedded in their management and use.

There is some good news though in recent years many promising initiatives have been launched around the world that aims to preserve and manage agricultural biodiversity. Small-scale family farmers often play a central role in these, acting as custodians of biodiversity. But other actors and institutions also play important roles. Producers, public and private institutions and consumers are reconnecting with each other through innovative market arrangements, many of them at local or regional level. Farmers and researchers are taking up joint research initiatives, and farmers' organizations are engaging in dialogues with policymakers, pushing for policies that enhance agro biodiversity.

As it is very clear that diversity exists in western Rajasthan is substantial and today need is to make it sustainable. A many constraints identified during survey conducted at farmer's field and need to rectify individually, socially and government part point of view. In this article we discuss the different constraints and measure to rectify them.

Types of Biodiversity

1. Genetic Biodiversity: Refers to the richness of variation in the species
2. Ecological Biodiversity: Existence of a variety of species in a geographical area.

Benefits of Biodiversity

Biodiversity directly and indirectly affected the different agro-ecosystems in various ways. It

- ✰ Provide food and feed.
- ✰ Supply of building materials, paper, fuel, wood etc.
- ✰ Providing raw materials such as rubber, ore, wax, oil etc for industrial use.
- ✰ Acting as a source of herbal medicines.
- ✰ Regulating global process such as water cycle and atmospheric gases.
- ✰ Nutrient recycling, nitrogen fixation.
- ✰ Soil and moisture conservation.
- ✰ Waste decomposition.

Objects of Biodiversity

- ✰ Widen the understanding of agricultural biodiversity.

- ☆ Role of biodiversity and its contribution in different production systems at all levels
- ☆ To improving integration and coordination of farm level activities and farming processes.

Dimensions of Agri-Biodiversity

- ☆ Production of food and other agricultural products sustainably along with genetic resources diversity (plant and animal) in different types of production systems.
- ☆ Conservation, sustainable use and enhancement of the biological resources to support sustainable production systems, particularly soil biota, pollinators and predators.
- ☆ Ecological and social services provided by agro-ecosystems such as landscape and wildlife protection, soil protection and health (fertility, structure and function), water cycle and water quality, air quality, CO_2 sequestration, etc.

Components of Biodiversity

The changes in crop pattern are the outcome of the interactive effect of many factors which can be broadly into the following groups;

- ☆ Resource related factors covering irrigation, rainfall and soil fertility.
- ☆ Technology related factors covering not only seed, fertilizer, and water technologies but also those related to marketing, storage and processing.
- ☆ Household related factors covering food and fodder self-sufficiency requirement as well as investment capacity.
- ☆ Price related factors covering output and input prices as well as trade policies and other economic policies that affect these prices either directly or indirectly.
- ☆ Institutional and infrastructure related factors covering farm size and tenancy arrangements, research, extension and marketing systems and government regulatory policies.

Constraints in Adoption in Arid Agro-ecosystem

- ☆ Harsh environmental conditions.
- ☆ Poor Education status
- ☆ Lack of proper media
- ☆ Desertification
- ☆ Lack of mechanization.
- ☆ Lack of suitable inputs for agro ecosystems.
- ☆ Suitable transportation

Threats to Biodiversity

Five hundred thousand years ago when human beings first appeared on this planet, the population was small and they had very little impact on environment. But then several major breakthroughs have occurred. Green revolution which took place in 1960's resulted in record food production, accelerating population increase.

Green Revolution - A Major Threat

This phenomenon combined with industrial revolution contributed to greater exploitation of natural resources leading to environmental degradation. The green revolution in particular resulted in the following undesirable consequences

a) Disappearance of indigenous varieties: About 75 per cent of diversity in agriculture crops is believed to have lost since the beginning of this century. In India due to green revolution rice varieties decreased from nearly one lakh to just less than one hundred.

b) Decrease in Pulse production and soil fertility: Due to mono-cropping of high yielding cereals, the pulse production in the country decreased and the soil became less fertile.

c) Pollution from fertilizers and pesticides: The excess of nitrogenous fertilizers applied to field crops were washed away in irrigation water and polluted drinking water. Application of pesticides not only pollutes atmosphere, water bodies and food products but also kill useful insects such as predators and pollinators.

d) Soil salinity: Due to continuous irrigation a large part of agricultural lands have become saline and unproductive.

e) Attack of invasive species from other regions due to their habitat disturbances.

f) Acid rain and ozone layer depletion due to over exploitation of fossil fuels resultant in production of CFC's. It also produces natural calamities like Hurricanes, melting of glaciers raising sea level, floods, drought, etc. disturbs biodiversity to a great extent.

Steps or Measures to Enhance Biodiversity

- ☆ To imparts training and conduct different capacity building programmes to awake awareness, technical know- how on agricultural biodiversity.
- ☆ Also spread and disseminate sustainable methods for agricultural biodiversity and its conservation vertically and horizontally through different means of mass media.
- ☆ Involve representatives of different farm families for decision-making and planning. Also involve women actively so that at least 60 percent farm activities conducted by them so educate them so they realise their responsibilities in agriculture production and biodiversity.
- ☆ To integrate agricultural biodiversity in national biodiversity programmes to make it as a part of agricultural strategies and plans and formulate

action plans to micro levels under strong guidance to local level workers or extension functionaries.

- ✰ To make rationale participation and proper coordination among experts of different streams that contributes to bio diversity *viz.*, agriculture, animal husbandry, forestry, watershed, social activists, etc. Also involve relevant organizations, ministries and sectoral bodies at all levels.
- ✰ Proper and sound marketing system so mitigate the influences of fluctuations of market, market forces and the existing economic framework because it have major impacts on agricultural biodiversity in terms of forcing to grow a particular crop(s), selling the products to get money for other purpose etc.
- ✰ Proper and affordable storage system for agriculture and related produce so they secure for fixed tenures.
- ✰ Impart knowledge about property rights and regulations so new innovation may be develop and rewarded.
- ✰ A proper Code of Conduct on Agricultural Biodiversity should be develop and implement which would assist private sector, government and civil society organizations to identify their rights and obligations and inform their policy-makers and programme developers.
- ✰ The program should be either started or strengthen that have effective linkages between research, production, extension, post-harvest management, processing, marketing and exports and bring about a rapid development of agriculture in the region.
- ✰ Implementing National Agriculture Insurance Scheme will cover food crops and oilseeds and annual commercial and horticulture crops. Some additional benefits given to Small and marginal farmers beyond the subsidy under the Scheme.
- ✰ A pilot scheme on Seed Crop Insurance has been launched which will cover the risk factor involved in production of seeds.
- ✰ Seed Bank Scheme: About 7-8 percent of certified seeds produced in the country will be kept in buffer stock to meet any eventualities arising out of drought, floods or any other form of natural calamities.
- ✰ An emphasizing greater coordination and information sharing between research and development programmes and better formal and informal sector linkages.
- ✰ Strengthening national agricultural research systems on agricultural biodiversity related issues.
- ✰ Furthering farmer-driven participatory research and technology development processes, for example through farmer field schools.
- ✰ Emphasizing three main issues: threats and positive incentives for agricultural biodiversity; ecosystem approaches and ecosystem functions; and specific research areas such as soil biota, pollinators and predator.

- ☆ Developing communication methods and facilitating the exchange of information on relevant scientific research and practical information between different actors and stakeholders, especially South-South.

Crop Diversification Impact

Crop diversification in the country is taking the form of increased areas under commercial crops including vegetables and fruits since independence. However, this has gained momentum in the last decade favouring increased area under vegetables and fruits and also to some extent on commercial crops like sugar cane, cotton and oilseeds crops specially soybean. The major problems and constraints in crop diversification are primarily due to the following reasons with varied degrees of influence:

- ☆ Over 117 m/ha (63 percent) of the cropped area in the country is completely dependent on rainfall.
- ☆ Sub-optimal and over-use of resources like land and water resources, causing a negative impact on the environment and sustainability of agriculture.
- ☆ Inadequate supply of seeds and plants of improved cultivars.
- ☆ Fragmentation of land holding less favouring modernization and mechanization of agriculture.
- ☆ Poor basic infrastructure like rural roads, power, transport, communications etc.
- ☆ Inadequate post-harvest technologies and inadequate infrastructure for post-harvest handling of perishable horticultural produce.
- ☆ Very weak agro-based industry.
- ☆ Weak research - extension - farmer linkages.
- ☆ Inadequately trained human resources together with persistent and large scale illiteracy amongst farmers.
- ☆ Host of diseases and pests affecting most crop plants.
- ☆ Poor database for horticultural crops.
- ☆ Decreased investments in the agricultural sector over the years.

Research Support for Diversification

Future agriculture will be much more knowledge and skill based rather than the raditional subsistence agriculture. In the wake of globalization and opening up of he global market, there will be much more opportunity for entrepreneurship levelopment in agriculture. This also calls for paradigm shifts in research and echnology development and also the transfer of technology for successful crop liversification. The research system not only needs to address the issues connected vith continuance and indulgence and knowledge in the areas of emerging echnologies but also create a cadre of scientists through the continuous upgrade of kills and human resource development. The researchers also need to popularize the echnologies, impart knowledge and skills to the extension functionaries for the transfer of technologies to the farmers. This knowledge-based farming will call for much more interaction between the researchers, extension workers and farmers. The fruits of the innovative technologies should reach the farmers at the earliest and also spread in the quickest possible time.

Conclusion

Farmers are required to implement a crop rotation that maintains or builds soil organic matter, works to control pests, manages and conserves nutrients, and protects against erosion. With the adoption of mixed cropping them also encourage organic farming and landscape diversity. Diversification provides more habitat for wildlife due to the resulting diversity of housing, breeding and nutritional supply but quantification depends on individual activities performed by farmers and substantial farming is also least detrimental for wildlife conservation and landscape and it is today's utmost need. Maintain the fields free from weeds will minimize the incidence of insecticides and diseases because these are secondary/alternate host for many insect and disease carrying pests. It is essential to develop sufficient bio-diversity especially at agro-ecosystem levels.

The need of the hour is transformation from chemical-based farming practices to eco-friendly alternatives, such as diversification in cropping patterns, crop rotation, intercropping, integrated pest and disease management, integrated nutrient management, and increasing use of green manure in fields. Mixed cropping will discourage monoculture without disturbing the yield or profits, by encouraging the activities of natural enemies of pests and also reduce the dependence on expensive and hazardous chemical inputs. (P.K. Shetty Socio-ecological Implications of Pesticide Use in India Economic and Political Weekly December 4, 2004)

References

B. S. Siddaramaiah. 2013. Conserving Natural Resources and Biodiversity: *International Conference on Extension Educational Strategies for Sustainable Agricultural Development - A Global Perspective: December 5-8, 2013, University of Agricultural Sciences, Bangalore, INDIA.*

FAO 1999: Opportunities, incentives and approaches for the conservation and sustainable use of agricultural biodiversity in agro-ecosystems and production systems 2–4 December 1998, FAO Headquarters, Rome, Italy.

2014, Sustainable Rural Development through Agriculture *Pages* ***149–158***
Editors: **Dr. Shobhana Gupta and Dr. S.S. Tomar**
Published by: **BIOTECH BOOKS, NEW DELHI**

Chapter 11

Combating Agricultural Risks through Knowledge Management in Rice Farming

R. Sendilkumar

ABSTRACT

Agricultural production is inherently a risky business and farmers face a variety of agricultural risks. The prevalence of risks in agriculture is not new and farmers have over generation developed and adapted ways of reducing, mitigating and coping with risks. Focusing this issue, study was carried out with an objective to account various adaptive strategies followed by the farmers to combat the agricultural risks. The 180 farmers from Nagapattinam District of Tamil Nadu were selected by adopting simple proportionate random sampling. The data related to risk management strategies adapted to tackle agricultural risks were collected though well structured interview schedule, focus group discussion and observation. The rice farmers were exposed to all kinds of agricultural risks *viz.*, production risks, market risks, financial risks and human resource risks. Some of the adaptive risk management strategies (*ex-ante* and *ex-post)* of farmers were selection of suitable and demand driven variety, Integrated Pest Management, registered for crop insurance, availing MSP facility, Crop loan, paying extra wages, adopting labour saving implements and representing issues to the Government in advance. Adequate empowerment mechanism may be taken up towards risk preparedness plan following ex-ante strategies and as a policy measure to develop individual risk management database supporting the issue at the micro level.

Keywords: *Agricultural risk management practices, Production risks, Market risks, Financial risks, Human resource risks, Policy risks.*

Introduction

India has been traditionally vulnerable to natural disasters on account of its unique geo-climatic conditions. Flood, droughts, cyclones, earthquakes and landslides have been a recurrent phenomenon (GOI, 2006). Agricultural production is inherently a risky business and farmers face a variety of agricultural risks *viz.*, production, market, financial, human resource and policy risks (World Bank Report, 2OO5). Management of risk at the farm and sectoral level is being recognized as a critical factor in achieving the Millennium Development Goals (MDGs) agreed by the nations/states in the year 2000 (India Country Report, 2005). The prevalence of risks in agriculture is not new and farmers have over generation developed and adapted ways of reducing, mitigating and coping with risk (Singh, 2005) and climate-resilient farming was a way of life in those earlier days (Swaminathan, 2011). Farmers have acquired different behavioural pattern to face various kinds of risks by adapting to formal and informal risk management strategies. However the diffusion of agricultural risk management knowledge among the rice farming community is very slow and not attracted much attention. Keeping this concept as a central theme, a study was carried with an objective to account various adaptive strategies followed by the rice farmers to combat the agricultural risks and suggest suitable extension strategies to foster the knowledge management of Agricultural risks.

Methodology

The study was conducted in six blocks of Nagapattinam District of Tamil Nadu with a sample size of 180 rice farmers distributed to 12 villages by adopting simple proportionate random sampling based on the secondary data served by the Primary Agricultural Credit Banks, Commercial Banks and Block level Agricultural Office. The data related to experiences in various kinds of agricultural risks and the strategies adapted to tackle were collected separately by using a well structured interview schedule, focus group discussion and observation method. The data obtained were analysed using percentage analysis for meaningful interpretation.

Results and Discussion

Agricultural Risks as Experienced by the Rice Farmers

It is seen from the Table 11.1 that vast majority (93.38 per cent) of rice farmers were subject to experienced the production risks, followed by human resource risks (88.34 per cent) financial risks (74.45 per cent), market risks (67.78 per cent) and policy risks (48.34 per cent). From the findings it could be concluded that the rice farmers were faced an array of agricultural risks and only that intensity of risks would vary according to the influence of external variables. Further it could be generalized that more than two-third of the rice farmers were exposed to all kinds of agricultural risks *viz.*, production risks, market risks, financial risks and human resource risks, whereas only less than a half (48.34 per cent) of the respondents had experienced the policy risks

Table 11.1. Proportion of Agricultural Risks Experienced by the Rice Farmers

n=180

Sl.No.	*Types of Agricultural Risks*	*Number*	*Per cent*
1.	Production risks	169	93.38
2.	Market risks	122	67.78
3.	Financial risks	143	74.45
4.	Human resource risks	159	88.34
5.	Policy risks	87	48.34

* Multiple responses.

Agricultural Risks Management Strategies Adopted by the Rice Farmers

The actual risk management practices adopted by the farmers physically to overcome each agricultural risk in rice cultivation are discussed below.

1. Actual Agricultural Risks Management Practices Adopted to Combat the Production Risks

It could be seen from Table 11.2 that there were ten strategies which have been found adopted physically by the rice farmers to combat the production risks. Among the ten, the widely adopted were choice or selection of variety, Integrated Pest Management, registered for crop insurance and adopted by 75.55, 71.11 and 66.66 per cent respectively. The reason attributed could be that the three strategies have direct effect on the yield attributes and showed visible results as per their perception. The other strategies *viz.*, adjusting sowing time (46.66 per cent) Integrated Nutrient

Table 11.2: Actual Agricultural Risks Management Practices Followed to Combat the Production Risks in Rice Cultivation

n=180

Sl.No.	*Strategies*	*No**	*Per cent*
1.	Diversification of crops (Mixed cropping/Intercropping)	32	17.77
2.	Diversification of enterprise	19	10.55
3.	Choice/Selection of variety	136	75.55
4.	Adjusting the sowing time	84	46.66
5.	Increased area under organic farming	12	06.66
6.	Adoption of Modern technology (SRI)	11	06.11
7.	Integrated Pest Management	128	71.11
8.	Integrated Nutrient Management	56	31.11
9.	Proper planning to stock inputs	21	11.66
10.	Registered for crop insurance	120	66.66

* Multiple responses.

Management (31.11 per cent), diversification of crops (17.77 per cent) proper planning to stock inputs (11.66 per cent), diversification of enterprise (10.55 per cent), increased area under organic farming (6.66 per cent) and adoption of modern technology (SRI) (6.11 per cent). Shiyani (1988) also reported that farmers had shifted from subsistence farming to commercial crops cultivation and exhibited their orientation towards crop diversification. It could be concluded that, no single agricultural risk management strategy alone would be sufficient to combat production risks. Hence the rice farmers could have adopted more than one practice to face production risks.

2. Actual Agricultural Risks Management Practices Adopted to Combat Market Risks

From the perusal of the Table 11.3 it could be seen that there were ten strategies found adopted by the rice farmers to combat market risks. Out of ten, the most important one was to grow demand driven or preferred variety (80.55 per cent), followed by availing the Governments' MSP facility (77.78 per cent), selecting and cultivate high quality seeds for production (67.22 per cent) and selling the product to private vendor who offers fair price (35.55 per cent). Chand (2003) also reported that price support mechanism had been the principle means by which Indian farmers had received some protection against market risks. The other strategies were storing the produce till they get increased price (30.00 per cent), reduced input cost through maximization of the owned resources (26.66 per cent), arrangement with private vendor who gives better price (17.77 per cent), selling value added products (4.44 per cent) marketing the produce through regulated markets and obtaining ware house receipt (3.33 per

Table 11.3: Actual Agricultural Risks Management Practices Followed to Combat the Market Risks in Rice Cultivation

n=180

Sl.No.	*Strategies*	*No**	*Per cent*
1.	Growing only demand driven or preferred variety	145	80.55
2.	Selecting and grow high quality seeds for seed production	127	67.22
3.	Storing the produce until the farmer gets increased price	54	30.00
4.	Reducing input cost through maximization of the owned resources	48	26.66
5.	Selling value added produces	8	4.44
6.	Growing unique variety, which was not grown in others regular season by others	3	1.66
7.	Marketing the produce through regulated markets and obtain warehouse receipt	6	3.33
8.	Availing the government's support through Minimum support Price	140	77.78
9.	Selling the produce to Pvt. Vendor, who offers higher price	64	35.55
10.	Arrangement with Pvt. Vendor, who promised to give better price on the event of maintaining stock over a reasonable period	32	17.77

* Multiple responses.

cent), growing unique variety, which was not grown in other regular season by others (1.66 per cent). It may be concluded that majority of the rice farmers adopted three strategies *viz.*, growing only demand driven or market preferred variety, availing the Government MSP facility, selecting and growing high quality seed for production. Through these strategies, farmers would have a belief that market risks could be managed to certain extent. However, little above one-third (35.55 per cent) of farmers would still resort to sold their produce to private vendor, who had offered higher price. Although there would be no guarantee and assurance on price due to uncertainty in output price, farmers would continue this arrangement based on confidence and trust.

3. Actual Agricultural Risks Management Prctices Adopted to Combat the Financial Risks

Rice farmers followed eight kinds of strategies to combat financial risks (Table 11.4). Among the eight, two-thirds (66.11 per cent) of rice farmers adopted the strategy of availing crop loan from PACBs/RRBs/CB in time, and about fifty per cent of the respondents resorted to the strategies such as availing credit facility from trader or commission agent and getting financial assistance farm private institutions or money lender. Almost every revenue village in the study district has been provided with either PACBs or Commercial Banks. Thus, the service of extending crop loan with low interest was readily available at their door step. The promotional efforts of formal institutions for extending credit and insurance were being yet another probable reason. Timeliness and easy dealing made the rice farmers to adopt the strategy like availing the credit facility from traders or commission agents. Another one-third (33.33 per cent) of farmers were followed the strategy of availing agricultural jewel loan (AJL) facility. The probable reasons would be that immediate liquidity of jewel, low interest rate for AJL, and banker's preference in giving loan in this strategy. The other operational strategies followed were, arrangement for immediate liquidity (12.77 per cent) availing the credit facility from input suppliers (12.22 per cent), ensuring enough

Table 11.4: Actual Agricultural Risks Management Practices Followed to Combat the Financial Risks in Rice Cultivation

n=180

Sl.No.	Strategies	No*	Per cent
1.	Availing crop loan from PACBs/RRBs/CB in time	119	66.11
2.	Ensuring enough owned funds at the time of monsoon	20	11.11
3.	Arranging financial help from relatives/friends/fellow farmers	16	8.88
4.	Getting financial assistance from Pvt. Institution/money lender	84	46.66
5.	Availing the Agricultural Jewel Loan(AJL) facility	60	33.33
6.	Availing the credit facility from input suppliers	22	12.22
7.	Availing the credit facility from trader/commission agent	94	52.22
8.	Arranging for immediate liquidity	23	12.77

* Multiple responses.

owned funds at the time of monsoon (11.11 per cent) and arranging financial help from relatives/friends/fellow farmers (8.88 per cent). The findings of Wenner and Arias (2002) also reported that maintaining financial reserves as one of the measures to mitigate financial risks. To conclude, the rice farmers adopted more than one strategy to face the financial risks aroused out of rice cultivation.

4. Actual Agricultural Risks Management Practices Adopted to Combat the Human Resource Risks

It is observed from the Table 11.5 that rice farmers were found to follow seven operational strategies to mitigate the human resource risks aroused out of farm operation from time to time. The list of strategies were paying extra wages with package to attract labour during high labour demand period (46.66 per cent) adopting labour saving implements (44.44 per cent) adjusting the time for farm operation so as to meet labour scarcity (21.11 per cent), inviting labourers from neighboring area where the human resource was plenty (17.77 per cent) empowering of feminization in agriculture (17.22 per cent), employing family labourers (14.44 per cent) and empowering the available human resources towards productive work (6.11 per cent). There was labour deficit especially during important farm operations like transplanting, weeding and harvesting. Among these operations, harvesting was the critical one and most farmers subjected to face this human resource risk. As a practical and on the spot solution, farmers immediately proposed extra wages with additional package (Rice @ 1/8 of gross yield with working lunch) to attract the available labourers in order to complete the operation in time. Thus it could be the possible reason. The availability of heavy machinery like rice thresher and combined harvester are so common in season time in the locale of research. The owners of combined harvester have preplanned their availability schedule to each and every packet of region ensuring their compact services for the whole block. Moreover, the attributes like the timeliness, easiness, comparatively less cost for the harvesting etc., have inclined the farmers to adopt the same. Because of these benefits, the farmers could have opted for adopting labour saving implement as one of the strategies to mitigate human resource risks. One-fifth (21.11 per cent) of the rice farmers used operational strategy "adjusting the time for farm operation so as to meet labour demand". Since, this strategy did not involve any cost; it was adopted by the considerable proportion of farmers. Inviting human resources from other neighbouring area where the human resource availability was plenty being yet another strategy adopted by less than one-fifth (17.77 per cent) of farmers. The male labour availability is so limited for agriculture due to many reasons. The secondary and territory sectors attracted lot of male labours and the urbanization circumvents the agricultural labours through migration. As a result, most of the farm works suffered due to these reasons. Therefore the only option before the farmers was using the female gender into the farming operations. However the wages paid for female labour was still low. This would have favoured the farmers to adopt this particular strategy empowering of feminization in agriculture. Employing family labour was also a strategy adopted for more than one-tenth (14.44 per cent) of farmers. The farmer cum agricultural labour mostly resorted to this type of strategy in order to face the labour shortage.

Table 11.5: Actual Agricultural Risks Management Practices Followed to Combat the Human Resource Risks in Rice Cultivation

n=180

Sl.No.	Strategies	No*	Per cent
1.	Employing family labour	26	14.44
2.	Empowering of feminization in agriculture	31	17.22
3.	Adopting labour saving implement(tractor, power tiller, thresher, and harvesting machine)	80	44.44
4.	Adjusting the time for farm operation so as to meet labour demand	38	21.11
5.	Inviting human resource from other area where the HR is in plenty	32	17.77
6.	Paying extra wages with package to attract labour during labour demand	84	46.66
7.	Empowering the available human resources towards productive work	11	6.11

* Multiple responses.

5. Actual Agricultural Risks Management Practices to Combat the Policy Risks

The results in Table 11.6 revealed that the rice farmers had resorted to five kinds of strategies to face the policy risks. They were listed in the order *viz.*, representing the issue to the Government in advance (10.55 per cent),organizing federation or congress to fight for right (6.66 per cent), representing the plea from bottom to top level through participatory approach, demonstrating the ill-effects of changes in policy through a non-violent mode and getting media support with 3.33 per cent each. The first strategy namely representing the issue to the Government in advance had been adopted by only one-tenth (10.55 per cent) of rice farmers. The periodical conduct of farmer's grievance and interface meeting by the district administration in the District Collectorate every month provide plate form for settling the issues could be the possible reasons for the reported findings. Organizing federation or congress to fight for the right was the next strategy to combat of policy risks. Cauvery Delta Farmers Association, Cauvery Farmer's Protection Group and Federation of Farmers Association are being some of the Farmers Interest Groups operating in the locale. The awareness on the advantages of formulating farmers interest group, farmers commodity group could be the probable reasons that, why farmers have had adopted this strategy. The remaining strategies were less pronounced even then it was believed that it could have yielded a positive response. Of course these strategies were time consuming one; whereas the strategy 'Getting media support' was a worth strategy would yield immediate response from the Government and public side. The powerful media magic could have helped the farmer to get relief from the clutches of policy risks.

Table 11.6: Actual Agricultural Risks Management Practices Followed to Combat the Policy Risks in Rice Cultivation

n=180

Sl.No.	Strategies	No*	Per cent
1.	Representing issues to the Government in advance	19	10.55
2.	Organizing federation/association/congress to fight for the right	12	6.66
3.	Representing the plea from bottom to top level through participatory approach	6	3.33
4.	Demonstrating the ill effects of Change in policy through non-violent mode	6	3.33
5.	Getting the media support	6	3.33

* Multiple responses.

7. Kinds of Risk Management Strategies Followed by the Rice Farmers

It is quite apparent from the Table 11.7 that more than two-third (67.22 per cent) of rice farmers had resorted to ex-post risk management strategies; where as little more than one-tenth (12.78 per cent) had followed the ***ex-ante*** risk management strategies. The possible reason for the adoption of ex-post risk management strategies that, most of the government interventions regarding the agriculture sector is of ex-post in nature. The relief package is one such example of these kinds. Hence, mostly the rice farmers adopted ex-post strategies. On the other hand, lesser proportion of farmers had resorted to ex-ante risk management strategies, because of their experience gained over a period of time might have instructed the farmers to adapt the ***ex-ante*** risk management strategies. ***Ex-ante*** or ***ex-post*** risk management strategies alone not able to manage adequately the agricultural risk and hence the farmers (20 per cent) could have adopted these two strategies combined.

Table 11.7: Kinds of Risk Management Strategies Followed by the Rice Farmers to Combat the Agricultural Risks

n=180

Sl.No.	Kinds of Strategies	Number	Per cent
1.	Ex- ante risk management strategies	23	12.78
2.	Ex-post risk management strategies	121	67.22
3.	Combined	36	20.00

Implications and Recommendations

On the light of above findings and discussions the following extension strategies could be suggested to promote the farmer's adaptive strategies to combat agricultural risks in rice farming.

1. Production risks, in spite of its good awareness and perceived causes, rice farmers were found still thrived hard to face the same. Hence an appropriate suitable technology transfer efforts like establishing Farm clinic and Agri. Clinic, Farmer Field School and Farmers advisory service on weather based information and market based information would empower the farmers to face the challenge.
2. Group farming and hiring of machinery could also be tried as interventions to handle human resource risk in farming. As mechanization is not fully possible, farmers club or FIG/CIG could be encouraged to buy labour saving machinery with a financial assistance from the bank with incentive linked package scheme.
3. Intensive educational efforts may be taken to popularize the advantages of futuristic risk management tools like crop insurance, contract farming, ware house receipts and commodity derivatives. Adequate knowledge empowerment mechanism may be taken up towards risk preparedness plan following ex-ante strategies and as a policy measure developing individual database supporting the issue at the micro level.
4. Risk management data base extension centre could be established under existing provisions of ATMA at block level. An integrated risk management database centre linked with the service of VKC/RKC could be better utilized for sharing risk related information.
5. Group marketing, value addition and allowing interstate sale of rice are suggested to tackle market related risks.
6. For the financial risks, dissemination of Kissan Credit Card Scheme (KCCS) may be made active and functional.

Conclusion

The suggested extension strategies may be incorporated in framing appropriate policy to address the various agricultural risks faced by the peasants through system approach model inclusive of policy system, research system, extension system and client system to integrate all the stake holders in a common platform.

Terminlogies

Production risks derive from the uncertain natural growth processes of crops and livestock. Weather, disease, pests, and other factors that affect both the quantity and quality of commodities produced cause production risk.

Market risks refer to uncertainty about the prices producers will receive for commodities or the prices they must pay for inputs. The nature of price risk varies significantly from commodity to commodity.

Financial risks results when the farm business borrows money and creates an obligation to repay debt. Rising interest rates, the prospect of loans being lent by lenders, and restricted credit availability are also aspects of financial risk.

Human resource risks refer to factors such as problems with human health or personal relationships that can affect the farm business. Accidents, illness, death, and divorce are examples of personal crises that can threaten a farm business.

Policy risks results from uncertainties surrounding government actions. Tax laws, regulations for chemical use, rules for animal waste disposal, and the level of price or income support payments are examples of government decisions that can have a major impact on the farm business.

Risk management is a scientific approach which deals with solving the problems of the pure risks faced by the individuals and business. It is a function of management in the same style as marketing management, financial management, or personnel management

References

Chand Ramesh 2003. Minimum Support Price in Agriculture - Changing Requirements, Economic and Political Weekly. 38(20): 3027-3028.

GOI. 2006. Towards faster and more inclusive growth: An approach to 11th Five- Year Plan, Planning Commission, Government of India, New Delhi.

India Country Report 2005. Millennium Development Goals. Ministry of Statistics and Programme Implementation, GOI, India.

Sendilkumar.R 2009. Risk management behavioural pattern of farmers - An empirical Study. UnPub. Ph.D thesis, Dept of Agricultural Extension and Rural Sociology, TNAU, Coimbatore.

Shiyani, RL, Pandya. H.L 1988. Diversification of Agriculture in Gujarat: A Spatial-Temporal Analysis. Indian J of Agricultural Economics 53(4): 627-639.

Singh, RP 2005. Yield Risk and its Management by the Farmers, Agricultural Economics Research Review 18(2): 163.

Swaminathan. M.S 2011. Document indigenous climate-resilient farming knowledge. The Hindu, dated 6th January.

Wenner, M, D. Arias. 2002. Agricultural Insurances-Where are we? US Agency for International Development (USAID) Agreement No. LAG-A-00-96-90016-00 through Broadening Access and Strengthening Input Market Systems Collaborative Research Support Program (BASIS-CRSP) and the World Council of Credit Unions, Inc.

World Bank 2005. Managing Agricultural Production Risk Innovations in Developing Countries. The World Bank Agriculture and Rural Development Department.

2014, Sustainable Rural Development through Agriculture Pages 159–167
Editors: Dr. Shobhana Gupta and Dr. S.S. Tomar
Published by: BIOTECH BOOKS, NEW DELHI

Chapter 12

Factor's for Improving Rice Quality

Smita Singh, A.K. Singh, U.N. Shukla, L.B. Singh, Anchal Sharma, Deepak Srivastawa, Satish Singh Baghel and M. Jerman

Asia's irrigated rice area of 73 M ha produces more than 70 per cent of the global rice production (Maclean *et al.*, 2002). About two-thirds of this area is located in the tropics and the remainder in the subtropics (Hossain and Laborte, 1993). To cope with the projected population growth and the associated increase in food demand, average irrigated rice yields in Asia must increase with 60 per cent from 4.9 t ha-1 to 8 t ha-1 in 2025 (IRRI, 1995). Yields in tropical areas are often below 5 t ha-1 (Dobermann *et al.*, 2003), while those in subtropical areas often exceed 6 t ha-1 (Dobermann *et al.*, 2003).Past research has compared the productivity of rice in subtropical and tropical areas to identify growth factors contributing to this difference. For example, (Ying *et al.*, 1998) found in two years that the same high-yielding genotypes yielded 33 and 62 per cent more under subtropical than under tropical conditions. Rice is one of the most important grain crops in the world. Obtaining both high yield and high-quality rice is becoming more and more important. Rice quality includes processing quality as well as appearance, cooking, eating, and nutritional quality factors. Many factors can affect quality. (Horie *et al.*, 1997) reported that irrigated rice yields were affected by genotypic characteristics, location and their interaction (G×E). However, they did not quantify the contribution of the individual factors to the yield variance. Remarkably, very few attempts have been made to analyze the performance of irrigated rice across subtropical and tropical areas in Asia. Where available, such reports are descriptive and do not explain yield differences (Ying *et al.*, 1998). Sound comparison among sites is only possible when experiments are carried out and data collected according to standard protocols. This is difficult to realize, which may explain the scarcity of systematically collected empirical information on the interaction between rice genotypes and environments across Asia. Better understanding of genotype-environment interactions facilitates design of new genotypes, identification of test conditions, and genotype selection for specific well defined conditions (Jackson *et al.*, 1996; Yan and Hunt, 1998). Yield is a function of biomass production and the proportion of that biomass invested in the grains (harvest index, HI) (Cock and Yoshida, 1972; Murata and Matsushima, 1975). Future yield increase must result from improvements in one or both components (Farrell *et al.*, 1998). Intercepted photosynthetically active radiation (IPAR) and crop-specific radiation-use efficiency (RUE, biomass produced per unit of IPAR) determine biomass production. Assimilates for grain filling originate from photosynthesis after flowering and from reserves stored in culms and leaves during the vegetative period. Longer crop growth durations, leading to higher radiation interception, result in higher yields if associated with favorable RUE and HI. High temperatures during vegetative growth stages accelerate crop development, shortening crop growth duration, resulting in lower grain yields. The photothermal quotient (PTQ, the ratio of average radiation intensity to average temperature during the grain filling period) has been proposed as a single indicator to capture the combined effect of the yield-determining factors radiation and temperature on yields (Nix, 1976). As basic food security in Asia improves, demand for rice with superior quality properties increases. Grain quality will become even more important in the future, when the economic situation of the very poor—many of whom depend on rice as their staple food—improves and they demand higher quality rice. The nutritional value of rice is related to its protein content, while its amylase content is an important indicator of cooking and consumption quality (Juliano, 1985; Lii *et al.*, 1996). Low amylose content increases water absorption, volume expansion, and stickiness of cooked rice (Delwiche *et al.*, 1996), while high amylose content is associated with hard grains after cooking (Juliano, 1998). Preferences for soft and hard rice grains vary widely across Asia. Quality characteristics of rice are formed during grain filling, and are determined by environmental and genotypic factors and their interactions. However, these interactions are poorly understood, which hampers breeding of rice varieties with targeted quality characteristics.

1. Challenges Facing Rice Sector Development

Challenges

A number of challenges that have been identified as limiting factors to the rice sector as follows:

1.1 Development and Availability of Improved Seeds

Availability of adequate varieties having tolerance to drought, cold weather, major insect pests and diseases are major challenges facing the rice sub-sector in the country. There are hundreds of local/traditional rice varieties grown by farmers in the rain-fed lowland, irrigated low land and upland ecosystems. Most of these varieties have, however, low yield potential, late maturing, and prone to lodging when improved management practices such as application of fertilizers are used. Improved seeds have been applied by only 10 percent of farmers. The use of self-saving seeds is common among small scale rice farmers and these seeds are of low quality.

1.2 Availability of Water is a Prerequisite for Increased Rice Production

Most of the rice production in the country depends on rainfall. Annual variation in the amount and distribution makes rain-fed rice production susceptible to flooding and/or drought, often within the same season. Drought risk impedes investment, causing production to stagnate at subsistence level. The deterioration of the drainage and irrigation facilities is posing a considerable constraint to the increased production of rice. The funding for the necessary rehabilitation of the irrigation infrastructure is beyond the capacity of small producers.

1.3 Development and Availability of Improved Post-harvest Processing Technologies and Addition Value Processes

Inadequate post-harvest technologies influenced limited use of the crop as opposed to Asian countries which can prepare many dishes from rice. Processing is done by inferior processing machines except for large scale production farms and hence the quality of milled rice is low.

1.4 Development of Labour Saving Technology

Ninety-five percent of farm operations in paddy production are done manually. These operations are coupled with intensive labour requirements. Planting is mainly done by hand (both during direct sowing/broadcasting, seeding and transplanting of seedlings), likewise harvesting, threshing and cleaning of paddy. Transportation of paddy from the field to storage (home/market) is by direct head loading and sometimes ox-carts or vehicles are used depending on availability. The labour input in paddling is high, requiring between 300 and 350 man hours/ha. Similarly manual transplanting and weeding are labour intensive, and each operation requires between 200 and 300 man hours/ha.

1.5 Improving Accessibility of Credit to Farmers

Rice producers and processors have restricted access to credit due to the reluctance and inability of commercial banking institutions to supply financial opportunities to the rural sector. The usual reason for the lack of credit is the insistence of commercial banks to have land as collateral, and their reluctance to accept leasehold land, especially short leases, as collateral. Farmers and processors lack collateral to be able to borrow from commercial banks. Also, loans from the banks have very high interest rates. The Government has been promoting Savings and Credit Cooperative Societies (SACCOS), but the members are still limited. Farmers are also decapitated by low producer prices.

1.6 Development and Rehabilitation of Communication, Transportation and Marketing Infrastructure

Most of the rural infrastructures are in poor state and in need of rehabilitation. It has been observed that rural feeder roads are impassable during the rain season when agricultural inputs are needed. This as a result contributes to the increased costs of production. Besides, inadequate storage and marketing infrastructures also deny the farmers access to sell their produce at convenient place compelling them to sell at lower price as they cannot bargain for better prices. In this regard, the development and rehabilitation of rural infrastructures can contribute positively to the reduction of production costs and fetch better prices of product at the market, to facilitate access in marketing information and to enhance quick transport of both paddy and milled rice to the central warehouses near major roads and to the markets in urban centers.

1.7 Improving Private Sector Participation

Private sector participation in rice value chain is limited due to inadequate financial capabilities and inadequate availability of information on markets. However, participation of private sector is noted in rice milling, supply of agro-inputs (fertilizers, pesticides) and trading.

2. Factor for Improving Rice Quality

2.1 Genetic Resources Conservation and Use

Collection and conservation of germplasm has been limited to a few crops due to poor finances and poor linkages. For this reason, germplasm collection and conservation for rice has been undertaken by the respective rice research programme in collaboration with interested international research institutions. A wide range of germplasm of about 400 genotypes has been collected within the country and from IRRI, IITA and WARDA. The materials are rejuvenated, field evaluated, characterized and conserved at KATRIN. Desirable genotypes are incorporated in rice improvement programme. NERICA lines are among materials introduced into the country and being evaluated in field. More collection missions are needed to be conducted and concerted efforts need to be put in place by the rice research programme in collaboration with National Plant Genetic Resource Centre (NPGRC) to collect, characterize and conserve germplasm for future mining of novel genes against biotic and abiotic stresses. In this regard, more funds are needed for collection and conservation and to build capacity in terms of facilities for both the rice research programme and the NPGR. In this endeavor; KATRIN which is the coordinating centre for rice research, will need to be provided with modern cold storage facilities and ensure that the same is conserved at the NPGRC. Participatory breeding methods enhanced through use of biotechnology tools will be used in developing high yielding varieties with desirable consumer/market qualities. These qualities include desirable post harvest and production attributes such as milling percentage, resistance to lodging, early maturity, and resistance to major biotic and abiotic factors. Developed and released varieties will be registered with the Registrar of Plant Breeders Rights at MAFC. Genetic engineering will be deployed in accordance to the national bio-safety framework.

2.2 Temperature

Of all the environmental factors, temperature, especially in the stage of grain filling, is the most important one affecting rice quality. High temperature in the tassel to maturity stage would speed up the grain-filling rate and shorten the grain-filling period, resulting in lower grain filling and decreased unpolished and polished grain ratios, with grain chalkiness becoming bigger and transparency lower. Low temperature will also result in bigger grain chalk spots and an increased percent unfilled grain. So, both high and low temperature would decrease the processing and appearance quality of rice (Sao Dong-sheng and Tang Jian 1987, Li jun and Gu De-fa 1995, Tang xiang-ru and Yu tie-qiao 1991). High temperature would increase the temperature of grain gelatinization and vice versa. The highest amylose content of rice grain will be obtained only in an optimum temperature condition. Both high and low temperature reduced the accumulation of grain amylose (Resurreccion 1997). It was reported that, after the tassel stage, high temperature would influence the nutritional quality (protein content) of rice. In the heading stage, if daily average temperature increased from 20 to 30 °C, the total polished grain ratio would decrease 24-35 per cent (He 1992). To keep a high polished grain ratio, japonica rice needs a lower temperature than indica rice (Yang Hua-long *et al.*, 2001). Compared with that of high temperature, the negative effect on rice from low temperature is much smaller. It was reported that, to obtain good rice quality, the optimum temperature should be in the range of 20-30 °C in the tassel and maturity stages. This is the main reason why the quality of late rice, ratooning rice, and rice planted in high-altitude regions is better than that of early rice, middle rice, and rice planted in low-altitude regions (Wu Guan-ting and Xia Ying-wu 1994).

2.3 Illumination

Illumination is an important factor related to rice quality, just after temperature. f illumination is low in the late growing stage, rice photosynthesis is restrained. In contrast, high illumination results in high temperature, thus shortening the heading stage. Therefore, both high and low illumination will make grain chalkiness larger and the ratio of chalky grain higher than normal. It was reported that there is a positive correlation between sunshine and gelatinization temperature and consistence, and a negative correlation between sunshine and amylose content. In tropical areas, rice protein content was reduced by strong radiation of sunlight (Li un *et al.*, 1997). The effect of different sowing times on rice quality is mainly caused y climatic factors such as temperature and illumination. The same rice variety at the ame location but sown at different times of the season will show significant differences ı quality. Usually, late-season rice has better quality than early-season rice (Zhu Xu-ong *et al.*, 1993).

.4 Soil Health and Soil Fertility Management

Research on soil fertility has been done to establish optimum rates of inorganic ertilizers for lowland rice at some areas. More fertilizer rates recommendations both or organic and inorganic fertilizers need to be established particularly in intensive ce producing areas. More work in revising fertilizers recommendations is required ı view of the increased prices of fertilizers and new brands of fertilizers being ıtroduced into the market. Packages for control of soil erosions are available and ill be adapted to conditions of the ecosystem. Mineral imbalances in rice have led to on sulphide, boron, manganese and aluminium toxicity. Integrated soil fertility and soil–water management options will thus be emphasized for sustainable natural resources. It was reported that total soil nitrogen (N) and available N content have a positive correlation with rice protein content and a negative correlation with amylose content, especially in late rice. Scented rice is usually grown on soils with higher phosphorus (P) content. Soil total and available P content also have a significant positive correlation with rice protein content. Soil available sulfur (S) content has a positive correlation with polished grain ratio and protein content. Soil available manganese (Mn) content has a significant negative correlation with rice quality. With higher soil-available Mn content, grain chalkiness size is bigger. Soil available zinc (Zn) content is significantly higher in scented rice-planting areas than that in nonscented rice-planting areas (Huang Shu-zhen 1990). Zn deficiency is the most widespread micronutrient disorder in rice. Its occurrence has increased with the introduction of modern varieties, crop intensification, and increased Zn removal (Dobermann and Fairhurst 2000). Rice soil and plant Zn deficiency will result in plant stunting, reduced tillering, severely affected growth, and some dusty brown spots appearing on upper leaves. This deficiency will not only reduce rice yield but also affect grain quality. Rice quality can be affected greatly by nitrogen and potassium (K). Other fertilizers such as P, Zn, S, silicon (Si), magnesium (Mg), and calcium (Ca) can affect rice quality as well, but not as significantly as N and K. N fertilizer applied at a rational rate increases polished grain ratio and protein content and decreases chalkiness ratio and size and amylose content. Split N applications can increase the unpolished and polished grain ratio, grain transparency, and protein content, whereas amylose content would be lowered when N is applied at one time. If N is applied in excess, especially at the rice grain-filling stage, taste would be negatively affected by increased protein content (Qian Qian *et a1* 1998). Research results indicated that topdressing N at the full heading stage for both early and late rice increased both grain filling and grain protein content. Potassium can promote N metabolization and the transportation of photosynthetic product to grain. It was reported that K application increased the ratio of winnowed paddy 1.3-1.7 per cent and increased the protein content of unpolished grain 2.8-6.8 per cent. Late rice has a tendency similar to that of early rice (He Dian Yuan *et a1* 1994).Potassium application can also reduce rice fungus diseases.

2.5 Crop Management and Protection Options

Rice production is affected by a wide range of diseases and their severity depend on location, season, variety, farming system, and weather. Rice yellow mottle viru (RYMV), which is indigenous in Africa, is a major scourge of lowland rice and ca sometime lead to a total crop failure. Rice blast caused by *Pyricula oryzae* is als another serious disease common in lowland rice. Pests are another biotic stresse that cause huge losses in rice production. Yield losses ranging from 30 to 100 percen have been recorded. Most damage to rice is caused by stem borers *(Chillo spp)*, Africa rice gall midge *(Orseolia oryzivora)*, rodents and birds. Integrated disease and pes management options developed or verified in the country will be disseminated t farmers. Also available improved crop management options for irrigated lowland rain-fed lowlands and rain-fed upland ecosystems will be disseminated an repackaged where necessary.

2.6 Processing

All the essential plant nutrients for rice growth are also essential nutrients for humans and animals, except boron. Cultivated crops provide humans with not only organic nutrition but also with many essential macronutrients (N, P, K), secondary nutrients (Ca, Mg, S, sodium Na), and micronutrients such as copper (Cu), iron (Fe), Mn, Zn, etc. Rice is the main grain food for humans and its nutritional quality can affect human health directly. Most of the nutritional value such as vitamins, minerals, and high-lysine-content protein are located in the exterior and embryo of rice grain, whereas the interior of the grain contains less nutrition. Thus, during processing of the grain, some nutrition will be lost. The finer the processing, the higher will be the loss of nutritional value, especially the vitamin-B family.

Conclusions

Rice and rice based cropping systems are cultivated widely and intensively in India under diverse soil and agro ecological conditions consuming major proportion of soil and water resources, agro chemicals and fertilizer inputs. The unique system of wetland rice cultivation provided many beneficial effects to the crop in terms of nutrient supply, weed and water control, and tolerance to specific soil stresses, the system also created problems of soil salinization and water logging of many fertile lands in the canal commands because of sharp rise in the water table. Similarly in areas of high productivity potential deficiency of many nutrients (Zn, Fe, S, K) have been reported after high intensity rice crop systems and high yielding varieties were introduced, which is further compounded by imbalanced and indiscriminate application of fertilizers nutrients. The impact on the productivity of the system has been perceptible with wide spread occurrence of multi-nutrient constraints besides low nutrient use efficiency and factor productivity, resulting in significant decline in yield growth under intensive agriculture. Compounded by this discouraging situation is the emerging problem associated with climate change influencing through its impact on land use, its quality, availability of irrigation water and use efficiency of resources and inputs, and crop growth and productivity. Adoption of precision technologies for more efficient use of resources and nutrients becomes more relevant in the current production scenario. While technological advancements, currently available, have the potential to address the issues when implemented in the right perspective of sustaining productivity of the soil system on a long term basis, the efforts also require addressing few issues connected with cataloguing of available information on soil variability systematically using modern tools of remote sensing and GIS. This provides opportunities to integrate crop based information for effective management of the field problems and dissemination. Some of the important strategies to minimize the existing abiotic stresses and sustain productivity growth without deteriorating soil quality are listed below.

References

Chemical Aspects of Rice Grain Quality. International Rice Research Institute, P.O. Box 933, Manila, Philippines, pp 251-260.

Cock, J.H., Yoshida, S., 1972. Accumulation of 14C-labelled carbohydrate before flowering and its subsequent redistribution and respiration in the rice plant. Proceedings Crop society of Japan 41, 226 – 234.

Delwiche, S.R., McKenzie, K.S., Webb, B.D., 1996. Quality characteristics in rice by near infrared reflectance analysis of whole-grain milled samples. Cereal Chemistry 73, 257– 263.

Dobermann A, Fairhurst T. 2000. Rice nutrient disorders and nutrient management. Singapore: Phosphorus and Potash Institute and International Rice Research Institute. p 84 – 86.

Dobermann, A., Witt, C., Abdulrachman, S., Gines, H.C., Nagarajan, R., Son, T.T., Tan, P.S., Wang, G.H., Chien, N.V., Thoa, V.T.K., Phung, C.V., Stalin, P., Muthukrishnan, P., Ravi, V., Babu, M., Simbahan, G.C., Adviento, M.A.A., 2003. Soil fertility and indigenous nutrient supply in irrigated rice domains of Asia. Agro. J. 95, 913-923.

Farrell, T.C., Williams, R.L., Reinke, R.F., Lewin, L.G., 1998. Variation in radiation use efficiency in temperate rice. The 9th Australian Agronomy Conference, (http: //www.regional.org.au/au/asa/1998/4/127williams.htm).

He Dian Yuan 1994. Soil fertility and planting crop fertilization in south China. Beijing (China): Scier.ce Publishing House. p 335-339. (In Chinese.)

He KY. 1992. Grain quality characteristics for brown rice. Korea J. Crop Sci. 39(1): 38-44.

Horie, T., Ohnishi, M., Angus, J.F., Lewin, L.G., Tsukaguchi, T., Matano, T., 1997. Physiological characteristics of high-yielding rice inferred from cross-location experiments. Field Crops Research 52, 55-67.

Hossain, M., Laborte, A.G., 1993. Asian rice economy: recent progress and emerging trends. Food and Fertilizer Technology Center, Taipei, Taiwan.

Huang Shu-zhen. 1990. Relationship between soil characteristics and rice quality in Hunan International Rice Research Institute, Los Baños, pp. 495-507.

International Trade Centre- UNCTAD/WTO Reports, 2008

IRRI, 1995. Rice facts. International Rice Research Institute, Los Baños, Philippines.

Jackson, P., Robertson, M., Cooper, M., Hammer, G.L., 1996. The role of physiological understanding in plant breeding: From a breeding perspective. Field Crops Research 49, 11-37.

Juliano, B.O., 1985. Criteria and tests for rice grain qualities. In: Juliano, B.O. (ed.), Rice Chemistry and Technology. American Association of Cereal Chemists, St Paul, MN, USA, pp 443–524.

Juliano, B.O., 1998. Varietal impact on rice quality. Cereal Foods World 43, 207-211, 214- 216, 218-222.

Li Jun, Gu De-fa, Li Lin-feng. 1997. Research advances on the effect of environment and planting factors on rice quality. J. Shanghai Agric. 13 (1): 94 – 97. (In Chinese)

Li Jun, Gu de-fa. 1995. Effect of different sowing time and fertilization rates on two high – quality rice varieties. J. Shanghai Agric. 13 (1): 63 – 67. (In Chinese)

Lii, C.Y., Tsai, M.L., Tseng, K.H., 1996. Effect of amylose content on the rheological property of rice starch. Cereal Chemistry 73, 415–420.

Maclean, J.L., Dawe, D.C., Hardy, B., Hettel, G.P., 2002. Rice almanac. IRRI, Los Baños, Philippines, pp 253.

Murata, Y., Matsushima, S., 1975. Rice. Cambridge University Press, London.

Nix, H.A., 1976. Climate, crop productivity in Australia. In: Yoshida S. (ed.) Climate, rice.

Qian Qian, Liu Xiou-yan, Zeng Da-li. 1998. Introducing research on the factors that affect quality. Hubei Agric. Sci. 6: 14-16. (In Chinese.)

Resurreccion AP, 1997. Effect of environment on rice amylose content. Soil Sci. Plant Nutr. 23 (1): 109 – 112.

Sao dong-sheng, Tang Jian. 1987. Relationship between grain milk filling and rice quality. Guizhou Agric. Sci. 2: 12-14. (In Chinese.)

Tang xiang-ru,Yu tie-qiao. 1991. Effect of temperature on rice quality and related biochemistry characteristics in milk filling and maturity stages. J. Hunan Agric. College 17(1): 1-8. (In Chinese.)

WARDA (2007) Africa Rice Centre. Medium Term Plan 2008-2010. Charting the Future of Rice in Africa.

WARDA (2007) Africa Rice Centre. WARDA Rice briefs 2007 edition

Wu Guan-ting, Xia Ying-wu. 1994. Effect of environment and cultivation on rice quality. Rice China 4 : 37 – 39. (In Chinese)

Yan, W., Hunt, L.A., 1998. Genotype-by-environment interaction, crop yield. Plant Breeding Reviews 16, 135–178.

Yang Hua-long, Yang Zhe-ming, Lu Bi-lin. 2001. Effect of ecological environment on rice quality. Hubei Agric.Sci. 6: 14 – 16. (In Chinese)

Ying, J., Peng, S., He, Q., Yang, H., Yang, C., Visperas, R., Cassman, K., 1998. Comparison of high-yield rice in tropical, subtropical environments - I. Determinants of grain, dry matter yields. Field Crops Research 57, 71-84.

Zhu Xu-dong, Xiong Zheng-ming, Luo Yu-kuen. 1993. Effect of different sowing seasons on rice quality. Rice Sci. China. 7(3): 172-174. (In Chinese.)

2014, Sustainable Rural Development through Agriculture *Pages* ***168–180***
Editors: **Dr. Shobhana Gupta and Dr. S.S. Tomar**
Published by: **BIOTECH BOOKS, NEW DELHI**

Chapter 13

To Conserve Resource Base in Rice through System of Rice Intensification

Smita Singh, L.B. Singh, Deepak Srivastava and Satish Singh Baghel

ABSTRACT

SRI is an acronym for system of rice intensification. This improved method of rice cultivation was developed in 1983 in Madagaskar and has now spread too many parts of the world. There is a notion that higher yields in rice come with high investments on seed, irrigation, high doses of fertilizers and pesticides. Contrary to this popular view, SRI method of cultivation produces higher yields with less seed and less water. SRI emphasizes on the need to shift from chemical fertilizers to organic manures. Increased soil aeration and organic matter helps in improving soil biology and thus helps in better nutrient availability.

Keywords: *Methodology, Resource conservation, SRI.*

1. Introduction

Water is the most precious and major factor in rice production. The demand for water is increasing by leaps and bounds due to increase in human population and urbanization. Therefore, the share of agriculture for this invaluable input for crop production would certainly decrease day by day. Rice is a water guzzling crop and requires about 3000-5000 liters of water for production of 1 kg grain. Thus, the global water crisis threatens the sustainability of irrigated rice production. Yet more rice needs to be produced with less and less water to feed the ever-growing population. Again the very belief by the farmers that application of more water would produce more yields is also a misnomer. Application of excess water is not only a waste, but sometimes harmful to the crop. Therefore, scientific and efficient use of water is necessary to enhance its productivity per unit area by increasing yield of crop with less use of water. In rice cultivation, system of rice intensification (SRI) is found to be very useful to produce more yields, improve nutrient use efficiency with almost one half of the quantum of water required under conventional method. The system of rice intensification, or SRI for short, is a fascinating case of rural innovation that has been developed outside the formal rice research establishment both in India and the rest of the world. This report documents the history of this practice in India in the last few years and presents some of the institutional changes and challenges SRI throws up. This report is in three parts. The first part looks at the complex and continuing evolution of SRI in India and presents SRI as an innovation in process and not as a completed product. Farmers and other actors are continuously shaping it through their practice. In part two use some of the insights of the innovation systems framework to understand SRI by looking closely at the nature and quality of linkages of the various actors and conclude by highlighting some features of SRI in India and its implications for pro-poor innovation. For the study the SRI crop was followed in two seasons, *Kharif* 2004 and *Rabi* 2004–2005 in a few southern states. The inputs and insights from the field were corroborated through detailed interviews with key stakeholders in SRI, involving structured and semi-structured surveys with farmers and other stakeholders. The study has relied on interviews with over 250 persons in India covering the southern states of Andhra Pradesh, Tamil Nadu and Karnataka as well as the union territory of Pondicherry and a diagnostic survey of SRI in Jharkhand. Along with these interviews and field visits, the has relied on extensive research of available material on SRI, primarily from the SRI website hosted by CIIFAD and Tefy Saina, and has followed the debates on SRI, placing it within the larger context of the International Year of Rice 2004 and SRI's neglect by the research establishment.

2. Origins of SRI

In 1983, a drought year, at the small work-study school that Laulanié established, young farmers reluctantly transplanted some rice seedlings that were much younger than what they had been using. They transplanted 15 day old seedlings, a quarter the age of those used in traditional cultivation. Yet, the plants were vigorous. The system of rice intensification or SRI emerged as a set of six practices Rabenandrasana (1999); Uphoff (2002), (2004); Berkelaar (2001); Stoop *et al.* (2002):

1. Transplanting of very young seedlings between 8 and 15 days old to preserve potential for tillering and rooting;
2. Planting seedlings singly very carefully and gently rather than in clumps of many seedlings that are often plunged in the soil, inverting root tips;

3. Spacing them widely, at least 25 x 25 cm and in some cases even 50 x 50 cm, and in a square pattern rather than in rows;
4. Using a simple mechanical hand weeder ('rotary hoe') to aerate the soil as well as to control weeds;
5. The soil moist but never continuously flooded during the plants' vegetative growth phase, up to the stage of flowering and grain production.
6. Use of organic manure or compost to improve soil quality

3. SRI in India: A Slow Start

India is one of the largest producers of rice in the world; however, rice cultivation in recent times has suffered from several interrelated problems. Increased yields achieved during the green revolution through input intensive methods of high water and fertiliser use in well endowed regions are showing signs of stagnation and concomitant environmental problems due to salinisation and water-logging of fields (the grain bowls of India Punjab and Haryana are some of the worst affected). In other parts there have been social conflicts between water users in several canal-irrigated areas due to the water intensive nature of the crop. However, unlike other rice-growing nations; India had a rather delayed start in SRI. T. M. Thiyagarajan of the Tamil Nadu Agricultural University, Coimbatore was the lone Indian representative at the 2002 international conference on SRI. He first heard about SRI in 2000 from Dr. Ten Berge of Wageningen's Plant Research International and was interested in the soil aeration aspect of SRI, and its water-saving potential. The 'modified' SRI practice that was evaluated by TNAU used three of the SRI principles (single seeding, wider spacing and use of weeder) but it used water and fertiliser in excess of normal SRI recommendations. The results indicated considerable water saving through modified SRI and a reduction of seed costs, but no significant increase in yields Thiyagarajan (2002). With less water, less seed, no fertilisers, no pesticides, more soil organic matter and more soil aeration, the productive potential of rice can be unleashed. As a new way of looking at rice cultivation and solely driven by the innovative farmers, System of Rice Intensification (SRI) is emerging as an alternative to conventional water and chemical intensive rice cultivation. Farmers across the country are adopting System of Rice Intensification (SRI), as it gives equal or more produce than the conventional rice cultivation; with less water, less seed and less chemicals. The net effect is a substantial reduction in the investments on external inputs. Conversely, increased labour needs for weeding and cultivation in saline lands are the two areas of major concern in SRI, on which innovations are forthcoming from various quarters. Farmers are leading the innovations and spreading that technology, while the scientific community is still to catch up with this emerging rice revolution.

SRI is not a new variety or a hybrid. It is only a method of cultivation. SRI is showing promising results in all rice varieties local or improved. Marking the plot before transplantation to ensure proper rows and spacing, and weeding are necessitating development of appropriate implements. SRI method is emerging as a potential alternative to traditional way of flooded rice cultivation and is showing great promise to address the problems of water scarcity, high energy usage and chemicalisation.

4. Important Features of SRI

- ☆ **Low seed requirement:** Since a single seedling is transplanted per hill at wider spacing, seed requirement is drastically reduced (6 kg/ha).
- ☆ **Low water requirement:** As there is no need to maintain standing water.
- ☆ **Transplantation of tender/young seedlings (8-12days):** Transplantation of young seedlings at shallow depth results in quick recovery and establishment and production of more tillers.
- ☆ **Transplanting at wider spacing (10 x 10 inches or 25 x 25 cm:** Wider spacing allows enough sunlight to reach the leaves of each rice plant thus reducing competition for water, space and nutrients resulting in the spread of roots and healthy growth of plants.
- ☆ **Incorporating weeds into the soil while weeding:** Weeding with a simple hoe helps in replenishing the nutrients in the form of green manure. Working with a hoe or weeder helps to aerate soil which in turn helps in vigorous root growth. (First weeding 10 days after transplanting and a minimum 3 weeding at 10-12 days interval).
- ☆ **Organic manures in place of chemical fertilizers:** Organic manures improve soil aeration and also microbial activity. This helps in decomposing organic matter into nutrients, essential for plant growth.
- ☆ **Pest management without chemicals:** Normally, the incidence of pests and diseases is low as the plants are widely spaced and are healthier in SRI. In case of incidence of disease, biological control methods/natural control measures can be applied to keep them under check.

5. Methodology of SRI

5.1 Suitable Soils

Unlike in the conventional method, saline or alkaline soils are not suitable for SRI cultivation. As water needs to be drained frequently in SRI, the salts come to surface, damaging seedlings or plants. It is therefore advisable to go for soil test before opting for SRI cultivation. Undulation in the plot causes water to stagnate in some parts of the land. So land should be properly levelled before transplantation to enable water to spread uniformly over the field. There should also be a provision to remove excess water *i.e.,* drainage facility.

5.2 Improving Soil Fertility

Yields in SRI respond well to organic manures than chemical fertilizers. This emphasizes the need to build up fertility through organic means from the beginning. There are some methods mentioned below to improve soil fertility. At least two such methods are to be followed.

5.3 Farm Yard Manure (FYM)

Application of well decomposed FYM (15 tons/ha or 3 tractor loads per acre) is quite beneficial for SRI cultivation. Vermi-compost can also be used in place of FYM or in combination with FYM.

5.4 Green Manure Crop

Green manures help in improving soil fertility to a great extent. Sunhemp and Sesbania are the commonly grown green manure crops that are incorporated into the soil at flowering time *i.e.*, around 45 days, when the nitrogen fixation is at its maximum. It takes nearly 10 days for decomposition. If the rice seeds are sown in the nursery on the day of incorporation of green manure, the field will be ready by the time the seedling are ready for transplantation.

5.5 Raising Nursery

In SRI method, utmost care should be taken in the preparation of nursery bed, as 8-10 days old seedlings are transplanted. 2 kg of seed (5kg/ha) is required to transplant in one acre of land. The nursery bed can be raised in a 48 square yard (40 Sq. Meters) plot. Depending upon the situation, two beds can also be raised each measuring 24 sq. yards (20 Sq. Meters) per 1 kg seed. Seeds should be thinly spread to avoid crowding of seedlings. Care should be taken that no two seeds should touch each other.

5.6 Bed Preparation

A bed with a width of 125cm or 4 feet is ideal. And length of the bed can be decided by the farmer depending on the ground situation. According to one's convenience either a single bed or several small beds (say, 4 beds measuring 4x28 feet or 1.25 X 8 meters each) can be prepared. As the roots of 8-12 day-old seedlings grow upto 3 inches (7.5cm) deep, it is necessary to prepare raised beds of 5-6 inches (12.5-15cm) height.

5.6.1 Nursery Bed Preparation

The nursery bed is prepared with application of farm yard manure (FYM) and soil alternately in 4 layers.

- ☆ **1st layer:** 1 inch (2.54 cm) thick well decomposed FYM
- ☆ **2nd layer:** 1 ½ inch (3.78 cm) soil
- ☆ **3rd layer:** 1 inch (2.54 cm) thick well decomposed FYM
- ☆ **4th layer:** 2 ½ inch (6.3 cm) soil

All these layers should be mixed well, as the FYM helps in easy penetration of roots. To prevent soil erosion, the bed on all sides should be made secure with wooden reapers/planks or paddy straw rope or anything of that sort. To drain excess water appropriate channels should be provided on all sides.

5.7 Preparation of Main Field

Preparation of the main field in SRI is the same as in conventional method. Field should be evenly levelled and there should not be standing water in the field during transplantation. In SRI method, seedlings are widely spaced (10 x10 inch or 25 x 25cm) and only one seedling is transplanted per hill (3-4 seedlings per hill in conventional system). SRI method can accommodate only 16 hills/sq. metre as against 33-40 hills/square meter in conventional method. Uniform spacing is also required for easy weeding by implements. To maintain uniform spacing, different methods can be employed. Small pegs can be tied to a rope at 25 cm or 10 inch distance and by using this rope, row after row transplantation can be done. Different types of 'Markers' are being developed for this purpose. These markers need to be run over the prepared field lengthwise and width wise. Transplanting at the marked intersection gives the required 25 x 25 cm spacing.

5.8 Transplantation

Young, 8-12 days old seedlings are transplanted in SRI method. Care should be taken to see that the plant does not experience shock during transplanting. The farmers and farm labour need to be educated on this aspect. Care should be taken to prevent any harm to seedlings while pulling them from nursery or at the time of transplantation. In the conventional method, the practice is to pull the seedling by holding the plant. In SRI method, a metal sheet is inserted 4-5 inches below the seed bed and the seedlings along with soil are lifted without any disturbance to their roots. Transplanting should be done as quickly as possible, preferably within half an hour to minimise trauma to the roots. Transplanting should be done with utmost care and concentration. Method of transplanting in the conventional method, 30 day-old seedlings are thrust into the puddle soil and the roots take 'U' shape *i.e.* the tips of roots face upward. Therefore the roots require time and energy to turn downward and establish in the soil. In SRI method, young seedlings are planted shallow and therefore establish quickly. Single seedlings with seed and soil are transplanted by using index finger and thumb and gently placing them at the intersection of markings. Light irrigation should be given on the next day of transplantation. Initially, it requires 10-15 persons to transplant one hectare due to its wider geometry (25 cm x 25 cm).

5.9 Irrigation and Water Management

Because of some special anatomical features, rice can grow well even in standing water; but it does not require standing water as a rule. The practice of growing rice in inundated condition is mainly to control weed growth. But such conditions result in lack of aeration and consequent stunted root growth. In SRI, irrigation is given to wet the soil, just enough to saturate the soil with moisture. Subsequent irrigation is suggested when the soil develops fine cracks. Irrigation interval depends on soil type and weather conditions. This method helps in better growth and spread of roots. Regular wetting and drying of soil results in increased microbial activity in the soil and easy availability of nutrients to plants. For a smooth weeding operation, the field should be irrigated maintaining a thin film of water. After the completion of the weeding, water should never be let out of the field. Once the tillering process is complete, standing water of one inch/2.5 cm height may be maintained.

5.10 Weed Management

Absence of standing water provides a congenial environment for weeds to proliferate in SRI. If these weeds are incorporated into the soil, they serve as green

Figure 13.1: Transplanting and Transplanted Field.

manure. Different models of manually operated weeders are being developed for effective weed management in SRI. The weeds in the vicinity of the hills that could not be reached by the weeder have to be removed by hand. Weeds can be incorporated

Figure 13.2: No Need to Inundate with Standing.

Figure 13.3: Water Incorporating Weed into the Soil with Cono-weeder.

by moving the weeder between the rows. First weeding should be done 10-12 days after transplanting. Later, depending on the need, weeding can be done once every 10 days.

5.11 Advantages of Weeder

- ☆ Controls weed
- ☆ Green manuring due to incorporation of weeds into soil
- ☆ Soil aeration
- ☆ Increased soil biological activity
- ☆ Increased nutrient availability and uptake

5.12 Management of Pests and Diseases

The incidence of pests and diseases is naturally low in SRI because of wider spacing and the usage of organic manures. Natural pest management methods and use of natural bio pesticides are recommended whenever necessary to keep pests under control.

6. Advantage and Disadvantage

6.1 Advantages of SRI

- ☆ Saving on seeds, as the seed requirement is less
- ☆ Saving of water, as the water requirement is less: 25-50 per cent when field are not continuously flooded.
- ☆ Withstands short gaps in water availability (like burning of transformers, delayed rains etc.)
- ☆ Saving on chemical fertilizers, pesticides
- ☆ More healthy and tasty rice due to organic farming practices: 30-50 or more tillers/plant, larger panicle with more and usually heavier grains, due to more growth of roots and soil organisms.
- ☆ Higher yields due to profuse tillering, increased panicle length and grain weight
- ☆ Easy and effective seed multiplication, as a small quantity is required as seed
- ☆ Resistance to lodging: Due to stronger root system and tillers.
- ☆ Shorter crop cycle: SRI crop mature 1-3 weeks earlier than conventional method.
- ☆ Lower cost of production: Lower by 20 per cent/hectare.
- ☆ Higher factor productivity

6.2 Disadvantages

- ☆ Higher labour costs in the initial years
- ☆ Difficulties in acquiring the necessary skills

7. SRI and Environmental Quality

- ☆ Reduced use of agrochemicals should enhance the health of soil and cultivators

- ✰ This should enhance groundwater quality by diminishing nitrate concentrations
- ✰ Reduced demand for water frees up water for other uses
- ✰ Soil that is not kept saturated has greater biodiversity.
- ✰ Un-flooded paddies do not produce methane, one of the major "greenhouse gases"

7.1 Reduced Need for Chemical Pest and Disease Control

Of special significance are the frequent reports by farmers in many countries that their rice crop grown with SRI has less damage from pest and diseases. SRI reduced need for chemical pest and disease control because due to greater abundance of insects that control of pests and crop predators. In Thailand, an on farm evaluation of alternative methods found not only less or nil white-ear head damage (per m^2) in Sri plot compared with nearby farmer practices plots, but there were higher numbers (per hill) of natural enemies on the SRI plants Salokhe *et al.* (2007).

7.2 Resource Base Conservation under SRI

7.3 Rate of Chemical Fertilizer Used in SRI *vs.* Conventional

The average use of organic matter for SRI was 4370 kg/ha compared to only 642 kg/ha. However, SRI Farmer still reports the inadequateness of organic fertilizer to apply to their land. The Study additionally found that, the average amount of various chemical fertilizer applied per hectare to SRI is considerably lower than the conventional practices. Kem Sothorn *et al.*,(2003) reported that the total number of chemical fertilizer for SRI is lower than half of the amounts apply for conventional. Similarly, comparatively high amount of urea, DAP and NPK were also used for conventional rather than SRI. This significance reveals that SRI is considerably low external input for rice production. Hence, the shift from conventional to SRI will lead to tremendously reduction of mineral fertilizer use.

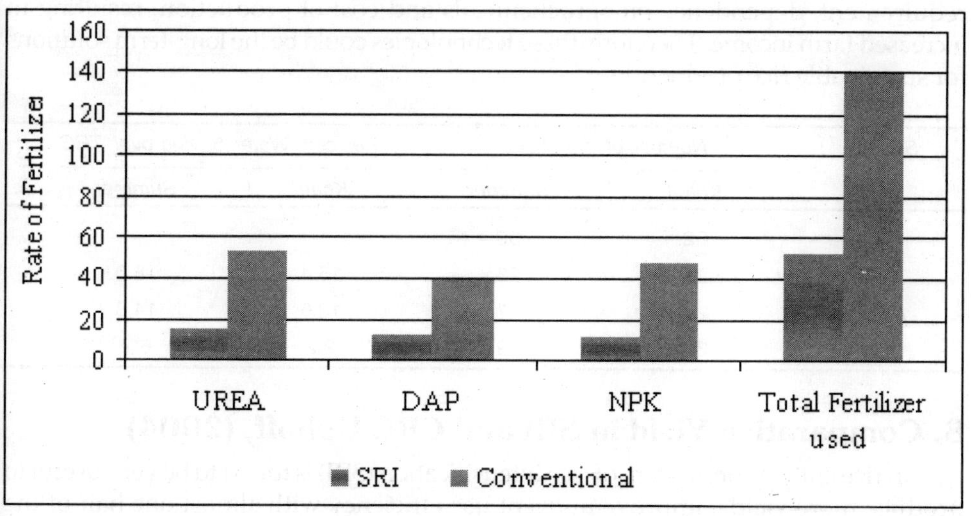

7.4 Effect of Nutrient Management Practices on Productivity (t/ha) of the Rice Under different Establishment Method

The SRI treatment showed a different pattern. The NPK-fertilized SRI plots showed a linear relationship for N uptake kinetics. This N uptake trend suggests that SRI and fertilized rice plants rely more on the remobilization of shoot N for their grain production. The lowest uptake occurred with the conventional treatment where N uptake not only varied very little from panicle initiation until anthesis, but also its increase at a later stage was relatively low compared to the other treatments. Nutrient uptake in plants cultivated with conventional methods may be constrained by the low root growth (RPR = 22 kg at panicle initiation) and the high root die-back (40 per cent RPR decrease between anthesis and maturity) Janssen, (1998).

Treatments	*Panicle Initiation*	*Anthesis*	*Maturity*
SRI with compost	62.38	95.32	176.74
SRI without compost	52.85	79.41	159.39
SRA with NPK and urea	53.01	77.38	133.63
SRA without fertilization	32.16	55.17	122.62
Conventional	20.18	27.87	62.95

7.5 Influence of different System of Rice Cultivation on Number of Irrigations and Water Saving

Uphoff (2004) studied the effect of SRI on yields, water saving, cost reduction and net income in rice cultivation from different countries, and concluded that SRI increases yield by 24 - 105 per cent, saves water by 40 - 50 per cent, reduces cost by 7 - 56 per cent and increases net income by 59 - 412 per cent. Therefore, it can be concluded that SRI methods of rice cultivation increases WUE, NUE, crop productivity and environmental sustainability and also significantly reduce the seed and water requirement, dependence on agrochemicals and cost of production, resulting in increased farm income. Therefore, these technologies could be the long-term solutions for sustainable rice production.

System	*Number of Irrigations*		*Per cent Water Saving over TRC*	
	Kharif	*Summer*	*Kharif*	*Summer*
TRC	32	33	–	–
SRI	24	23	15.4	18.0
AWD	29	27	12.6	14.8
DSR	37	39	2.2	6.7

8. Comparative Yield in SRI and CRC Uphoff, (2004)

In rice cultivation, system of rice intensification (SRI) is found to be very useful to produce more yields, improve nutrient use efficiency with almost one half of the

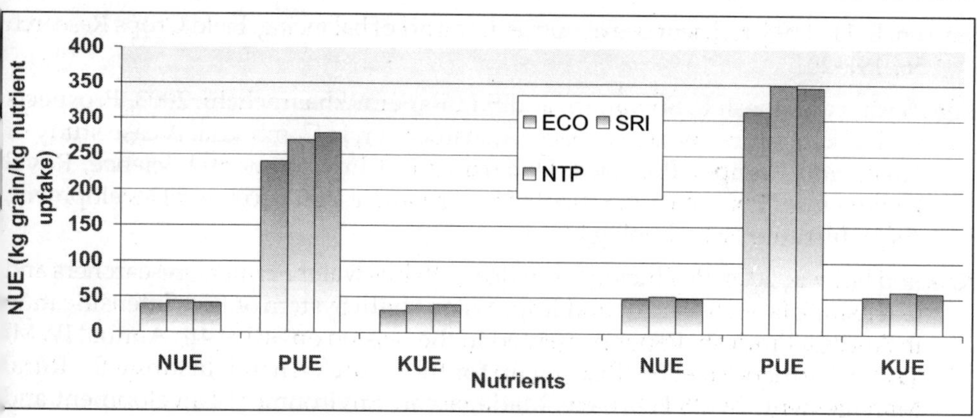

Nutrient Use Efficiency (NUE) under different Treatments.

quantum of water required under conventional method. It was observed in multi-locational trials under AICRIP that grain yield increased by 5 - 65.2 per cent under SRI compared to normal transplanting (NT) at 19 locations. Grain yield also increased by 5 - 10 per cent under SRI compared to ICM at 17 locations but reverse was true at 5 locations (Viraktamath, 2008).

In General, SRI Recorded Higher Nutrient Use Efficiency

Thiyagarajan. (2002) observed the significantly higher values of root length, root volume and root dry weight was recorded with SRI compared to control but remained at par with ICM practice. SRI roots were found more active and healthy at harvest, whereas conventional practice was week, thin and degenerated. Satyanarayana, (2004) conducted on farm trial in Andhra Pradesh and found that all the growth (plant height, no. of tillers) and yield parameters (tillers/plant) were significantly higher in SRI method of rice cultivation as compare to farmer's practices.

9. Conclusion

Rice is the staple food for >70 per cent Indians, and it holds the key for food security.SRI method of rice cultivation increase WUE, NUE, crop productivity and improve environmental quality and significantly reduce the seed and water requirement, dependence on agrochemicals and cost of production, resulting in increase farmers income. Further research is needed to understand the factors contributing to higher yield, soil health parameters, and various aspects of sustainability. Therefore, these technologies could be the long-term solutions for sustainable rice production

References

Berkelaar, Dawn. 2001. SRI, the System of Rice Intensification: Less can be More, ECHO Cornell International Institute for Food, Agriculture and Development, Ithaca, NY.

GOI, 2007-08. *Economic Survey*, Government of India, New Delhi.

Janssen, B. H., 1998. Efficient use of nutrients: an art of balancing. Field Crops Research 56, 197-201.

Kem Sothorn, Ganesh P. Shivakoti, and S.L. Ranamukhaarachchi, 2003. Prospects for Disseminating System of Rice Intensification in Cambodia; A case study of Takeo and Kampot Province. Department of Environmental Science, Royal University of Phnom Penh School of Environment Resources and Development, Asian Institute of Technology.

Krik and Solivas, 2004. Producing more rice with less water Farmers, researchers and extensionists' experiences and impressions with System of Rice Intensification in Andhra Pradesh. Paper presented in the session on SRI at 4th Annual IWMI TATA Partners Meet on 'Bracing up for the future', Anand, Institute for Rural Management, 24–26 February. Madagascar. Environment Development and Sustainability, 1(3/4), pp. 297–31. Newsletter, Vol. 15 No. 3/4 (Dec.).

Rabenandrasana, Justin 1999. Revolution in Rice Intensification in Madagascar, ILEIA.

Salokhe, V. M., Kumar, P., Mishra, A., 2007.Increasing water use efficiency by using mulch under SRI management practices in northeast Thailand. Second progress report: Participatory action research phases, 1st July to 30th December, 2006 Bangkok: Asian Institute of Technology.

Satyanarayana, A. 2004. The System of Rice Intensification: Evaluations in Andhra Pradesh. Presented at the Panel assembled for the World Rice Research Conference, Tokyo Tsukuba, 4–7 November.

Stoop, W.A., N. Uphoff and A. Kassam. 2002. A review of agricultural research issue raised by the system of rice intensification (SRI) from Madagascar: opportunitie for improving farming systems for resource-poor farmers, Agricult. Syst, 71: 212 217

Thiyagarajan. 2002. Experiments with a Modified System of Rice Intensification i India. pp. 137–139. In Uphoff *et al.* (eds). Assessments of the System of Ric Intensification (SRI): Proceedings of an International Conference held in Sanya China, 1–4 April, 2002.

Uphoff, N. 2002. 'Opportunities for raising yields by changing management practice The System of Rice Intensification in Madagascar' in N. Uphoff. (ed Agroecological Innovation: Increasing Food Production with Participator Development, Sterling, Virginia, Earthscan.

Uphoff, N. 2004. 'Development of the System of Rice Intensification in Madagasca in Participatory Research and Development for Sustainable Agriculture an Natural Resource Management: A Sourcebook', pp. 13–19.

Viraktamath, B. C. 2008. Evaluation of System of Rice Intensification (SRI) Under A India Coordinated Rice Improvement Project (AICRIP). Extended Summarie Second National Symposium of System of Rice Intensification (SRI) in Indi Progress and Prospects. October 3-5, Agartala, Tripura, pp. 40.

2014, Sustainable Rural Development through Agriculture *Pages* ***181–187***
Editors: **Dr. Shobhana Gupta and Dr. S.S. Tomar**
Published by: **BIOTECH BOOKS, NEW DELHI**

Chapter 14
Genetic Purity and Varietal Maintenance

Birendra Prasad[1] and Shambhoo Prasad[2]
[1]*Senior Research Officer, Genetics and Plant Breeding*
[2]*Ph.D. Scholar, Seed Science and Technology*
G.B. Pant University of Agriculture and Technology, Pantnagar

The economic importance of genetically and physically pure seed of high-yielding varieties and hybrids has been, of late, receiving increased recognition. It is an established fact that good quality seed is a pre-requisite for optimum return of the crop. In fact, few other agricultural inputs meet with as ready a response and adoption from farmers as high quality seed. It is the cheapest input and forms only a small fraction of cultivation expenses. Further, improved seed encourages the use of modern production technology. Continuous supply of good quality seed, therefore, is essential for maintaining tempo of revolution in agriculture.

Genetic purity refers to trueness to type, or the degree of contamination by seeds or by genetic materials of off-type varieties or species. Since the success of hybrid seed production is dependent on the genetic purity of parental lines and both out crossing and the inadvertent mixing of seed can compromise seed quality, genetic purity tests are critical tools for seed producers and plant breeders. A high level of genetic purity in crop varieties must be achieved and maintained for agronomic performance as well as to encourage investment and innovation in plant breeding and to ensure that the improvements in productivity and quality imparted by breeders are delivered to the farmer and, ultimately, to the consumer. Traditionally, morphological comparisons have formed the basis for genetic purity evaluations. Production of genetically pure and otherwise good quality pedigree seed is an exacting task requiring high technical skills and comparatively greater financial investment. During seed production strict attention must be given to the maintenance of genetic purity and other qualities of seeds in order to exploit the full dividends sought to be obtained by introduction of new superior crop plant varieties.

Causes for Deterioration of Genetic Purity

The genetic purity of a variety or trueness to its type deteriorates due to several factors during the production cycles. Kadam (1942) listed the following important factors responsible for deterioration of varieties.

1. Developmental Variations

When seed crops are grown under environments with differing soil, fertility, climate photoperiods, or at different elevations for several consecutive generation's developmental variations may set in as differential growth responses. It is therefore, preferred to grow the varieties of crops in the areas of their natural adaptation to minimize developmental shifts.

2. Mechanical Mixtures

Mechanical mixtures, the most important reason for varietal deterioration, often take place at the time of sowing if more than one variety is sown with the same seed drill, through volunteer plants of the same crop in the seed field, or through different varieties grown in adjacent fields. Two varieties growing next to each other field is usually mixed during harvesting and threshing operations. The threshing equipment is often contaminated with seeds of other varieties. Similarly, the gunny bags, seed bins and elevators are also often contaminating, adding to the mechanical mixtures of varieties. Roguing the seed fields critically and using utmost care during seed production and processing are necessary to avoid such mechanical contamination.

3. Mutations

Mutations do not seriously deteriorate varieties. It is often difficult to identify or detect minor mutations occurring naturally. Mutants such as 'fatuoids' in oats or 'rabbit ear' in peas may be removed by roguing from seed plot to purify the seeds.

4. Natural Crossing

Natural crossing can be an important source of varietal deterioration in sexually propagated crops. The extent of contamination depends upon the magnitude of natural cross-pollination. The deterioration sets in due to natural crossing with undesirable types, diseased plants, or off types. In self-pollinated crops, natural crossing is not a serious source of contamination unless variety is male sterile and is grown in close proximity with other varieties. The natural crossing, however, can be major source of contamination due to natural crossing.

Extent of genetic contamination in seed field due to natural crossing depends up on

- ✰ The breeding system of the species
- ✰ Isolation distance
- ✰ Varietal mass
- ✰ Pollinating agent.

The isolation of seed crops is the most important factor in avoiding contamination of the cross-pollinated crops. The direction of prevailing winds, the number of insects present and their activity, and mass of varieties are also important considerations is contamination by natural crossing.

5. Minor Genetic Variations

Minor genetic variations can occur even in varieties appearing phenotypically uniform and homogenous when released. The variations may lose during later production cycles owing to selective elimination by the nature. The yield trials of lines propagated from plants of breeder's seed to maintain the purity of self-pollinated crop varieties can overcome these minor variations. Due care during the maintenance of nucleus and breeder's seed of cross-pollinated varieties of crop is necessary.

6. Selected Influence of Pest and Diseases

New crop varieties often are susceptible to newer races of pests and diseases caused by obligate parasites and thus selectively influence deterioration. The vegetative propagated stock also can deteriorate quickly if infected by virus, fungi or bacteria. Seed production under strict disease free conditions is therefore essential.

7. The Techniques of the Plant Breeder

Serious instabilities may occur in varieties owing to cytogenetic irregularities in the form of improper assessments in the release of new varieties. Premature release of varieties, still segregating for resistance and susceptibility to diseases or other factors can cause significant deterioration of varieties. This failure can be attributed to the variety-testing programme.

In addition to these factors, other heritable variations due to recombination's and polyploidization may also take place in varieties during seed production, which can be avoided by periodical selection during maintenance of the seed stock.

How to Maintain the Genetic Purity

Hartmann and Kester (1968) and Agarwal (1995) described steps to maintain the genetic purity of variety during seed production.

The following measures have been suggested to safeguard the genetic purity

1. Adoption of Crop: Growing crops only in areas of their adaptation to avoid genetic shifts.
2. Approved Class of Seeds: Use of only approved class of seed in seed multiplication and adopt generation system.
3. Preceding Crop Requirement: Inspection and approval of seed plots prior to planting
4. Isolation: Isolation of seed crops from various sources of contamination by natural crossing or mechanical mixtures.
5. Roguing: Roguing of off- types differing in characteristics from those of the seed variety.

6. Field Inspection: Qualified and experienced personnel of seed certification agency should inspect seed crops at all appropriate stages of growth and verify seed lots or purity and quality.
7. GOT: Periodic testing of varieties for genetic purity.

Varietal Maintenance

Varietal maintenance implies to keep the variety purity level as it was developed, tested for a specific agro-climatic area and accepted at the time of release and notification. The main aim is to provide true to type seed of high physical purity, good health and high germination ability to the farmers. In other words careful management of seed stock in beginning of seed production chain *i.e.*, nucleus seed is referred as Varietal Maintenance. Maintenance of varietal purity, uniformity and stability are main important characters for any variety. The complete yield potential of a variety can be realized if the variety is genetically pure.

At the time of variety release a small quantity of seed is available with the breeder. The relatively small amount of seed of improved cultivar needs to be multiplied and made available to farmers as quickly as possible. During seed multiplication, varietal purity and identity needs to be maintained. Each multiplication cycle starts from the 'breeder seed'. If the breeder seed is not of high purity, the contaminants present get multiplied several times in the succeeding generations of foundation and certified seed production. The presence of contaminants may even lead to complete loss of the improved features of the cultivar. Prevention of contamination is at the heart of a successful breeder seed production programme. However, plant breeders develop a variety he must ensure that the characteristics he has bred into variety are genetically stable.

Maintenance of a Variety

Once a variety is released for cultivation, the breeder usually supplies a small quantity of seed for further multiplication and maintenance. The responsibility of breeder seed production centre is to produce breeder seed and varietal maintenance. In order to release seed of an improved variety to farmers, it has to be multiplied. Each multiplication cycle has to start from its basic seed stock, 'Nucleus Seed'.

Our basic objective of varietal maintenance is to maintain the purity and identity of a cultivar. The maintenance procedures are in fact the extension of the normal breeding process. The difference is that during maintenance breeding, selection process is relatively mild and our aim is not improvement but to keep the identity unchanged. Selection should maintain the plant type, its uniformity and freedom from diseases. The maintenance procedures for self pollinating crops with a substantial amount of out-crossing are slightly different from cross-pollinating crops. In cross pollinating crops the important characters are assessed before flowering. The fields where plants and progenies are to be assessed should be uniform. Essentially these should be grown under optimal growing conditions.

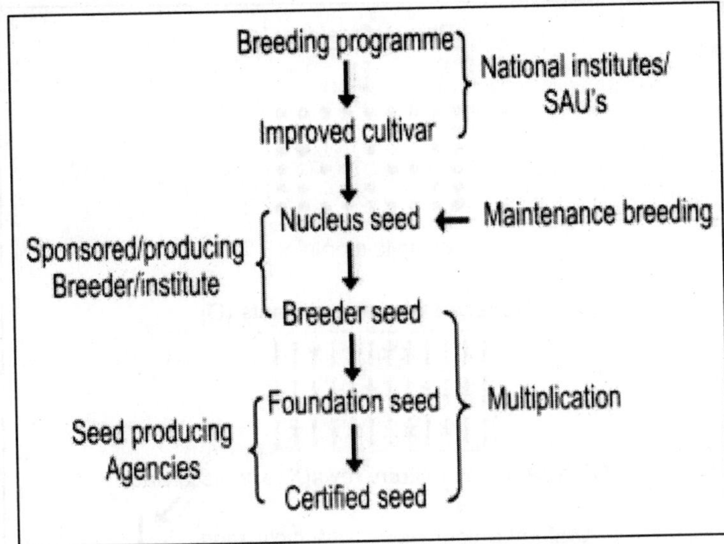

Scheme Showing Cultivar Breeding, Maintenance and Multiplication in India.

Varietal Maintenance in Self-Pollinated Crops

The maintenance procedure starts with a small plot raised from the parental material received from the breeder or uniform seed multiplication field in case of established cultivars. A fair number (300-500) of healthy plants typical of the cultivar are selected and marked for progeny testing. The seeds of the marked true- to- type plants are harvested separately. The seeds of each ear/head/plant are planted in a 3m long progeny row. These progeny rows are assessed critically several times during the growing season. Progeny rows that deviate in one or other characteristics are discarded and entire plant progeny rows is rejected. The plant progenies that are uniform and true to type are selected and bulked together as nucleus seed stage-I. This nucleus seed is used for planting larger breeder seed plots. If the breeder seed requirement of a particular cultivar is more, then another cycle of nucleus seed production is followed.

In this case the true- to- type individual plant progenies are harvested and thrashed separately. Seed of each selected plant- row progeny is now sown in a small plot called plant- row- progeny plot. The second cycle of nucleus seed production provides another opportunity to eliminate any plant row progeny showing segregation or off-type plants. The plots with required uniformity and plant type are bulked together to produce nucleus seed stage-II.

Isolation distance should be maintained appropriately between varieties.

Varietal Maintenance in Cross Pollinated Crops

The maintenance of varieties in cross fertilized crops probably more complicated than self pollinated crops. The following methods used for maintenance of nucleus seed of inbred lines.

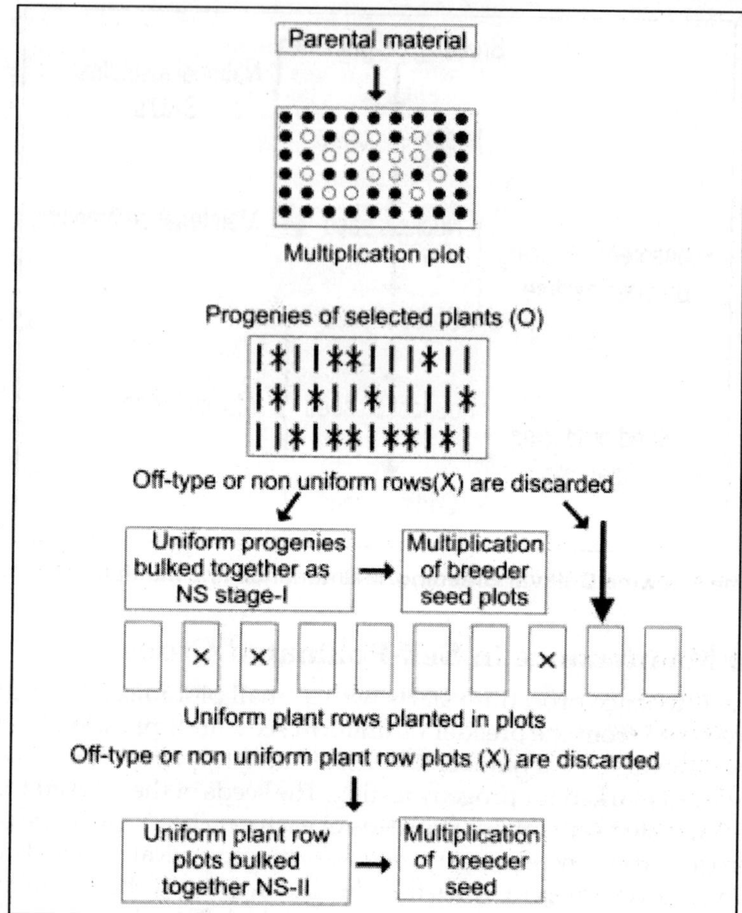

Maintenance Scheme for Self-Pollinating Cultivars.

Hand Pollination

Maintaining nucleus seed of inbred lines involves self pollination, sib pollination or a combination of two procedures. Generally maintenance by sibbing is preferred by some breeder because it does not reduce the vigour excessively. However, if a change in breeding behavior is noted and then selfing should be used as a means of stabilizing the inbred lines. It is preferable to maintain some parental lines by alternate selfing and sibbing from one generation to the next. The individual selfed or sibbed ears should be examined for different traits. Those ears which are not in uniformity to the varietal characteristics should be discarded. Each selfed or sibbed ear should be thrashed separately. The hand pollinated seed should be sown on fertile land where same kind or variety has not been sown in previous season. The crop specific isolation requirement and other seed production measures will be adopted as such.

These separately threshed ears should be planted in ear to row as following the stringent isolation distance. This ear to row planting system helps in easy identification of off-type plants. Careful inspection is needed to detect the off-type plants. The off-type plants should be checked prior to pollen shedding stage. It is very easy to recognize the out-crossed rogue because they are normally more vigorous. Harvest these rows separately and careful examine for the ear characteristics. All off colored, textured of diseased or otherwise undesirable ears should be sorted out. After that, the remaining ears may bulked.

References

Agrawal R L 1995 Seed Technology. Oxford and IBH Pub. Co. Ltd. New Delhi.

Agrawal R L 1998 Fundamental of Plant Breeding and Hybrid seed Production. Oxford and IBH Pub. Co. Ltd. New Delhi.

Hartmann H T and Kester D E 1968. Plant Propagation: Principle and Practices. Prentice Hall, New Delhi.

Kadam B S 1942 Deterioration of Varieties o Crops and the Task of the Plant Breeder. Indian J. Genet. And Pl. Br. 2: 159-172.

Singhal C 2001 Concept of variety maintenance. Seed Tech. News 3(1) : 3-5.

Singhal N C 2010 Variety: definition, characteristics and maintenance *In* : N C Singhal (Ed.) Seed Science and Technology.Kalyani Publishers New Delhi PP 217-234.

2014, Sustainable Rural Development through Agriculture *Pages* ***188–210***
Editors: **Dr. Shobhana Gupta and Dr. S.S. Tomar**
Published by: **BIOTECH BOOKS, NEW DELHI**

Chapter 15
Orchard Management and Varietal Narrative of Mango

Abhishek Bahuguna, Sandhya Bahuguna and Birendra Prasad

Mango (*Mangifera indica* L.) is the most important fruit of India. The genus *Mangifera,* which belongs to the family Anacardiaceae, originated in South - East Asia at an early date. According to Mukherjee (1958), the natural spread of the genus is limited to the Indo-Malaysian region, stretching from India to the Philippines and New Guinea in the east. The mango (*Mangifera indica*), sometimes called the 'king of fruits', by volume is the second largest tropical fruit crop in the world after bananas and fourth in total fruit after bananas, citrus and apples. It is native to north-eastern India and Burma. India, the main producer, accounts for 65 per cent of the world's mango crop, which is estimated at 16 million tonnes (FAO 1995).

Common Name	:	Mango
Botanical name	:	*Mangifera indica* Linn
Latin name	:	*Mangifera Indica*
English name	:	Mango
Sanskrit	:	Amrah
Hindi	:	Aam
Marathi	:	Amba
Tamil	:	Mamaram

Telgu	:	Mamidi
Malayalam	:	Mavu
Kannada	:	Mavu

It is very well adapted to tropical and subtropical climates. There are mainly two types of mangos classified according to the place of origin, mono-embryonic mangoes or Indian varieties of subtropical origin and poly-embryonic mangoes, varieties of tropical origin. Mono-embryonic varieties were evolved in subtropical India while the poly-embryonic mangoes were originated in wet tropical South East Asia. Over the years, these varieties have been moved from place to place around the world and cultivated. Because of the cross pollination nature of mango flowers, new varieties have been evolved as a result of natural and man-made crosses within and between these two groups of mango. Thus ranges of widely differing varieties have been appeared over the last few decades. Out of these varieties, superior candidates giving high yields and quality fruits were adopted as cultivars (Cultivated Varieties) in mango growing regions of the world. The best of these have been mass propagated with vegetative means and well established and famous cultivars have been born in different regions of the world. These cultivars show very distinct differences in relation to yields, bearing habits, fruit size, color, flavor and aroma etc. Way back in 327 BC, Alexander the Great too had its view on the Indus Valley when he was there on his conquering spree of the world. No other fruit, excepting banana, is so closely associated with the history of agriculture, nay, the very history of civilization itself, as the mango is records suggestthat it has been in cultivation in the Indian subcontinent for well over 4000 years now (De Candolle 1904) and has been since time immemorial, the favorites of the kings and commoners alike because of its luscious taste and captivating flavor.

Uses

The mango, because of its great utility, occupies a pre-eminent place amongst the fruit crops grown in India and is acknowledged as the king of Fruits of this country, Amir Khusrau (1330 AD) has stated (Popenoe 1920)-'The mango is the pride of the garden, the choicest fruit of Hindustan, other fruits we are content to eat when ripe, but the mango is good in all stages of growth' Young and unripe fruits, because of their acidic taste, are utilized for culinary purposes as well as for preparing pickles, chutneys and amchoor etc. Ripe fruits are utilized in preparing squash, nectar, jam, cereal flakes, custard powder and baby food, mango leather (am-papar) and toffee. Besides, fruits of some cultivars like Alphonso and Dashehri are sliced and canned for catering to the needs of consumers during the off- season.

Composition

The mango is a good source of sugars, vitamins A and C and minerals. The composition in general differs with the cultivar and the stage of maturity. The unripe green mangoes are reported to have 90 per cent moisture, 0.7 per cent protein, 0.1 per cent fat, 8.8 per cent carbohydrates, 0.01 per cent calcium, 0.02 per cent phosphorus, 4.5 mg/100g iron, 150 IU carotene (vitamin A), 30μg/100g riboflavin, 3mg/100g ascorbic acid (Anon. 1962). Consumption of a medium size mango could provide the daily requirement of Vitamins A and C. The level of various nutrients may vary depending on the cultivar, ripeness of the fruit and area of cultivation.

Mango	One mango without peel contains 1.06 grams of protein, 135 calories and 3.7 grams of dietary fiber.	Potassium - 323 mg Phosphorus - 23 mg Magnesium - 19 mg Calcium - 21 mg Sodium - 4 mg Iron - 0.27 mg Selenium 1.2 mcg Manganese - 0.056 mg Copper - 0.228 mg Zinc - 0.08 mg	Vitamin A - 1584 IU Vitamin B1 (thiamine) - 0.12 mg Vitamin B2 (riboflavin) - 0.118 mg Niacin - 1.209 mg Folate - 29 mcg Pantothenic Acid - 0.331 mg Vitamin B6 - 0.227 mg Vitamin C - 57.3 mg Vitamin E - 2.32 mg Vitamin K - 8.7 mcg

Cultivars

Mango is the most important fruit covering about 35 per cent of area and accounting of 22 per cent total production of total fruits in the country, which is highest in the world with India's share of about 54 per cent. India has the richest collection of mango cultivars. Major mango growing States are Uttar Pradesh, Bihar, Andhra Pradesh, Orissa, West Bengal, Maharashtra, Gujarat, Karnataka, Kerala and Tamil Nadu. The main varieties of mango grown in the country are Alphanso, Dashehari, Langra, Fajli, Chausa, Totapuri, Neelum etc.

The popular commercial cultivars of different regions in India are-

Northern region	:	Dashehari, Langra, Chausa and Bombay Green.
Eastern region	:	Himsagar, Langra, Fazli, Zardalu, Krishnabhog and Gulabkhas
Western region	:	Alphonso, Pairi, Kesar, Rajapuri, Malkurad and Jamadar
Southern region	:	Bangalora, Neelum, Swarnarekha, Pairi (Peter), Banganpalli, Mulgoa and Badami (Alphonso)

Characteristics of Important Indian Varieties

1. Alphonso

This is the leading commercial variety of Maharashtra state and one of the choicest varieties of the country. This variety is known by different names in different regions, *viz.* Badami, Gundu, Khader, Appas, Happus and Kagdi Happus. The fruit of this variety is medium in size, ovate oblique in shape and orange yellow in colour. The fruit quality is excellent and keeping quality is good. It has been found good for canning purpose. It is a mid season variety

2. Bangalora

It is a commercial variety of south India. The fruit size is medium to large, its shape is oblong with necked base and colour is golden yellow. Fruit quality is poor. Keeping quality is very good. It is widely used for processing. It is a mid season variety.

3. Banganpalli

It is a commercial variety of Andhra Pradesh and Tamil Nadu and also known as Chapta, Safeda, Baneshan and Chaptai. Fruit is large in size and obliquely oval in shape. The colour of the fruit is golden yellow. Fruit quality and keeping quality are good. It is a mid season variety and is good for canning.

4. Bombai

It is a commercial variety from Bihar state. It is also known as Malda in West Bengal and Bihar. Fruit size is medium, shape ovate-oblique and colour yellow. Fruit quality and keeping quality are medium. It is an early season variety.

5. Bombay Green

It is commonly grown in north India due to its early ripening habit. It is also called Malda in Northern India. Fruit size medium, shape ovate oblong and fruit colour is spinach green. Fruit quality is good and keeping quality is medium. It is a very early variety.

6. Dashehari

This variety derives its name from the village Dashehari near Lucknow. It is a leading commercial variety of north India and one of the best varieties of our country. The fruit size is medium, shape is oblong to oblong oblique and fruit colour is yellow. Fruit quality is excellent keeping quality is good. It is a mid season variety and is mainly used for table purpose.

7. Fajri

This variety is commonly grown in the states of Uttar Pradesh, Bihar and West Bengal. Fruit is very large, obliquely oval in shape. Fruit colour is light chrome. Fruit quality and keeping quality are medium. This is a late season variety.

8. Fernnadin

This is one of the oldest varieties of Bombay. Some people think that this variety originated in Goa. Fruit size is medium to large, fruit shape is oval to obliquely oval and fruit colour is yellow with a red blush on shoulders. Fruit quality and keeping quality are medium. It is a late season variety mostly used for table purpose.

9. Himsagar

This variety is indigenous to Bengal. This is one of the choicest varieties of Bengal and has gained extensive popularity. Fruit is of medium size, ovate to ovate oblique. Fruit colour is yellow. Both fruit and keeping quality are good. It is an early variety.

10. Kesar

This is a leading variety of Gujarat with a red blush on the shoulders. Fruit size is medium, shape oblong and keeping quality is good. It is an early variety.

11. Kishen Bhog

This variety is indigenous to Murshidabad in West Bengal. Fruit size is medium, fruit shape is roundish oblique and fruit colour is yellow. Fruit quality and keeping quality are good. It is a mid season variety.

12. Langra

This variety is indigenous to Varanasi area of Uttar Pradesh. It is extensively grown in northern India. Fruit is of medium size, ovate shape and lettuce green colour. Fruit quality is good. Keeping quality is medium. It is a mid season variety.

13. Mankurad

This variety is of commercial importance in Goa and in the neighboring Ratnagiri district of Maharashtra. The variety develops black spots on the skin in rainy season. Fruit is medium in size, ovate in shape and yellow in colour. Fruit quality is very good. Keeping quality is poor. It is a mid season variety.

14. Mulgoa

This is a commercial variety of southern India. It is quite popular among the lovers of mango owing to high quality of its fruit. Fruit is large in size, roundish oblique in shape and yellow in colour. Fruit quality is very good. Keeping quality is good. It is a late season variety.

15. Neelum

This is a commercial variety indigenous to Tamil Nadu. It is an ideal variety for transporting to distant places owing to its high keeping quality. Fruit is medium in size, ovate oblique in shape and saffron yellow in color. Fruit quality is good and keeping quality is very good. It is a late season variety.

16. Chausa

This variety originated as a chance seedling in the orchard of a Talukadar of Sandila district Hardoi, U.P. It is commonly grown in northern parts of India due to its characteristic flavour and taste. Fruit is large in size, ovate to oval oblique in shape and light yellow in colour. Fruit quality is good keeping quality is medium and it is a late variety.

17. Suvarnarekha

This is a commercial variety of Visakhapatnam district of Andhra Pradesh. Other synonyms of this variety are Sundari, Lal Sundari. Fruit is medium in size and ovate oblong in shape. Colour of the fruit is light cadmium with a blush of jasper red. Fruit quality is medium and keeping quality is good. It is an early variety.

18. Vanraj

It is a highly prized variety of Vadodra district of Gujarat and fetches good returns. Fruit is medium in size, ovate oblong in shape and colour is deep chrome with a blush of jasper red on the shoulders. Fruit quality and keeping quality good. It is a mid season variety.

19. Zardalu

This variety is indigenous to Murshidabad in West Bengal. Fruit size is medium, oblong to obliquely oblong and golden yellow in colour. Fruit quality is very good. Keeping quality is medium. It is a mid season variety.

Hybrid Varieties

1. Amarapali

This hybrid is from a cross of Dashehari x Neelum. It is dwarf, regular bearing and late maturing variety. The variety is suitable for high density planting as about 1600 plants may be planted in a hectare. It yields on an average 16 tonnes/hectare.

2. Mallika

It is from a cross of Neelum x Dashehari. Its fruit is large in size, oblong elliptical and in shape cadmium yellow in colour. Fruit and keeping quality are good. It is a mid season variety.

3. Arka Aruna

It is a hybrid between Baganpalli and Alphonso. It is dwarf regular bearing, precocious. Fruits are large having attractive skin colour with red blush free from spongy tissue.

4. Arka Puneet

It is a hybrid between Alphonso and Banganpalli. It is regular and prolific bearer. Fruits are medium sized having attractive skin colour with red blush and free from spongy tissue. Excellent keeping quality.

5. Arka Anmol

This hybrid is from a cron of Alphonso and Janardhan Pasand. It is regular bearer and good yielder. Fruits are medium sized having uniform yellow peel colour, excellent keeping quality and free from spongy tissue.

6. Arka Neelkiran

It is a hybrid between Alphonso and Neelum. It is, regular bearering late season variety with medium sized fruits having attractive red blush free from spongy tissue.

7. Ratna

This hybrid is from a cross of Neelum x Alphonso. Tree vigorous, precautions, fruits are medium sized, attractive in colour and free from spongy tissue.

8. Sindhu

It is from a cross of Ratna x Alphonso. It is regular bearer, fruits medium sized, free from spongy tissue with high pulp to stone ratio and very thin and small stone.

9. Aurumani

It is from a cross of Rumani x Mulgoa. It is precocious, heavy and regular bearing with large fruits having yellow cadmium skin colour.

10. Manjeera

This hybrid is from a cross of Rumani x Neelum. It is dwarf, regular and prolific bearer with firm and fibreless flesh.

How to Select a Mango Variety

Since the profitability of an orchard depend on the right variety and the performance of a variety is greatly affected by the environment, selection of varieties to be adopted should be done very carefully. For intermediate and dry zones of Sri Lanka, where much of the lands are available for commercial agricultural activities varieties Willard, Karthacollomban and Vellaicollomban are more suitable. When authenticated planting materials are used and appropriate sites for orchards are selected these varieties may be ideal because under these conditions, varieties mentioned above starts to bear at 3 - 4 years after planting and bear continuously every year. If good cultural management practices are adopted growers can expect reasonable high and profit able yields from these varieties every year. If growers are interested in going for new varieties, especially exotic varieties which have good market demands, it is advisable to analyze the growing environment of those varieties with that of the local environment where the orchard is going to be established. Since mango is very sensitive to environment, particularly to air temperature, temperature profiles may be helpful to decide the appropriateness of such varieties to local environment. To illustrate how this could be done, the temperature profiles for Homestead, Florida is given in comparison with temperature profiles at Aralaganwila, Mahawali System B. If a grower in Aralaganwila is interested in going for variety Haden or Tommy atkins, it is better to find answers to the following questions before desiding to adopt the variety. If this variety is successfully grown in Florida, what effects the low temperatures experienced in Florida during September - February have on the fruit yield in following season. What will happen to yields if the winter temperatures are higher than the average in a particular season? If you look into the answers for these questions you will get to know that cool temperatures during the period from September to February is an essential prerequisite for those varieties to produce good crops in the following summer and above average temperatures during winter will produce more leaf flushes and fruit yields will be lower in the following season. Then you have to think, what effects the temperatures in Aralaganwila, which are considerably higher than that in Florida, will have on the productivity of these varieties. It should be mentioned here that most of these varieties are very sensitive to temperatures and a little difference as much as 5ºC accumulated over seasons would make a significant difference in tree productivity. Again as another example, if you think of growing Kensington, Australia's major variety, just consult an Australian mango grower and ask what happens to there crop if their winter temperatures in June -July go above average in a particular season. Or if you meet a Thai mango grower who had visited Australia, ask him why they do not grow Kensington in Thailand. If you intelligently analyze these facts you can make wise decisions on adoption of those new varieties. This kind of environmental analysis, not only with temperature but also with rainfall, relative humidity etc., is very helpful in deciding what varieties to grow in a particular area if growers are interested in going for new exotic varieties.

5oil

The mango can grow well in all types of soil from alluvial to lateritic, except ·lack cotton soils which are considered to be poor. The only prerequisite is a deep (2 o 2.5m) and well drained soil. However, it grows successfully in soft rocky areas of he west coast. Like most other fruit crops, it prefers a slightly acidic soil. It does not lo well beyond a pH of 7.5 (Singh 1960). Soils with an appreciable amount of gravel r kankar ($CaCo_3$) in the profile too can grow good mangoes provided they are not ery alkaline. Saline and alkaline conditions are not conducive to profitable mango ultivation.

:limate

Mango is very well adapted to tropical and subtropical climates. The mango is usceptible to cold. Young trees may be killed by temperatures below 0.5°C. Older ees will survive a few degrees of frost, but may be severely damaged. In areas isceptible to frosts such as Gingin, selection of planting sites is critical. Ideally, a orthern facing slope is the most suitable. Adequate frost protection is also essential. langoes will tolerate temperatures up to 48° Celsius, without serious damage to stablished, irrigated trees, however sun damaged fruit and fruit drop can be caused y excessively high temperatures combined with low humidity. It thrives well in most all the regions of the country from sea level to an altitude of 1500m, *i.e.*, from ape Comerin to Himalayas. However, it cannot be grown commercially in areas oove 600 m. Temperature, rainfall, wind velocity and altitude are the main climatic ctors which influence its growth and fruiting. It cannot stand severe frost, especially hen the tree is young. High temperature by itself is not so injurious to mango, but in ombination with low humidity and high winds, affects the trees adversely. Most of e mango varieties thrive in places with good rainfall (75 to 375 cm per annum) and y season. The distribution of rainfall is more important than its amount. Dry weather fore blossoming is conducive to profuse flowering. Rain during flowering is trimental to the crop as it interferes with pollination. However, rain during fruit velopment is good but heavy rains cause damage to ripening fruits. Strong winds d cyclones during the fruiting season can play havoc as they cause excessive fruit op.

election of Orchard Site

The orchard performance and profitability to a greater extent depend on the lection of planting sites. In selecting the site factors like soil type, soil depth, soil tility, drainage, water table etc. must be carefully studied. Mango is not too particular to soil type providing it has good drainage. Mangoes grow in most soils, but for mmercial production the poorer, shallow soils are better. On these soils, the trees y smaller and more manageable. On the other hand, rich deep loam soils certainly ntribute to maximum growth, but if the soil is too rich and moist and too well tilized, the tree will respond vegetatively but will be deficient in flowering and iting. Mango performs very well in sand, gravel and even on oolitic lime stones as Southern Florida and the Bahamas.

The water table must be well below one to one ant will affect root functioning and ultimately will affect the tree productivity. When trees grow bigger, if water tabl is high, it will affect flowering and fruit set also because one of the prerequisites fo good flowering is a dormant period of 2-3 months prior to flowering with less water

If inappropriate sites such as places with high water table and with very rich fertile, deep moist soils are selected, it will affect the productivity of orchards. Therefore selection of the site must be very carefully done especially because if something i wrong it cannot be corrected easily after the orchard is established.

Field Preparation

Mango is cultivated both as a home garden crop and a commercial scale crop Before establishment of a commercial cultivation, clear the land and plow and harrow At the same time, take steps to adopt appropriate soil conservation measures.

Preparation of Pits and Planting

Planting should be done when the growth flushes in the plants have hardenec Leaves should be cut to half its size to minimize transpirational water loss and t reduce transplanting shock. It is better to establish plants in the orchard after heav rains are over. In case appropriate drainage and irrigation facilities are availabl planting may be done any time of the year. It is advisable to paint the stems of plant with a white water based paint to protect plant stems from direct sunlight and hea

Plant spacing may be adjusted depending on soil type, variety and managemer systems to be adopted. For variety Willard a spacing of 7 x 10 meters is recommende while for Karthacollomban and Vellaicollomban, recommended spacing is 10 x 1 meters. If the soils are sandy or tree dwarfing methods such as regular pruning or us of chemical growth retardants are to be used, high density planting may be done wit narrower spacing. However, when spacing is manipulated, it should be kept in min to use with in raw spacing to adjust plant density. It is always better to have 10 mete of inter raw spacing to facilitate orchard management practices when trees gro bigger.

Planting designs such as rectangular planting, triangular planting etc. may b used. However, rectangular planting designs are commonly used in commerci orchards. Planting holes are made about 2 - 3 months priors to planting and fille with topsoil, compost and basal fertilizer mixture and heap up to allow the soils establish. In loose soils 60 x 60 x 60 cm planting holes are sufficient while in ha soils 90 x 90 x 90 cm holes are recommended. By the time of transplanting soils mu have well compacted in heaped planting holes. Planting should be carried out such a way that the stock-scion union is about 4 - 6 inches above the ground lev Scion should never touch the ground at all.

Time of Planting

Planting can be commenced with the onset of Maha rains in the dry zone. intermediate and wet zones, planting is possible with onset of Maha or Yala rai For a home garden, planting is possible at any time of the year except during perio of heavy rains. If a prolonged dry condition exists, plants must be irrigated as a when necessary. Use only very vigorous plants for field planting. Minimize the stress during field planting by hardening plants exposing to direct sunlight and with less water application. This hardening held improves the success rate of field establishment. At planting remove the cover. Cut around the edge of the bottom of the pot and remove the intermingled roots by pruning tap root.

- ✩ Place the plant in such a way that the base of the plant in the pot is aligned with the ground level. Then remove the polyethylene bag with two longitudinal cuts from bottom up.
- ✩ After removing the polyethylene cover, fill the planting hole with soil and slightly tighten the soil. These steps help reduce root damage due to breaking and splitting of potting media block.
- ✩ Allow the plant to grow directly up. Use a stick closer to the plant and tighten it into the stick carefully. - To minimize water loss under dry weather conditions, remove half of each mature leaf.
- ✩ Use mulch around the plant using easily available mulching material such as dry grass or salvenia. Mulching helps to reduce soil temperature in the root zone. Weed control also become easy. It also reduces drying of soil and wind erosion of soil.
- ✩ After planting watering is an essential requirement. Construct a basin around plants to control runoff of applied water.
- ✩ Provide shade appropriately to protect plants from heavy sunlight.

Protection of Plants

In the initial 2 to 3 years, it is advisable to protect plants against low temperature injury by covering plants with some sort of cover, leaving the eastern side open for entrance of light. Building up slow fires, which emit smoke, and resorting to flood irrigation may also be essential to ward off the ill effects of frost.

Irrigation

Water requirements for tree fruits, in general, have not been worked out so far. The mango is relatively tolerant of dry conditions due to its deep root system. However, most feeder roots still are concentrated in the top 75 to 80 cm of soil. Mangoes are able to tolerate water logging and moderately saline conditions better than most tropical tree crops, but for best results, irrigation water should contain less than 1000 ppm total salts. Plant physiology studies indicate that a period of cold stress before flowering should promote flower bud production, instead of shoot growth. In Carnarvon and Gingin trees are stressed enough by cold conditions. To save water, Carnarvon growers often do not water mature trees from April to June inclusive. In all areas, it is essential to provide adequate water from the first sign of flower spikes, until harvesting. In Kununurra, Kensington Pride only crops well after a cold winter. Under normal winter or dry season temperatures, imposing water stress does not necessarily lead to the promotion of good flowering. However, it can reduce the chances of a vegetative flush just prior to flowering which reduces the tree's potential for a good flowering. The critical time for irrigation is during fruit development.

Water stressing the trees at this time of year can lead to reduced yield. Trees up to four years old should be watered every one to three weeks throughout the dry months. In most growing regions under-tree sprinklers provide the most efficient irrigation delivery system. Local irrigation consultants are able to advice on specific sprinkler types for individual situations. In Kununurra on the heavy clay soils, furrow irrigation is often used. Although this system is not as efficient as under-tree sprinklers, it does offer significant cost advantages.

The irrigation system is one of the most important resources in a commercial orchard. The type of irrigation system may be decided on the capital budget for orchard development. For mango orchards flood, basin, and furrow or under tree sprinkler irrigation systems are commonly used. Under tree sprinkler system is the best out of these systems and it is the most expensive system as well. With this system it is possible to enhance the water use efficiency of the orchard by cutting down the water requirements significantly as the conveyance losses are very low compared to other systems. It also has an added advantage that trees can be manipulated with greater efficiency with irrigation, a powerful management tool in commercial orchards, in controlling tree productivity to its optimum

Manuring and Fertilization

Only generalized recommendations of mangoes' fertilizer requirements can be made. It is important that for good nutritional management regular leaf samples are taken. Nutritional levels will change in the tree during the season. It is therefore important to sample the trees at the same time each season. The most stable time of year for sampling is just prior to flowering. Take leaf samples from the last mature flush, ensuring that the shoot is dormant at that time.

Application of plant nutrients economically at correct time with right amounts in a way that nutrients could be taken up by plants efficiently with minimum losses cover the whole issue of nutrient management. Proper nutrient management has significant effects on tree productivity. The purpose here in nutrient management is to keep mineral nutrient levels in the tree with in the desired range to have the growth and development effects and fruiting of trees as desired by the grower.

It is time to just review the functions of major plant nutrients in mango. Mango needs annual applications of Nitrogen, Phosphorus and Potassium. Nitrogen is needed at the time of vegetative flushing to support leaf growth. Nitrogen is utilized to synthesize chlorophyll molecules found in green sub cellular organs known as chloroplasts where photosynthesis occurs. In addition to this, N is used as components in many enzymes, co-enzymes, hormones, nucleic acids and many other plant proteins. N is highly mobilized and if there is a deficiency, N in older leaves may be transported to new developing leaves. Therefore, deficiency symptoms appear in older leaves becoming yellowish in color and early senescence.

Phosphorus is required to synthesis ATP, cellular molecules engaged in respiration process to support plant life. P is also found in other bio-molecules such as nucleic acids, enzymes, certain proteins and in cell membrane phospholipids.

Potassium is essential mainly for food translocation within the plant. This nutrient is not generally found in plant cells as integral components of bio molecules. Potassium is found in cellular sap as K^+ and are facilitating many plant physiological processes such as stomatal opening, activation of enzymes and charge balancing agent in transport of anionic nutrients such as nitrates and sulphates etc.

The most important thing in nutrient management for bearing trees is to analyze the importance of timing of nutrient application in relation to tree phenology or growth cycle. The recommended fertilizer mixture for bearing mango trees is 12 – 8 - 34 for Dry and Intermediate Zones and 11 – 10 - 25 for wet zone. Based on these mixtures, annual requirement of each nutrient given in Table 15.1.

Table 15.1: Nutrient Requirements (g/tree) of Bearing Mango Trees.

Nutrient	*Age of Tree*						
	4	*5*	*6*	*7*	*8*	*9*	*10*
A. For dry and intermediate zones							
N	108	162	216	270	324	378	432
P_2O_5	72	108	144	180	216	252	288
K_2O	306	459	612	765	918	1071	1224
B. For Wet Zone							
N	100	150	200	250	300	350	400
P_2O_5	90	135	180	225	270	315	360
K_2O	225	337	450	562	675	787	890

With the exception of N, the essential nutrients for growth can be applied to the soil at any time without forcing competitive, non productive vegetative flushing. This does not mean to say that some rationalization of fertilizer application is not needed. However, looking at tree nutrient requirements, application timing can be programmed in relation to growth cycle.

Nitrogen is one of the most important and powerful management tools currently available and learning to use it effectively is very essential to improve orchard productivity. Application of N stimulates vegetative flushing. At the same time withholding N temporarily favors fruitfulness, but we have to be careful as overall productivity declines if trees become N deficient. The timing in N application is also critical in balancing the tree between vegetative and reproductive growth. Therefore, all N requirements should be given just after harvesting to induce growth flush immediately. Never after the growth flush are mature until the next harvesting time.

P may be added at any time of the year. Its best, however, to apply it in rainy season under rain fed conditions. Otherwise it may be applied at any time of the year with another fertilizer to cut d requirement of plants are high during the period from flowering to fruit maturity for during this period trees are very active in translocating food reserves to maturing fruits. Trees need K during flush development also. Therefore K may be applied after harvesting and again during flowering to fruit maturity.

In addition to above major nutrients, other nutrients that are important include Calcium, Magnesium, Sulphur, Zinc, Boron, Iron, Manganese and Copper. Under high tree management, the requirements of these nutrients may be increased as most of these nutrients are removed from the field with harvested fruits. Under such circumstances external applications may be required. Requirements of those nutrients may be decided by leaf analysis. Leaves are usually sampled during flowering to assess micro-nutrient requirements.

It is imperative to see that trees are not over fertilized with any of the major or minor nutrient as excess amounts may cause problems like excessive vegetative growth, low yields, inferior fruit quality and nutrient toxicity.

The recommended fertilizer mixture for non bearing mango trees has a higher N content than that recommended for bearing trees. This is because plant needs more N at young stage to promote tree growth. DOA recommends 16-20-12 mixture for Dry and Intermediate Zones and 12-14-14 for Wet Zone. In wet zone mixture, as the phosphorus source less expensive rock phosphate is used instead of Concentrated Supper Phosphate (CSP).

Fertilizer requirements for plants increase annually. It is always better to split the annual quantity as far as possible and this is very important under sandy soil conditions. Also fertilizer dose for the rainy months in the Dry Zone should be applied after the peak high intensity rains are over to minimize leaching losses of nutrients.

Fertilizers can be placed as a band spread approximately 30 cm from the base of the plants and should be incorporated to the soil mechanically without severely disturbing the roots. Irrigation is necessary after specially N if urea is used in the fertilizer mixture. Soil incorporation is not necessary if under tree sprinklers are used for irrigation of trees. Then fertilizers can be effectively incorporated by irrigation.

Fertilizer Placement

In addition to timing, the way fertilizers are placed for the trees to take up effectively is also important in nutrient management. For efficient utilization by trees, fertilizers must be equally distributed on the root zone area. Active roots are scattered 30 cm from the trunk to drip line in small trees and in well grown trees this may extend up to about 1 - 1.5 m away from the drip line. After broadcasting equally on the root zone area, fertilizers should be incorporated to the soil by cultivating the top soil. Care should be taken not to damage the roots too much. If thick mulch with under tree sprinklers is available for irrigation such mechanical incorporation is not required as a number of frequent irrigations may slowly incorporate the fertilizers to root zone.

Especial care should be taken when applying urea as the source of N. Urea must only be applied when the soils are moist. If not trees must be irrigated immediately after urea application.

Fertilizers may also be applied as bands or in few spots around the tree to reduce application costs. However, equal broadcasting around the tree and incorporation to the root zone by mechanical ways or by irrigation is the best technique of application. Band and spot application methods are less efficient than broadcasting on the root zone.

Intercropping and Cover Cropping

Inter Cropping

Inter crops such as vegetables, legumes, short duration and dwarf fruit crops like papaya, guava, peach, plum, etc. depending on the agro-climatic factors of the region can be grown. The water and nutrient requirements of the inter crops must be met separately.

Cover Cropping

Besides the intercrops one may also grow some crops like sunnhemp (in light soils) or 'dhaincha' (for heavy soils) to protect the crchard soil from erosion and also for enriching the soil fertility. These crops are sown with the beginning of the rainy season and are ploughed before the stems become woody (towards the end of monsoon). One can choose any other crop like cowpea, pea, berseem, 'mung', 'urid' and 'guar'.

Pruning and Training

The mango, being an evergreen plant, hardly needs any pruning. The only pruning that it requires is periodic removal of the dead and diseased branches, as and when one notices them. It was found that in 17th year old trees of cvs. Himayuddin, Rumani, Kalepad and Potalima, when the criss cross branches were pruned to open the centre and the terminal whorl of shoots were thinned to retain only one or two healthy shoots, gave greater yield than in previous years (Shanmugavelu and Selvarajan 1985). Training, however, is an essential practice in the initial 2-3 years. This is done with a view to provide a good framework for the future so that the branches are spaced properly and these do not break with the crop load at the bearing stage. The branches are not encouraged too low on the truck or too high from the ground level.

Training of Trees

Training gives a tree good appearance, management of the tree becomes easy, high yields with quality fruit is possible and pest and disease incidence minimized. Training of trees must be started right from the early stages of growth. Pay special attention to train trees from the time of planting. Allow a plant to grow as a single stem up to about 1/2 M. Let the first branch form at 1/2 M height. Then at about 15-20 cm spacing allow growing 3-4 branches around the tree. Let these branches to grow in opposite directions to give a tree a good appearance. This is also important to minimize break of branches at latter stages of growth. Natural shading of branches also minimized when branches are equally well distributed around the tree.

Shoots that do not receive sufficient sunlight do not produce enough food reserves for the tree. Thus, fruit set in such branches are not satisfactory. Such branches must be removed. Also diseased, dead and intermingling branches must be removed. In removing branches the cut must be very close to the main stem or limb when pruned. Prune trees under dry weather conditions. Apply a paint mixed with a fungicide to the cut surface.

Pruning

When a young Kensington Pride tree is about one metre tall, remove 20 to 30 millimetres of the terminal shoot to promote branching. If this is not done, the tree will grow four to five metres tall as a single stem and will be subject to wind buffeting. Three side branches should be allowed to grow from beneath the point of topping. When these have made 40 to 60 centimetres of growth, repeat pruning of terminals to induce further branching. Ensure that your pruning maintains wide branch-crotch angles - up to 90 to minimise wind damage. This structural pruning is essential in the early years of tree development to assure a strong structured tree and to maximise the number of terminals (potential fruiting sites). Significant yield increase can be achieved on young trees if they are intensively pruned.

Fruit is borne on new season's growth and usually on the tips of the outer branches of the tree. Therefore, it is only necessary to lightly thin trees by removing weak, overcrowded or broken branches, keeping the centre of the tree open. Cut off branches which are too near the ground.

Pruning in the Tropics

In the tropics mango trees trend to be extremely vigorous. The tendency is to want to reduce the size by heavy pruning but this can be detrimental to yield for several seasons after pruning. This is primarily due to the lack of a growth check, *e.g.* cold temperatures. Light pruning of trees in the tropics just prior to flowering (see tip pruning: flowering and fruiting) has shown promise. Research is currently under way in the tropics to address these issues.

Pruning in the Subtropics

Cool winter temperatures in the subtropics generally impose a growth check on the trees. Postharvest pruning is a common practice in these regions. The frequency and severity of the pruning will be a matter of judgement depending on the timing and extent of the previous season's growth. For cropping not to be detrimentally affected by maintenance pruning, a following growth flush must occur.

Interculture

Maintenance of good sanitary condition is a must for keeping an orchard in healthy and diseases free condition. In the initial years of establishment, when one has to go in for intercrops or cover crops, one normally gives preparatory tillage to the plots. However, when the mango trees are full grown and intercropping is neither feasible nor desirable, it is necessary to plough the soil to a shallow depth at least twice a year, *i.e.*, in June and October. Also, the tree basins must be kept free from weeds at all times of the year by shallow hoeing and weeding. The best control of weed in the mango bud wood nursery was reported by pre-emergence application of atrazine at 4 kg/ha and post-emergence of gramoxone at 3 l/ha (Pawar *et al.*, 1985).

Flowering, Pollination and Fruit Set

Greenish white flowers, tinged with pink are borne on large panicles which develop from terminal buds on the outer stems. A tree may bear over a million flowers but most of the flowers do not have functional female parts, so only a few fruits are borne on each panicle. The flowers are self- and cross-pollinated. Various insects such as flies, bees and ants transfer the pollen in cross-pollination. Flowering in the Kimberley occurs from June through to September. In Carnarvon, flowering usually starts in mid-June and continues through to early October. Kensington Pride mango trees generally crop lightly on windward-side branches and also on heavily shaded branches.

Promotion of better flowering is an area which has received large amounts of attention over recent years. The use of Paclobutrazol has shown to generally improve flowering in poor seasons. Methods and accuracy are critical in applying this chemical. Contact your local horticulturist for latest recommendations on rates. In tropical areas a pre-flowering tip pruning has been shown to improve flowering on Kensington trees. This is where the trees are lightly pruned back to mature wood just prior to flowering.

During the rest period tree carbohydrate reserves levels increase as against tree N level. Thus C: N ratio of the trees increase. Also at the same time, shoots bolecules required to make flower primordial on the terminals of mature shoots. When all these conditions are accomplished the mature shoots become physiologically ready for flowering. After this stage is achieved trees are ready to flower when the environment is appropriate for flowering. The flower bud differentiation in India in most cases occurs between October December (Singh 1960), although Musahibuddin and Dinsa (1946) reported it to occur in August which is rather too early in the light of the present state of knowledge. The flowering period of mango is usually of a short duration of 2 to 3 weeks, low temperatures may extend it, whereas higher temperatures may shorten it. The mango inflorescence or panicles bear mainly two types of flowers male and perfect, though neutral flowers are also encountered occasionally. The number of flower per panicle variety between 1000 and 6000, depending upon the cultivars (Mukherjee 1953).

After flowering, fruit set takes place as a result of pollination of complete flower. Mango has two types of flowers, Complete Flowers or hermaphrodite flowers having both male and female organs *i.e.* anthers and ovary, and male flowers having only anthers producing pollen. Soon after pollination and fertilization of the ovary with pollen fruits start to develop. Then the plant become active in translocation carbohydrate and other food reserves from the storage tissues such as roots and trunks to the developing fruits for their growth.

Though mango flower panicles produce thousands of flowers and a lot of these pollinate, only a few, in some varieties one fruit per panicle while in other varieties several fruits per panicle and commence further development. Then the size of the crop a tree can bear is primarily determined by the tree reserve level. Excess fruits drop and this may happen at various stages of fruit development. Fruit drop is governed physiologically by the size of reserve food base in the tree and also by climatic conditions prevailing during this time. Even if trees have sufficient reserves, lack of K or soil moisture deficits during fruit development phase or pest and disease attacks on flowers and developing fruits may act as limiting factors controlling the size of the final crop. Therefore, at this stage trees have to be managed very well providing all necessary inputs correctly.

Fruit Growth and Development

This aspect has been studied in Dashehari, Langra and Chausa (Singh 1954b) and the growth curve is of the sigmoid type (Singh 1978). Development of the fruits of Langra and Dashehari starts in the last week of March and is completed by the second week of June. Percentage increase in growth is highest in April, followed by that in May, least growth being in the month of June, however, maximum increase in weight and volume occurred in May, followed by April and June (Saini *et al.*, 1971). The period of rapid growth in mango is directly related to the contents of auxin and gibberellins-like substances in the seeds (Chacko *et al.*, 1970). Natural parthenocarpy is not very common in mango. Mango fruits are usually flat, mostly undersized and seedless and in some cases a rudimentary cotyledon is also present. Thimmappaiah and Singh (1983) reported that in Dashehari, on an average, each tree produced 5.44 per cent parthenocarpic fruits of which 3.94 per cent was completely seedless and 1.43 per cent fruits had rudimentary cotyledons showing sign of degeneration.

Fruit Drop

The natural fruit drop in mango is rather too high amounting to about 99 per cent at various stages of growth, more especially during the initial four weeks in the cultivars Bombai, Langra and Fazli, about 13-28 per cent of the perfect flowers sets fruit and only 0.1 – 0.25 per cent reaches maturity stage (Sen, 1939).the extent of fruit drop appears to be a varietal characteristic. Amongst the causes attributed to this phenomenon are lack of pollination, low stigmatic receptivity, defective perfect flowers, poor pollen transference, occurrence and extent of self-incompatibility, competition between developing fruitiest and drought or lack of irrigation. Some other contributing factors are unfavorable climatic conditions during fruit development period, high incidence of serious diseases like powdery mildew and anthracnose and pest like hopper and mealy bug (Hayes 1957). The extent of fruit drop in mango can be reduced significantly by regular irrigation during the fruit development period. Timely and effective control measures against major pests and diseases can also be of great help. Subsequently, 2, 4-D (30 ppm) has given good results in controlling fruit dropped in Neelum, without having any adverse effect on fruit size or TSS (Rao and Subba Rao 1963). Gill (1966) tried a number of growth regulators including 2,4-D, NAA and 2,4,5-T on Bomby Green, Dashehari, Langra and Chausa of these, 2,4-D appeared to be the best at concentrations lower than 20 ppm about six weeks after fruit set. Exogenous application increased the fruit retention by increasing their levels or antagonizing adverse effects of endogenous growth hormones like ethylene and ABA (Singh and Ram 1983).

Biennial Bearing

This is one of the most burning problems, since it renders mango cultivation less remunerative to the growers. This term is synonymous to alternate bearing which denotes yield variation in alternate years, *i.e.*, a year of optimum or heavy fruiting is followed by a year of little or no fruiting. The terms periodicity of cropping and irregular bearing are sometimes erroneously used to describe the phenomenon of biennial bearing. The terms irregular bearing and periodicity of cropping imply that cropping does not follow a systematic pattern, *i.e.*, an optimum crop is obtained only once in a number of years. Such a behavior is largely due to lack of proper orchard management practices. On the other hand, the alternate or biennial bearing habit is of a genetic nature since this trait can be identified amongst the mango plants right from the initial years of fruiting. This is contrary to the viewpoint prevalent in the literature (Singh and Khan 1940), wherein it is mentioned that the biennial bearing habit sets in only after the age of 10-12 years. The problem of biennial bearing has been studied in great depth by a number of workers during the last 4 decades or so several aspects like climatological factors, age and size of shoots, C/N ratio and hormonal balance have been investigated into with a view to arriving at a better understanding of the problem and for providing a solution to this problem.

Climatological Factors

Adverse climatic conditions like rain, high humidity and low temperature sometime convert an on year into an off year directly or by promoting the incidence of diseases like powdery mildew and anthracnose. Climatological factors, such as, do not form the basic causes of biennial bearing.

Age and Size of Shoots

This necessarily means that a certain amount of physiological maturity is prerequisite to flowering in the case of cultivars which are prone to this phenomenon. This has also been established in another regular bearing cultivar Neelum, wherein shoots emerging in October are in a position to flower by the following March (Singh *et al.*, 1962a). In an on year, shoots of any size or maturity differentiate flower buds where as in an off year, even if the shoots of requisite size and maturity are available, and they fail to flower.

Carbon Ratio

Subsequently, Sen (1943a) reported that irregular bearing in mango was caused by nutritional deficiency, especially that of nitrogen. Studies on this aspect have since been conducted by a number of workers (Sen *et al.*, 1963) and it is indicated that higher starch reserves, total carbohydrates and C/N ratio favour flower bud formation in most of the cultivars studied except Baramasi and regular bearing cultivars. The available evidence suggests that in fruit plants, nitrogen and carbohydrate reserves play important roles in flower bud initiation, even if these do not form the primary cuase of the phenomenon of biennial bearing (Singh 1978).

Hormonal Balance

The problem of biennial bearing in mango appears to be closely associated with the fruit development process and the inhibitory influence of the developing fruits on vegetative growth (Singh 1960). Choudhuri and Rudra (1971) found an inverse relationship between the level of endogenous inhibitor in the shoot and vegetative growth and suggested that higher inhibitor content promotes flowering in mango. The distinct difference in the bearing behavior of the regular and biennial bearing cultivars leaves hardly any doubt that problem is an inherent one and a lasting solution can only be obtained through genetic engineering.

Control of Biennial Bearing Problem

The biennial bearing problem has been a major bottleneck in the expansion of the mango industry. Obviously, this problem has, therefore, posed a major challenge to the horticultural scientists during the last seven decades. Consequently a number of remedial measures have been proposed from time to overcome this limitation in mango production. These include:

- ✰ Proper upkeep and maintenance of orchards
- ✰ Deblossoming
- ✰ Pruning
- ✰ Growing regular bearing cultivars

Hedging

Hedging is a management system whereby trees are planted at a close spacing growing into each other and maintained as a hedge rather than as individual trees. Trees may be mechanically pruned annually to maintain size. This system has proven to be successful in Carnarvon, giving very high yields. However, fruit colour may not develop as well due to high amounts of shading.

Harvesting and Packing

Mango trees grow broad and relatively tall. The foliage is dense, and the fruit difficult to pick. To harvest most of the fruit, traditionally ladders and hydraulic platforms have been used. This method, whilst convenient, is expensive and has led to the recent development of semimechanised harvest aids. Harvest aids are either self-propelled or tractor-pulled trampolines whereby fruit is harvested onto canvas whilst being sprayed with a detergent solution to minimise sap burn.

Maturity Testing

Determining the mature green stage for harvesting mangoes is a difficult task. Often the temptation is there to pick the fruit too early due to high market prices at the beginning of the season. Several maturity indices are currently used by industry but none of the indices is completely accurate.

1. **Flesh colour.** This method is derived from the basis of internal flesh colour and fruit shape. Fruit are said to be mature when the internal flesh colour is light yellow. Colour charts are available for this. This is the most commonly used in-field method. It is, however, destructive and is not useful at the market as immature fruit will still colour internally after harvest.
2. **Dry matter.** Dry matter at harvest is a commonly used maturity indices in Queensland and the Northern Territory. It has been found that there is a correlation between dry matter and Brix (sugar levels). Dry matter is calculated by measuring a fresh sample of fruit then oven drying it and calculating the remaining dry matter as a percentage of the total weight. The recommended dry matter for Kensington is 14 per cent. This method can be used on the market floor, however, there is some concern over the

accuracy of this method as dry matter can be manipulated by irrigation management.

3. **Specific gravity.** Although this method has been well researched, it is currently not used commercially in Australia. Fruit with an SG greater than one are said to be mature; this means they will sink in water, while immature fruit will float. This method has the potential to be used for sorting the fruit on a grader once it has been harvested.
4. **Heat sums.** This method being developed by the NTDPI is still in its early stages however, it appears to have great potential as an accurate non-destructive method for determining maturity. The method is a calculation of the daily accumulated heat taken from the time of floral emergence.

   Heat sum = (max °C + min °C/2) - 12°C. (Diczbalis *et al.*, 1997). For Kensington an accumulation heat sum of 1600 hours has been found to indicate optimum maturity.

Preventing Sap Burn

Sap burn is the largest single quality problem with mango. Poor harvesting and handling techniques will result in a high incidence of sap burn which will severely downgrade the fruit quality. The sap in Kensington Pride consists of two components, oil and a protein component. The oil component of the sap is mostly responsible for the burning of the skin. The oil component is at its highest concentration in the spurt sap which is exuded immediately the stem is snapped. All harvesting and handling techniques used should be aimed at eliminating the potential for the sap to come into contact with the skin of the fruit.

Skin Browning

Harvest aids and the use of detergents have greatly reduced the incidence of sap burn on mango. However, a new problem has become increasingly more noticeable, being brown markings on the surface of the skin of the fruit. This is commonly called skin browning. Several forms of skin browning have been identified as well as their causal agents. Hygiene is the single greatest method of minimizing skin browning. Care needs to be taken that equipment used in the harvesting and handling procedure is maintained clean at all times. Fruit should never be packed when wet as this is a primary cause of skin browning.

Grading

Fruit are graded into lines according to their blemish level. Charts are available to assist in determining the amount of blemish level for each grade of fruit. Some growers participating in group marketing have already set standards for each grade.

Ripening and Storage

Removal of field heat is critical in maintaining fruit quality and for achieving maximum storage life. For this reason forced air cooling is recommended for producers in Carnarvon and the Kimberley. For maximum storage life the fruit should be stored at 13° Celsius. This is the most suitable temperature for transportation without the risk of causing chill damage to the skin. Fruit is generally put into ripening rooms at the market and gas ripened. This gives more uniformed ripening throughout the tray. It is currently not recommended to pre-ripen Kensington before leaving for market, although this is commercially practiced in some parts of Queensland. Optimum temperatures for ripening mangoes range from 20 to 23°C. If temperatures are high, fruit will not develop yellow background colour. Rather it will tend to ripen in a less attractive dull yellow/green colour. If temperatures are too low then sometimes fruit flavour can suffer.

Commercially, ripe Kensington Pride mangoes may be stored for three weeks by dipping in a prochloraz (Sportak) solution for controlling anthracnose and other postharvest rots then maintained at 13° Celsius. Both in storage and transport, the temperature must not fall below this level, unless under strictly controlled conditions where the temperature is gradually reduced. The critical minimum storage temperature for most other varieties has not yet been determined. Processed mangoes or mango pulp can be stored at 0° to 1° Celsius for up to six weeks. In the home, sliced mango can be stored in a freezer at minus 18° Celsius for up to 18 months.

Medicinal Properties

Both unripe and ripe mango has medicinal properties. A drink made out of unripe mango is used as a remedy to prevent various body ailments caused by a raise in ambient air temperature. Unripe mangoes are also used in treating stomach problems and to stimulate bile formation and in treatment of blood related diseases. Ripe mango has many medicinal properties. Consumption of ripe mango is useful to overcome night blindness and to protect health of skin. There is a common belief that consumption of ripe mango with cow's milk helps gain weight. Mango seeds, leaves and bark are used in the treatment of diarrhea and disorders in reproductive system of women.

Health Benefits of Mangoes

1. Fresh mangoes are a very rich source of potassium. Potassium is an important component of cell and body fluids that helps controlling heart rate and blood pressure.
2. Mangoes are useful to children who lack concentration in studies as it contains Glutamine acid which is good to boost memory and keep cells active.
3. According to new research study, mango has been found to protect against colon, breast, leukemia and prostate cancers. Several trial studies suggest that poly-phenolic anti-oxidant compounds in mango are known to offer protection against breast and colon cancers.
4. Mangoes are rich in pre-biotic dietary fiber which helps in digestion, vitamins, minerals, and poly-phenolic flavonoid antioxidant compounds and high level of soluble dietary fiber, Pectin and Vitamin C present in mangoes helps to lower serum cholesterol levels specifically Low-Density Lipoprotein (LDL) Cholesterol.

5. Mango leaves help normalize insulin levels in the blood. Boil a few mango leaves in water and allow it to saturate through the night. Consume the filtered decoction in the morning for diabetic home remedy. The glycemic index of mango is low, ranging between 41 -60. So, mango does not have any significant effect in increasing blood sugar levels.
6. Mangoes are beneficial for pregnant women and individuals suffering from anemia because of their iron content, mangoes can be beneficial for people wanting to gain weight. A 100 gram of mango contains about 75 calories and mango is effective in relieving clogged pores of the skin, thus helping in curing acnes
7. One cup of sliced mangoes supplies 25 percent of the needed daily value of vitamin A, which promotes good eyesight. Eating mangoes regularly prevents night blindness, refractive errors, dry-ness of the eyes, softening of the cornea, itching and burning in the eyes.

References

Anonymous (1962). Wealth of India, Raw Materials, CSIR, New Delhi.

Chacko, E.K., Kachru, R.B. and Singh, R.N. (1970). J Hort Sci. 45: 341-349.

Chudhuri, J.M., Rudra, P. (1971). Indian Agric, 15: 127-135.

De Candolle (1904). Origin of Cultivated Plants, Kegan Paul, London.

Diczbalis. Y., Wicks, C. and Landrigan, M. (1997). Heat sums to predict fruit maturity in mango (cv. Kensington Pride). Draft report for HRDC FR605 NTDPI and F.

Food and Agricultural Organizing (1995). Production Year Book, 1979, Rome, 33.

Gill, A.P.S. (1966). Ph D Thesis submitted to the PG School, IARI, New Delhi.

Hayes, W.B. (1957). Fruit Growing in India, Kitabistan, Allahabad.

Mukherjee, S.K. (1953). Economic Botany, 7: 130-162.

Mukherjee, S.K. (1958). Indian J. Horti, 16: 129-134

Musahibuddin and Dinsa, H.S. (1946). Punjab Fruit J, 10: 35-42.

Pawar, S.S., Gunjate, R.T. and Lad, B.L. (1985). Pesticides, 19: 32-35.

Popenoe, W. (1920). Manual of Tropical and Subtropical Fruits, MacMillan, New York.

Rao, S.N. and Subba Rao, C.H. (1963). Punjab Hort J, 3: 205-208.

Saini, S.S., Singh, R.N. and Paliwal, G.S. (1971). Indian J Hort, 28: 247-256.

Sen, P.K. (1939). Annual Report, Fruit Research Station, Sabour (Bihar), India.

Sen, P.K. (1943a). Indian J Horti, 1: 48-71.

Sen, P.K., Sen, S.K. and Guha D. (1963). Indian Agric, 7: 133-138.

Shanmugavelu, K.G. and Selvarajan, M. (1985). 2nd Int Symp on mango, India, Abst No 4.18, p. 37.

Singh, L. and Khan, A.A. (1940). Indian Fmg, 1: 380-383.

Singh, L.B. (1960). The Mango: Botany, Cultivation and Utilization, Leonard Hill, London.

Singh, R.N. (1954b). Indian J Hort, 11: 69-88.

Singh, R.N. (1978). Mango, Low-priced Book Series No 3, ICAR, New Delhi.

Singh, R.N., Majumder, P.K. and Sharma, D.K. (1962a). Sci and Cult, 28: 484-485.

Singh, R.S. and Ram, S. (1983). Indian J Horti, 40: 188-194.

Thimmappaiah and Singh, H. (1983). Indian J Hort, 40: 195-198.

— *Theme III* –

New Initiatives

2014, Sustainable Rural Development through Agriculture *Pages* ***213–230***
Editors: **Dr. Shobhana Gupta and Dr. S.S. Tomar**
Published by: **BIOTECH BOOKS, NEW DELHI**

Chapter 16

Intensifying Smallholders' Income through Profitable Enterprises by KVKs: Case Studies

S.R.K. Singh, Anupam Mishra, Prem Chand and A.P. Dwivedi

Krishi Vigyan Kendra (KVK) is an innovative institution of Indian Council of Agricultural Research (ICAR) play vital role in the application of technology by the farmers. Since 1974, KVK has grown as a largest network in the country with presence of 637 at present. Small farming in resource poor areas must be sustainable, economical, and intensive in order to provide long-term support for livelihood and income to the rural households. To achieve these capabilities, farmers must have access to sustainable technology suitable and available in agriculture and allied sectors. Zonal Project Directorate, Zone VII, Jabalpur coordinates, monitors and evaluates the mandated activities of 100 KVKs spread across the three states namely Madhya Pradesh, Chhattisgarh and Odisha.

KVK is playing very crucial role in assessment and dissemination of the agro-technologies to the farmers through frontline extension system. They show the pathway to the extension functionaries for further dissemination of the tested technologies at wider scale for bringing the prosperity and equity in the society through higher productivity, income and happiness. All these are carried out by the activities like – On-farm testing, frontline demonstrations, capacity building of farmers and extension personnel, extension activities, SAC Meeting, seed and planting material production, etc.

Small holders needs mostly low-cost technology having minimum initial investment and less risk of the climatic and marketing factors. Vegetable cultivation and small scale income generating activities offers opportunity to these farmers to get the assured income in limited time period protecting them from climatic as well as other vulnerability. Some case studies are given in brief as testimony for the same:

Case Studies

1. Cultivation of Spine Gourd for Additional Income: A Case of KVK Ganjam, Odisha

Background

During Kharif farmers usually cultivate short duration paddy in the uplands which gives less yield and more investment. In district Ganjam vegetable has much importance than other field crops. Due to less yield from short duration paddy the farmer interested to divert from paddy cultivation to vegetable cultivation (Spine gourd) as it fetches more market value and gives more yield which increases the income of the famers.

Technology

Interventions	*Details of Technology*
Variety	Desikankad.
Nutrient management	Application of bio-fertilizers (azotobacter + phosphate solubulizing bacteria @ 6 kg/ha) +RDF(70:40:60kg/ha)

Success Points

It provides better income over paddy cultivation in upland as it fetches high demand and much profit in the local market.

Outcome

Crop	*Technology*	*Yield (q/ha)*	*Income (Rs./ha)*	*B:C Ratio*
Spine gourd Var.-Desi Kankada	Application of bio-fertilizers (azotobacter+ phosphate solubulizing bacteria @ 6 kg/ha) +RDF(70:40:60kg/ha)	117.73	66500	1.61

Impact

The technology widely accepted by the farmers and interested for future adoptions in large scale.

2. Water Melon Cultivation for Prosperity in Odisha: Case of KVK Boudh

Background

- ☆ Paddy-Paddy and Paddy-Greengram are two major cropping sequences in the irrigated land.

- ☆ Sri.Manoj Kumar Pradhan of Vill- Lambakani, Block- Harabhanga was cultivating Paddy in 6 ha during Kharif and Greengram in 2 ha during Rabi season.
- ☆ He was getting a net profit of Rs 60,000 from Paddy and Rs 12,000 from Greengram which is insufficient for his livelihood support.
- ☆ He was in search of cultivating some profitable crop during Rabi season.
- ☆ After being trained on Watermelon cultivation he was motivated and started Watermelon cultivation in Rabi season since 2010.

Technology Used

- ☆ Use of hybrid seed Black Magic
- ☆ Transplanting of 15 days old seedlings raised in poly bags
- ☆ Foliar application of Ethrel @ 4 ml/10 lit in alternate row at 2 and 4 true leaf stage
- ☆ Application of RDF, NPK@ 120:60:100 kg/ha

- ☆ Application of herbicide quizalofop ethyle@ 1000 ml/ha at 15 days after transplanting.
- ☆ Foliar application of Boron @ 1000 gm/ha at flowering and fruit setting stage
- ☆ Need based application of insecticide and fungicide
- ☆ Judicious and timely irrigation to prevent fruit cracking

Dissemination Process

- ☆ Training on improved package of practices of watermelon
- ☆ FLD on transplanting technique in watermelon
- ☆ FLD on application of ethrel in watermelon
- ☆ Field day, Field visit, Diagnostic visit
- ☆ Agro- advisory services and extension literature on watermelon cultivation

Institutes Involved

Sl.No.	Name of Institute	Role
1.	KVK, Boudh	Technological support through training, FLD, field day, field visit and diagnostic visit
2.	Horticulture Dept, Boudh	Supply of micro nutrient and pesticides
3.	Jai Matadi Farmers Producer Company Ltd. Boudh (NGO)	Supply of seed and marketing of produce

Success Points

- ☆ Transplanting of watermelon reduced plant mortality, weed infestation and increased yield.
- ☆ Ethrel application resulted in more female flower and more yield.
- ☆ Uniform fruit size and more yield due to Boron application.
- ☆ Herbicide application save labour cost in weeding.

Outcome

By investing Rs 86,200 in 2 ha of watermelon cultivation, he got gross return of Rs 1,72,900 with net profit of Rs 86,700. So his net income in Rabi season is increased from Rs 12000 in green gram to Rs 86,700 by cultivation of Watermelon.

Year	Cost of Cultivation (Rs.)	Production (qt)	Gross Return (Rs.)	Net Return (Rs.)	BC Ratio
2010-11	68200	474	142200	74000	2.08
2011-12	68600	468	140400	71800	2.04
2012-13	86200	494	172900	86700	2.00

Impact

☆ His social standard has been Improved. He set an example as a progressive farmer in his locality.

- His economic condition has been improved. Now he has constructed a Pucca house, bought a motorcycle and is able to provide better education to his children.
- Seeing the success, other farmers of his village started cultivation of Watermelon.
- Now Watermelon is cultivated in 30 ha area in his village and around 800 ha area in the district.

3. Pointed Gourd Cultivation for Crop Diversification

Background

- Puri, a coastal region of Odisha having total area under vegetable is 16,980 ha.
- Major vegetables grown are okra (3865 ha), cauliflower (2356 ha), brinjal (2312 ha), tomato (1272 ha).
- Among the cucurbitaceous crop - pumpkin, cucumber, bitter gourd, pointed gourd are grown by the farmer of Puri district
- Farmer are growing pointed gourd, local variety gedi potal and mainshia potal in ground trailing system but they are not getting high profit due to adoption of local varieties and root rotting, stem rotting and fruit rotting during Kharif
- Pointed gourd is a perennial crop and have high market price and demand throughout the year

Description of Technology

Pointed gourd (var. -Swarna Alaukik)is cultivated in triangular staking system

- Two bamboos were tied in a triangular shape and placed in a row with spacing of 5 ft(within row)X 5 ft (between the row). Bamboos stripes and G.I wires were tied horizontally between two triangular bamboos and in between bamboo stripes locally available staking materials were given. Planting was done on both the sides of staking system at spacing of 5 ft X 5 ft.

Institute Involved in Dissemination

- Agriculture Department, Govt. of Odisha
- Horticulture Department, Govt. of Odisha
- ATMA, Puri
- Central Horticultural Experimentation Station
- Doordarsan All india Radio
- OTV Electronic media

Success Points

- It is low cost as compared to other trailing system
- In this system farmer can harvest 256.2 q pointed gourd per ha with a net income of Rs 2,19,510/- with an investment of 87,800 per ha.

- ✰ Increased yield upto 38.1 per cent over farmers' practice
- ✰ Easy plucking of fruits after maturity
- ✰ Improved quality of fruits as there is no rotting during rainy season.
- ✰ Suitable for high rainfall area.
- ✰ Suitable for small and marginal farmers
- ✰ It can be substituted for other crops like spine gourd, bitter gourd, ridge gourd etc.

Outcome

- ✰ In the system he harvested 256.2 q pointed gourd per ha with a net income of Rs 2,19,510/- with an investment of 87,800 per ha.
- ✰ Increased yield upto 38.1 per cent over farmers' practice
- ✰ B:C ratio-3.5

Impact

- ✰ The variety Swarna Aloukik has been accepted by the farmers due to pleasing appearance, softness, better taste, high market price than the local variety due to its taste and good keeping quality
- ✰ The success story of the variety Swarna Aloukik published in ATMA Newsletter for its horizontal spread and it has been recommended to the line department for large scale cultivation in whole district due to its performance and wide acceptability

- QRT team also appreciated the variety. PD, ATMA, Project Officer ATMA also have seen the performance of the variety Swarna Aloukik in the field and appreciated during the field visit
- The variety is now has been cultivated in an area of 20 ha of land in Puri district

	Farmers Practice	*Recommended Practice*
Technology	Cultivation of local var. Gedi potal	Var. Swarna Aloukik with FYM application 225q/ha, NPK@90:60:60 kg/ha
Yield (q/ha)	185.4	256.2 (38.1 per cent increase in yield)
Gross return	1,87,120	3,07,310
Net return	1,22,600	2,19,510
B:c ratio	2.9	3.5
Horizontal spread		22ha.

4. Sweet Corn Cultivation for Getting more Money: KVK Rayagada

Background

Pitambar Bhalu, a resident of village Gunupur having cultivated area of 5 acres of irrigated land. He is a Graduate and having 6 family members. Pitambar Bhalu is a hard working farmer. Out of 5 acres land, 3 acres cover vegetables (Tomato, Brinjal and Pointed gourd) throughout the year and rest 2 acres for paddy cultivation during kharif only. He has 60- 80000/year income previously and always worried about more income from any source. He came to KVK in search of job giving the lands on lease to Andhra farmers. Pitambar Bhalu, Programme Co-ordinator and scientists visited the Sweet corn Plot of KVK and he directly observed the number of brokers coming for buying Sweet corn.

Description of Technology

- Variety- Madhuri
- Seed rate- 10 kg/ha
- Spacing- 60 x 30 cm
- Fertilizer dose – 80:40:40
- Spraying of Boron and zinc twice during taseling and 7 days after tasseling

Dissemination Process

- Training and demonstration conducted at Limapadar, Khilapadar, Nilamguda and Nuagaon of Ramnaguda, Padmapur and Gunupur block.
- Seeds of Sweet corn provided to the farmers along with 50 per cent of RDF fertiliser and micronutrient. Scientists visited frequently to the demonstrated plots and compared the income with Maize cultivation.
- Farmers satisfied with the high income 75, 000/ha within 3 months.
- Success story of Pitambar Bhalu is now in more than 10 villages of Gunupur, DAO circle.

Economics of Sweet Corn Cultivation

Crop	Variety	Yield		Cost of Cultivation		Gross Return		Net Return		Benefit : Cost ratio	
		Demo	Check	Demo	Check (HYV maize)	Demo	Check	Demo	Check	Demo	Check
Sweet corn	Madhuri	38000 cobs	44000 cobs	37000	27500	114000	45000	77000	17500	3.08	1.63

In Sweet corn average cost of cobs Rs. 3- 5/- Maize: average cost of cobs Rs. 1/-

Outcome

Due to high market demand and good transportation facility in urban area like Gunupur and Rayagada, marketing is not a problem. Pitambar Bhalu got a profit of Rs. 77, 000/- per ha in 2012-13 and interested to continue throughout the year. It is expected more number of farmers will divert their uplands for sweet corn cultivation instead of paddy.

Impact

ATMA, Rayagada demonstrated 5 ha of land on Sweet corn cultivation near Gandhi Institute of Engineering and Technology after listening a success story from Pitambar Bhalu in KVK SAC meeting. More than 50 no. of farmers, teachers from different schools, Govt. officials coming to KVK, Rayagada for the learning of package and practices of Sweet corn cultivation. Around 100 hectares of land will be covered under sweet corn cultivation in Rayagada district during 2013-14. Villagers of Limapadar are always pressurising Mr. Pitambar Bhalu for the Seeds of Sweet corn to cultivate during Kharif 2013-14.

5. Kharif Onion Cultivation for Higher Income in Madhya Pradesh

Background

- Dewas district of Madhya Pradesh comes under Malwa Plateau agro-climatic zone of Madhya Pradesh.
- The diversified soil and agro-climatic condition of the district offer a vast scope for cultivation of all kind of vegetables.
- In the district farmers are growing onion in large scale during rabi season but farmers are not growing onion in kharif season, though there is a ample scope for cultivation of this crop in the district.
- During participatory Rural Appraisal (PRA) survey of the villages, farmers and scientists comes to the conclusion that kharif onion may be more remunerative over traditional crop soybean which covers a larger area under kharif season.

Technology

- Varieties taken for trials : AFDR
- Preparation of raised bed during the month of May-June.
- Soil solarization
- Sowing of seeds in lines at 10 cm apart during June-July.
- Seedlings are ready for transplanting in 45 days during kharif season.
- Soil test based manures and fertilizers were applied alongwith micro nutrients especially Zinc.
- Need based application of insecticides especially Fipronil @ 1.5 ml/litre water and fungicides *i.e.* Mancozeb were applied to control the sucking pest and fungal diseases.

- Bulbs are ready for harvesting in 110-120 days after transplanting.
- Average yield of Kharif onion varieties are about 250-300 q/ha.

Dissemination Process

- Participatory Rural Appraisal (PRA) survey, general discussion with farmers, officials of Horticulture department and NHRDF were undertaken continuously since 2007 in various villages for crop diversification by Kharif onion.
- 18 trainings were imparted to the farmers.
- On farm trials on improved variety of Kharif onion *i.e.* Agrifound Dark Red (AFDR) were conducted by the KVK for 2 consecutive years (2007-08 and 2008-09).
- After taking interventions in HYV of Kharif onion, KVK had conducted On farm trials on Integrated Nutrient Management as per soil test value.
- Afterwards OFT were converted to front line demonstrations (FLD) on improved variety of Kharif onion *i.e.* AFDR in the operational areas of KVK with full package.

Institutes Involved

- Krishi Vigyan Kendra, Dewas
- Department of Horticulture, Dewas
- National Horticulture Research and Development Foundation, Sub centre-Indore.
- Farmers Welfare and Agriculture Development Department, Dewas.

Success Points

- Shri Radheyshyam Bhagat was a very poor farmer of Village Narana, Block-Sonkutch, Dist. Dewas. He has only 1.0 ha. cultivable land. He was given a trial on improved variety of kharif onion during the year 2007-08. In this trial he had given 500 g of ADR variety seed for 500 m^2 area. From this 500 m^2 area, he earned a net profit of Rs. 16803.00. In the next season he has taken loan of Rs. 4000 from SHG constituted by the KVK in the year 2008. Then he grow kharif onion in the area of 0.5 ha and obtained 109 q yield. He sold his produce in Dewas mandi for Rs. 1,03,550.00 @ Rs. 950 per quintal. Now Shri Radheyshyam Bhagat is living a properous and happy life.
- Therefore, as an off season crop, kharif onion may prove a bright prospect to uplift the socio-economic status of the small and marginal farmers.

Impact

- Before the inception of KVK, area under kharif onion in the district was very limited (about 12-15 ha). Now the area under this crop is increased to *1500* hectare.

Outcome

Year	No. of Farmers	Area (ha.)	Demo. Yield (q/ha)	Yield of local Check (q/ha)	Increase in Yield (per cent)	Average Net Return (Rs./ha)		B:C Ratio	
			A			Demo	LC	Demo	LC
2007-08	20	1.0	103.02	75.25	37.25	51542.93	35097.50	3.00	2.64
2008-09	18	0.9	190.37	151.45	25.70	40778.50	31450.25	2.58	2.46
2009-10	18	0.9	184.20	140.85	30.78	117296.65	80079.50	3.03	2.49
2010-11	38	1.90	187.14	144.82	29.22	230997.10	167329.95	4.91	3.93
2011-12	10	0.5	167.96	131.29	27.93	87719.49	63038.23	3.29	2.78
	05	1.0	170.84	150.23	13.71	89879.49	77243.23	3.35	3.18
2012-13	05	1.0	217.34	188.92	15.04	200147.51	170792.17	6.1	5.6

Trainings imparted to the Farmers — Seedlings in nursery stage — Uprooting of Seedlings
Transplanting of seedlings — Standing crop — Produce of Kharif onion

Different stages of crop diversification by Kharif onion

- ✰ Under technological dissemination to 5000 farmers, Horticulture Department give responsibility for taken kharif onion in all the 25 villages *i.e.* for 5521 Farmers.

6. Enhancing Income by Chilli Cultivation in Nimar Valley: KVK Dhar

Background

- ✰ Cotton is used to be grown as dominant crop in Nimar region of Dhar district in five blocks *viz.* Dharampuri, Manawer, Kukshi, Gandhwani, Nisarpur and covering an area of 1.30 lakh ha with an average productivity of 1370 kg/ha.
- ✰ During past four to five years the area and productivity of Cotton is decreasing due to heavy weed infestation, lack of quality Bt seeds, problem of wilt, heavy incidence of sucking pests and fluctuation of market price of cotton.
- ✰ Looking to above problems and economic losses the interest of farmers is reduce to grow cotton and they are shifting to grow chilli as a alternative crop. In this way the area of hybrid chilli cultivation is popularizing among farmers due to high yielding and profit.
- ✰ Now the chilli cultivated in about 15,000 ha area and present area of cotton is reduced to about 1.14 lakh ha.

Technology

- ✰ Interventions taken through OFTs and FLDs for selection and promotion of hybrid chilli cultivation using hybrid Charmi G303, US-11, US-14 and Ujala.

- ☆ Proper spacing (2′x1′), staking, drip irrigation, INM (FYM @ 20 tones/ha + 120:60:60 NPK, Full dose of P and K as basal, ½ dose of N after 15 days and another ½ dose of N after 40 days of Transplanting)
- ☆ Fertigation of 19:19:19 and micro nutrients was given through drip.
- ☆ Pest and disease management practices were also applied by spraying of Imidachlorprid followed by COC and Streptocycline.

Dissemination Process

- ☆ On the basis of survey and group discussion the Chandawada village was selected during 2010-11.
- ☆ Several farmers group meeting were organized to cultivate chilli using drip irrigation system.
- ☆ Farmers were also motivated through capacity building, interaction, regular field visits, repo building, etc.

Institutes Involved

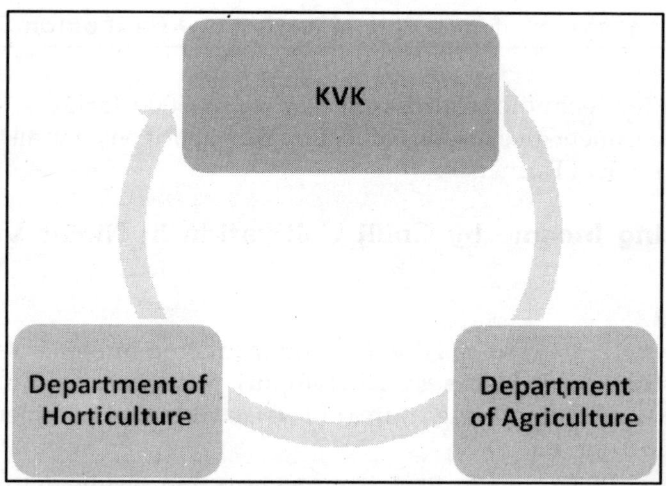

Outcome

- ☆ Five farmers of Chandawad village Block Dharampuri District Dhar are successful chilli var. Ujala growing farmers.
- ☆ Previously they were growing cotton crop only. By growing cotton they were earning only **Rs.**80000/ha **as** gross income.
- ☆ They sold **about** 44 q green chilli and 28 quintal dry chilli/ha. They sold the produce in the Manawer and Dhar market and earned Rs.170000/ha as gross income and B:C ratio 2.74 from chilli cultivation.

Impact

- ☆ Due to diversification of Cotton by Chilli farmers are getting more remunerative than the Cotton.
- ☆ Chilli cultivation is spread over 15000 ha area.
- ☆ Some other farmers of the Nimar region are also adopting Chilli cultivation.

7. Tomato Cultivation by Tribes

Background

Jhabua is a tribal district of Madhya Pradesh. It represents the 11th agro-climatic zone *i.e.* Jhabua Hills. The climate is semi-arid with annual average rainfall 828 mm.

Krishi vigyan kendra, Jhabua had identified the problem of lower cropping intensity due to more than 70 per cent fellow during Rabi season in Jhabua district. Scientists of KVK also correlated the problem with lower income of small and marginal farmers. Farmers of district usually practiced traditional cultivation of maize, blackgram, cotton and soybean during Kharif season yet, they could not get better monetary return due to low productivity and lower prices during the harvest stage of crop.

To uplift the poor farmers and to increase the cropping intensity with introduction a cash crop during kharif season under rain fed agro eco system, KVK had started with extension activities, training and on farm trials on tomato cultivation.

Results of On Farm trials were very much encouraging and with tireless efforts of KVK, farmers of different villages were going to adopt the new technology especially in Petlawad and Thandla block.

Technology

- ☆ Crop: Tomato
- ☆ Variety: Hybrid (Red Gold, Namdhari,Chamtkar, Avinash)
- ☆ Planting Method: Raised bed method, Plastic Mulching, Stacking
- ☆ Spacing: RXR-90 cm and PXP-60cm
- ☆ Irrigation: Drip irrigation, fertigation
- ☆ INM: 15-20 t/ha FYM + 150:75:75::N:P:K kg/ha

Dissemination Process

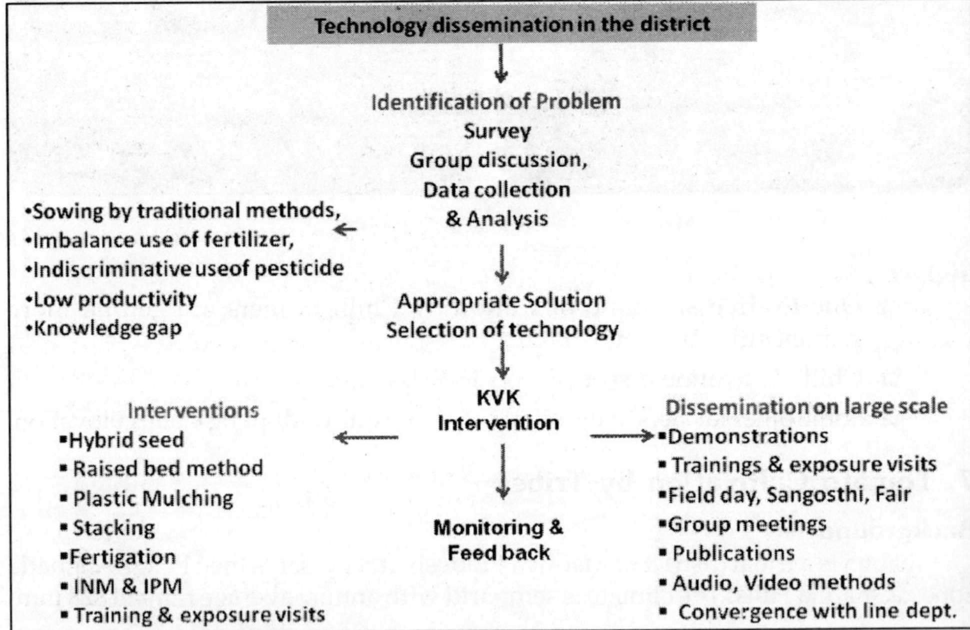

Institutes Involved

- ☆ Problem identification and solution: KVK, Jhabua (M.P.)
- ☆ Seed and input Supply: Private dealers
- ☆ Technology Dissemination (large Scale):
    1. Dept. of Horticulture
    2. NGO
    3. ATMA
    4. Service Provider

5. Publication
6. Audio, Video methods

- ☆ Monitoring and Feedback: KVK, Jhabua (MP)

Success Point

- ☆ Suitable under Changing Climatic Conditions (Temperature – Maxi. 45 °C and Mini. 7 ° C)
- ☆ Suitable in rain fed agro eco system.
- ☆ Higher monetary return due to Higher Yield
- ☆ Employment generation-180 man days/ha.
- ☆ Higher Benefit cost raitio
- ☆ No need to storage
- ☆ Produce sold at Delhi, Rajathan, Gujrat, Maharastra through cooperative societies and brokers

Outcome

- ☆ Average Productivity (q/ha): 800.00
- ☆ Income generation (Rs/ha): 300000
- ☆ Additional Employment Generation (Mandays/ha): 135
- ☆ The productivity of Petlawad and Thandla block is 1200q/ha

Impact

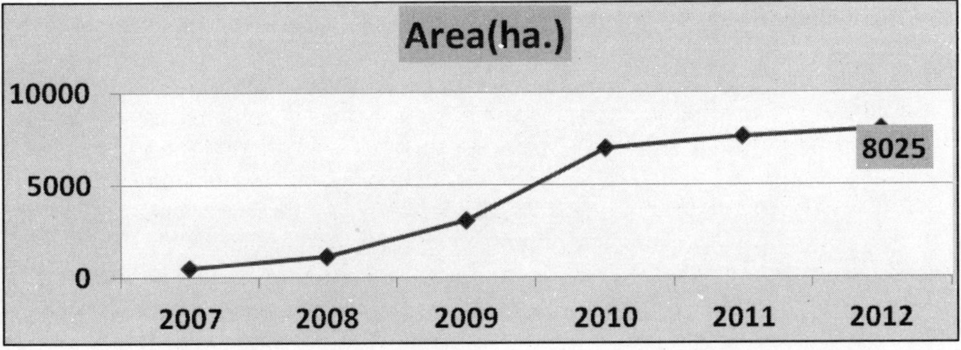

Tomato production is increasing gradually in district. Farmers are regularly asking the technology from KVK, other line department and input suppliers.

Three Tomato Processing Unit to be established by

1. KVK, Jhabua
2. Progressive farmer
3. Private firm

2014, Sustainable Rural Development through Agriculture Pages 231–246
Editors: Dr. Shobhana Gupta and Dr. S.S. Tomar
Published by: BIOTECH BOOKS, NEW DELHI

Chapter 17
Strengthening Extension System: Issue and Options

Sarju Narain, Vikas Kumar,
Sudhir Kumar Rawat and R.R. Kushwah

Introduction

Over the years, India has made various experimentations to enhance its reach to the farmers by initiating a number of projects and programmes in Agriculture sector. The organized agricultural extension date back to early part of 19th century started with the inception of sporadic individual efforts by legendry social reformers such as Gandhian Experiment (1920) to Nilokheri Project (1947). Post independent initiatives include Etawah Pilot Project (1948) to Krishi Vigyan Kendras (since 1974 to till now).

These programmes/projects increased production and productivity of food crops, improved food security and raised rural incomes. Expansion of Farm income is still the most potent weapon for reducing poverty. The strategy behind increasing Farm productivity depends upon research and extension. Thus, research and extension both are the cardinal pillars for agriculture growth and development. In India, there are four major organizational streams devoted to extension work for agriculture and rural development, the first and most important is (i) First line Extension system includes Indian Council of Agricultural Research (ICAR) - Institutes and Agricultural Universities; (ii) Extension system of Ministry of Agriculture and State Departments of Agriculture; (iii) Extension system of ministry of Rural development and state development departments and (iv) Development work by the voluntary organizations, business houses, etc. (**Prasad *et al.*, 1987**).

Public research and extension played a major role in bringing about the Green Revolution. In the post-Green Revolution era, however, extension faces important challenges in the areas of relevance, accountability and sustainability. The changing economic scenario in India and the need for appropriate agricultural technologies and agro-management practices to respond to food and nutritional security, poverty alleviation, diversifying market demands, export opportunities and environmental concerns are posing new challenges to the technology dissemination systems. It is expected that future agricultural growth would largely accrue from improvements in productivity of diversified farming systems with regional specialization and sustainable management of natural resources, especially land and water. Effective linkages of production systems with agro-processing and other value added activities including marketing would play an increasingly important role in the diversification of agriculture.

It is becoming increasingly evident that public extension by itself can no longer respond to the multifarious demands of farming systems. There is need for appraisal of the capacity of agricultural extension to address contemporary farmers requirements effectively. Public funding for sustaining the vast extension infrastructure is also under considerable strain. In response to market demand, the existing public extension network is being complemented, supplemented and being replaced by private extension. As the nature and scope of agricultural extension undergoes fundamental changes, the outlook is for a whole new policy mix nurturing a plurality of institutions.

Indian agriculture is essentially small farm agriculture with the majority of farmers owning less than 1 ha. of land. Small and marginal farmers now constitute over 80 per cent of farming households in India. They are facing lot of problems including poor access of technology. The commercialization of agriculture also requires specialized knowledge and new skills regarding various practices. The varied agro-ecological cropping situations, socio-economic and political issues, alteration in demand and supply of inputs and outputs, need and belief of farming population make a diverse scenario for diversified demand of agriculture technology. It means agriculture technology requirements/demands vary among farmers/locations/pockets/regions/states type and system of farming, etc. Hence, it is clear no one uniform extension system will serve as panacea to all situations or all farmers. In India, where agricultural technology transfer activities are mainly in the hand of public sector which could not capable to fulfill the diverse and specific needs of all section of farmers. Except above issues sustainability, diversification and to increasing productivity are the major challenges for the same.

Changing Agricultural Development Goals

To rejuvenation/reorientation/refreshing/strengthening the extension system several steps were taken during 11th Five Year Plan and their outcomes seeing slowly. But for more outcomes it takes more time because extension is really very challenging and complex. It is very difficult to get the results, output, in a manner that people should appreciate (**Kokate, 2011**). The changing scenario of agricultural, changes

Table 17.1: Types of Organizations Providing Extension Services in India

Sl.No.	*Types*	*Example*
1.	Line departments	Department of Agriculture, Horticulture, Animal Husbandry etc. Department of various state governments.
2.	University	Directorate of Extension and other colleges of State Agricultural University and Agricultural Schools of Open University (YCMOU Nasik)
3.	Krishi Vigyan Kendra	Krishi Vigyan Kendras sponsored by the ICAR in various districts.
4.	Farmers Associations	Maharashtra Grape Growers Association, Kerala Mushroom Growers Association, etc.
5.	Producers' Cooperatives	Milk and Sugar Co-operatives in different parts of the country.
6.	Research Institutions	Various research units of ICAR and SAUs mainly through the outreach programmes.
7.	Input Industry	Seed companies such as Ankur Seeds, Nagpur, Kumar Gente Pune Messina Beej Pvt. Ltd. Bihar, Fertilizer Companies as IFFCO, KRIBHCO, FACT, Indo-Gulf etc. Pesticides and machinery firms.
8.	Consultants	Individual consultants and consultancy firms such as Green Plus, Nasik.
9.	Non-Governmental Organisations	NGOs working in the area of agriculture and rural development such as BAIF, Pune, PRADAN, Bihar, GSSS, Udaipur etc.
10.	Community Boards	Spices board, Rubber Board, etc.
11.	Marketing Boards	Maharashtra State Agricultural Marketing Board
12.	Media Print	News papers (Agricultural pages of most language dailies) and farm magazines
13.	Media-Audio and Visual	All India Radio and Doordarshan through its farm programmes and satellite channels such as E-TV (Telugu) through their farm programmes.
14.	Others	Autonomous agencies in specific areas (Command Area Development Agency) and crops (Kerala Horticultural Development Programme); Banks through their field officers, Panchayat Samit through their agricultural wings, Farmers Training Centres (DOA); Irrigation Management Training Institutes under the Irrigation Departments, agro-processing companies such as Pepsi Foods (Punjab) and ITC (Andhra Pradesh) for its contract growers and internet.

Source: Dimensions of Agricultural Extension by A.K. Singh *et al.*, 2006.

the demand of information and advisory support (**Van den Ban, 1998**). So gear up accordingly to meet the demands of farmers for successful technology transfer. Extension should engage with a wide range of issues related to agriculture. These includes market, credit, insurance, weather forecasting in addition to technology and research services and making arrangements for supply of inputs. The field of extension now needs to address a wider range of activities *viz.*, linking farmers to market (**Neuchatel group, 2002**), reducing vulnerability and empowering the rural poor (**Farrington *et al.*, 1998**), developing microenterprizes (**Rivera *et al.*, 2001**), poverty reduction and environmental conservation (**Alex *et al.*, 2002**) and strengthening and supporting farmers organizations. The conventional mode of technology transfer through personal contact is no longer the sole or redominant mode of extension. There are more diverse needs of information by farmers not only by production procedures, but also for quality certification, reporting, grading, packaging, storage, transportation and other requirements of marketing (**Kokate *et al.*, 2011**). To achieve more rapid productivity growth, both research and extension need to be strengthened. The draft approach paper to the 12th Five year plan argues the need for ensuring a minimum of 4 per cent growth in agriculture during XIIth plan (2012-17). There are no silver bullets or unique models that will lead to enhance extension performance but following reforms in agricultural extension already initiated/and proposed to be undertaken for accelerating, strengthening, revitalizing the extension system.

Reforms in Agricultural Extension

Ground realities as discussed above demands broadbased and holistic extension services which requires following reforms: policy reforms, institutional restructuring, management reforms, strengthening research extension linkage, capacity building and skill upgradation, empowerment of farmers especially farming women, use of print media and Information and Communication Technology (ICT), financial sustainability, changing role of government, etc.

1. Policy Reforms

This type of reforms refers to replacement of public sector monopoly in Extension to Multiagency extension service and single discipline based commodity approach to holistic approach *i.e.* farming system approach. On the basis of extension service providers the extension agencies can be further divided into two main system *i.e.* Public extension system and Private extension system. These bath extension system commonly known as Pluralistic extension.

1.1 Pluralistic Extension Approach

Future demands a different set of approaches building on pluralistic extension approaches with appropriate mix of public and private funding and delivery mechanism. The extension approaches focus on 'Cost reduction without yield reduction' (KV Thomas, 2011). Extension has to address the issue of 'how to bring together all the stalk holders because technology is the hand of ICAR, University and some private players but implementation goes in hand of others. On these lines several public and private extension service providers including farmers organizations and NGOs are providing several type of extension services as discuss under.

1.2 Public Extension Service Providers

- ICAR institutes, Centers, Directorates, Stations, etc. through their schemes/ projects/programmes like Krishi Vigyan Kendras (KVKs), Agriculture Technology Information Centers (ATICs), Institution Village Linkage Programme (IVLP), National Agricultural Technology Project (NATP), National Agricultural Innovation Project (NAIP), etc.

- ✫ Agricultural Universities based Extension activities.
- ✫ State government line departments like Agriculture, Horticulture, Livestock, Sugarcane, etc.

1.3 Private Extension Service Providers

- ✫ Cluster/community based Organizations like Farmers' Organizations, Farmers Cooperative, Self Help Group (SHG), Commodity groups, etc.
- ✫ Para Extension Workers like *Kisan mitra, Gopal*, etc.
- ✫ Agriclinics and Agribusiness Centers
- ✫ Input marketing Companies and their local dealers
- ✫ Commercial crop based Corporate sector

Except above both system, now a days Mass Media and Information and Communication Technology (ICT) also played very significant role in transferring agricultural technology. Both Public and Private Extension System also take help to this new generation ICT system.

2. Institutional Restructuring

Under this Public Extension System decentralize and develop demand driven, farmer accountable, bottom-up and broadbased aspects. For this purpose Agriculture Technology Management Agency (ATMA), Single window broadbased extension model, Panchayati Raj Institutions (PRIs) operated extension and State Agricultural Universities (SAUs) operated extension model started for fulfilling increasing demands of extension services.

2.1 Strengthening to Krishi Vigyan Kendras (KVKs)

Based on the recommendation of the Education Commission (1964-66), discussion by the Planning Commission and Inter Ministerial Committee and further recommendation by the committee headed by Dr. Mohan Singh Mehta appointed by ICAR in 1973, the idea of establishment Farm Science Center (Krishi Vigyan Kendra) give the birth first KVK in 1974 at Pondicherry. KVK is designed to impart need based and skill oriented vocational training to the practicing farmers, inservice field level extension workers, and to those who wish to go in for self employment (Prasad *et al.*, 1987). Since the birth of one KVK (1974) to 632 KVKs (2011), it proved to continue a technology transfer engine at district level. KVK has evolved its role and undergone several changes in its mandate over the years, that being a vocational training institution to Technology Assessment and Refinement (TAR) and to work as a Resource and Knowledge Center (RKC) of agriculture technology for supporting initiative of public, private and volunteer sector for improving the agricultural economy of the district.

Agriculture technology is the complex blend of materials, process and knowledge and can be classified into two main group, (i) material technology and (ii) knowledge based technology. KVK are best positioned to convert the information into knowledge and knowledge into practical application with situational refinement. Material technology refers to knowledge embodied into technological products such as tools, fertilizers, seeds, etc. are the components of technology capsule required by farmers, knowledge based technology refers technical knowledge, management skill and other processes that farmers need for better production in their enterprizes. The KVK has to convert the bits and pieces of information/knowledge into technology at district level (Kokate, 2012). Out of 632 KVKs, the e-connectivity provided to 192 KVKs and 8 ZPD units for strengthening connectivity within system (upto 2011).

- ☆ Now days KVK work as fountainhead of technology at district level with good co-ordination with Agriculture Technology Management Agency (ATMA).
- ☆ Now need to KVK became a brand name of technology transfer.
- ☆ Resource generation for KVKs is a important issue for sustaining this model.

2.2 Decentralization of Agricultural Extension Management through Agriculture Technology Management Agency ('ATMA')

ATMA is a registered body/society at district level work on the concept of decentralize decision making power upto farmers for their agricultural programme planning and resource allocation at block level and increasing accountability of stalkholders. Under ATMA, grass root level extension mainly channelized through the involvement of Block level Technology Teams (BTTs) and Farmers Advisory Committees (FACs), Farmer groups, Self help groups, etc. ATMA also provide a plateform for interaction between line departments and farmers and brought some new concepts such as bottom up planning and commodity interest groups into field extension practices.

In June 2010, the Central government issued revised guidelines on ATMA implementation (DAC, 2010) mainly to address the constraints associated with the national implementation during the past five years. The revisions included hiring exclusive staff for ATMA at the district and block levels, inclusion of farmer advisory committees at the block, district and state levels and greater emphasis on ATMA's links to the KVKs. ATMA is now operational in 603 districts of India spread over 28 states and three Union Territories. Provision of separate staff for ATMA has brought improved attention to ATMA. With improved links to KVKs, better convergences among different schemes/departments, and greater focus on commodity interest groups, ATMA is expected to strengthen Indian extension system during the XIIth Plan (2012-2017).

2.3 Agricultural Technology Information Center (ATIC)

This is initiated within NATP (1999) as a competent and implemented by Agricultural Universities and ICAR Institutes. Now days some successful ATIC(s) are popular among farmers like 'Dharwar' ATIC but most of them could not achieved their objectives. Therefore, an urgent need of revitalization of such innovation center.

Objectives

- ☆ Single window system of technology delivery from SAUs/ICAR for farmers.
- ☆ Provide advance input delivery, advisory and diagnostic services to farmers.

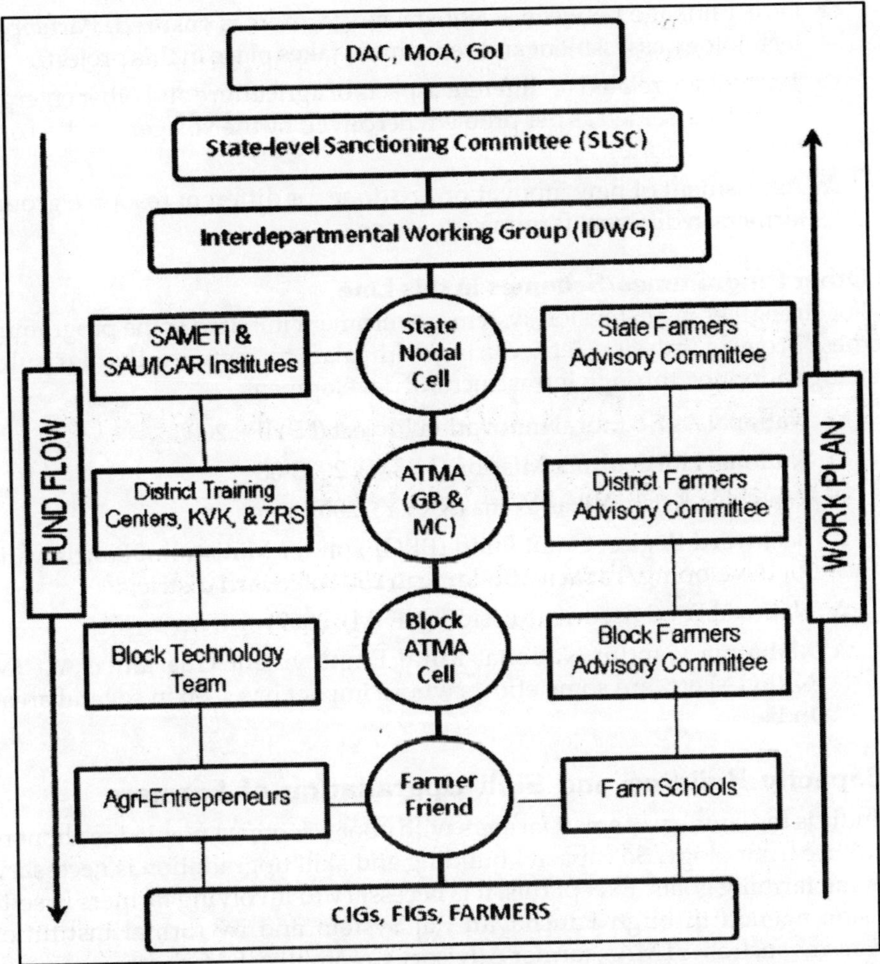

Figure 17.1: ATMA Model as per the Revised Guidelines (2010).
***Source*: DAC (2010).**

- ☆ To overcome technology losses and generate resource for host SAUs institutes
- ☆ Strengthen linkage between farmers and SAUs/institutes.

3. Strengthening Research Extension Linkage

For strengthening research – extension linkage Technology Assessment and Refinement (TAR) is necessary, therefore, ICAR started Institute Village Linkage Programme (IVLP) in 1995.

3.1 Institute Village Linkage Programme (IVLP)

This was implemented through ICAR institutions, SAUs, Zonal Research Stations (ZRSs) and KVKs following basket approach in cluster of Villages. The specific objectives of technology assessment and refinement programme were :

- ☆ Direct linkage between scientists and farmers is ensured. Participatory technology assessment and refinement takes place in this project.
- ☆ Technology related to different aspects of agriculture and other enterprizes are to be assessed as per problem perceived by the villagers and according to their priorities.
- ☆ Assessment of new innovation are done for different resource groups of farmers in different farming situations.

3.2 Other Programmes/Schemes in this Line

For strengthening extension system, government initiated some programmes/ schemes/projects/mission, etc. during XIth plan to increase the agricultural technology adoption through infrastructural development.

- ☆ National Agricultural Innovation Project (NAIP), 2006
- ☆ National Horticulture Mission (NHM), 2005-06
- ☆ Rashtriya Krishi Vikas Yojna (RKVY), 2007
- ☆ Backward Region Grant Fund (BRGF) under Ministry of Panchayati Raj for developing/capacity building in 250 backward districts.
- ☆ National Food Security Mission (NFSM), 2007
- ☆ Mahatma Gandhi National Rural Employment Guarantee Act (MG-NREGA) etc. are some efforts whose impact has seen in several areas of India.

4. Capacity Building and Skill Upgradation of Farmers

India is the country of small farmers with poor scientific knowledge about new agriculture technology. So capacity building and skill upgradation is necessary to achieving farming goals. Except this, it is necessary to involving farmers in setting extension agenda through Panchayati Raj System and by formal institutional mechanisms such as ATMAs, farmer Advisory Committee (FACs), etc.

For effective technology dissemination skill upgradation of farmers is an alarming issue. To improve skills of farmers several training cum skill development programmes are the central part of all programmes. But availability of able extension machinaries is the key constraints at local level. Therefore several training and skill upgradation programmes are not to effective as required. For removing this constraint use of ICT as miracle tools and proved as landmark network.

For capacity building and skill upgradation of extension functionaries and empowerment of farmers State Agricultural Management Extension Training Institute (SAMETI) is upgraded, increases farmers participation in setting extension agenda at different levels, implementation of programmes through farmers' user groups, acquisition of skill by farmers, etc. steps are taken.

Majority of the government plans, proposal, schemes, programmes, etc. are planned and implemented through rural peoples/farmers group, such as watershed associations, vegetable growers, producer and marketing co-operatives/societies etc.

Government decentralized policy increases the role of Panchayati Raj Institutions in implementation of several programmes.

5. Empowering Women

Gender issue is also important in extension. The present Public and Private extension system is male dominated. All the technology, processes, inputs, etc. are male driven in nature while women play greater role in Agriculture. For this purpose it is necessary to (i) improving access to extension and training; (ii) Redesign of extension services to reach woman farmers; and (iii) Expanding the sphere of woman extension workers.

To empowering women several steps to be taken by state and central government level at different areas which can be made landmark in extension strengthening.

6. Role of Mass Media and ICT in Technology Transfer

This is the era of globalization, liberalization, privatization and commercialization, where information and communication technology reach upto grass root level. Therefore, it is necessary to well utilize the technology for agricultural extension purposes. On this line Radio, T.V. (both public and private operated FM, AIR, TV channels), internet, cellphone, etc. have to be harnessed for enhancing communication.

6.1 Need to Harvesting these Options

Except above, several pilot based communication centers, information Kiosks, portals with high integration of print media playing greater role in this line. Geographical Information System (GIS), interactive multimedia with web connectivity, e-governance, e-extension or virtual extension, Gyandoot (community owned rural internet Kiosks), e-choupal (farmer friendly networking technology), Kisan Call Centers (KCCs), etc. initiatives are proved milestone in transferring technology to farmers. Kisan-Mobile Advisory Services (K-MAS) is a new initiative launches by KVK as a magical tool of technology transfer to small and marginal farmers. IFFCO Kisan Sanchar Limited (IKSL), Reuters Market Light (RML) and Tata M-Krishi are three successful examples of mobile information services for the farmers. The 'Digital Green' is an another initiatives uses video for agricultural development.

6.2 Kisan Call Center (KCC)

This is a good initiative taken by government from 2004 but our farmers could not well utilized this service. Government provide one toll free number (1800 180 1551) within nation to solve advisory and diagnostic problems of farmers. The call center operated at three levels *viz.* (i) providing immediate reply to the querry raised by the farmers; (ii) transferring the unanswered question to the specialist located at second level in agricultural universities or other technical institutes for immediate answer and (iii) in case of question unanswered at the 2nd level also it is referred to the 3rd level where the expert will examine and detailed answer could be provided to the farmer. The call centers operate all the seven days of the week.

Table 17.2: Paradigm Shift from Production-Led to Farmers-Led Extension System

Components	*Production-led*	*Farmers-led*
Purpose/Objective	Transfer of production technologies	Capacity building (especially farmers, extensionist), create para- professional extension workers, creating or strengthening local institutions
Goal	Food self-sufficiency	Livelihood security including food, nutrition, employment to alleviate poverty, sustainability and conserving bio-diversity
Approach	Top-down, commodity and supply driven	Participatory, bottom-up and demand driven
Actors	Mostly public institutions	Pluralistic with public, private, non government and farmers organizations as a partner rather than competitors
Mode	Mostly interpersonal/individual approach	Integration of clients oriented on-farm participatory/experiential learning methods supported by ICTs and media
Role of extension agents	Limited to delivery mode and feedback to research system	Facilitation of learning, building overall capacity of farmers and encouraging farmers experimentation
Linkages/liaison	Research-Extension-Farmers	Research-Extension-Farmers Organizations (FIGs, CIGs, SHGs)
Emphasis	Information management, Production "Seed to Seed"	Knowledge management and sharing
Nature of technology	Input intensive, crop based and general recommendations as per agro-climatic zone, fixed package of information	Knowledge intensive, broad based, farming system perspective and blending with ITKs.
Critical areas	Improvement, production and protection	Decision support system, integrated farming system approach, natural resource management, clients group formation and community empowerment
Critical inputs	Money and material	Access to Information, building human and social capital
Accountability	Mostly government	To farmers rather than donors

Source: Technology Dissemination in Indian Agriculture : Lessons learnt and Future Strategies by Dr. K.D. Kokate, DDG-Agril. Extension, ICAR, New Delhi.

7. Changing Role of Government

In changing regime the role of government is also changing from service providers to regulation and inforcement of legislation, which ensure quality control of inputs as well as assuring accountability of all service providers to the farmers and insure transparency through provision of information.

8. Financial Sustainability and Resource Mobilization

Government increases the role of private sector in technology generation and extension, therefore several public good and services converted into semi-private or private goods and services like Artificial Insemination (AI) services, soil testing, fertilizer testing and advice etc. Several cases/models are also emerging in the area of 'contract farming' through involvement of private sector particularly in the area of high value/export oriented agriculture.

8.1 Business driven Extension Approaches

These approaches would be discussed under following heads:

8.1.1 Agri-business Orientation

Country produce surplus number of agricultural graduates every year. To generating employment and utilizing their potential as agricultural experts and supplementing/augmenting extension services Ministry of Agriculture in Association with the National Bank for Agriculture and Rural Development (NABARD) and *MANAGE* is implementing the agri-cilinics and Agri-Business Centers (ACABCs) schemes. For this purpose a two month training (free of cost) is provided to agriculture graduates for opening ACABCs at local level to serve the farmers on business basis. Interested candidates can received bank finance from this purpose. These ACABCs generally provides input delivery, advisory and diagnostic services to the farmers.

8.1.2 Agro-input Providers

Now a days farmers dependency increases on several market agro-input providers. Agro-input providers includes input manufacturing/and marketing companies, input agencies and their dealers, etc. are also playing significant role in production system. Now days 2.82 lakh agro-input dealers covering parts of India providing input delivery and advisory services to input users. They work as crop doctors at local level even they are untrained or sometimes not equipped with knowledge. The importance and role of input dealers also revealed through the NSSO (2005).

Tata Kisan Kendra, Khushhali and Hariyali-Kisan Bajar (H.K.B.), etc. are new initiates taken by private sectors. These are agro-input retail stores to provide end to end support to farmers. HKBs also provides credit and market access through by back opportunity. This type of initiatives limited upto prosperous state/belt of farming. On these lines 'Dhanuka' and 'DSCL' group also providing some extension services. Several private sugar mills also having sugarcane seed multiplication farm, advance lab of plant protection, Bio-fertilizers etc. facilities for their sugarcane farmers under sugarcane mill jurisdiction to increase productivity. In the field of Animal husbandry and dairying *'Namaste India'* taken several initiatives to increase milk productivity among farming group at Kanpur belt (U.P.).

8.1.3 Public Private Partnership (P-3)

Public and private organizations having different mode of action, purpose, methodology, strategy, network, management and controlling authority but they could come together for mutual gains. P-3 initiative is the true picture of new world. This strategy more helpful in research/development, processing, marketing and extension field where much investment and infrastructure is required such as Hashangbad model (Dhanuka joint hand with *MANAGE*). Under this system 'private' word used for corporate, private businessmen, organizations like community based or NGOs.

Now a days PPP/P-3 is the reality for growth and development. On these pattern concept of 'Contract farming' is blooming in several places.

8.2 Farmer Driven Extension Approaches

Farmer organizations are relatively strong in most industrially developed countries. In recent years, public support for extension has declined with the transformation of agriculture sector. The future strength and structure of farmer controlled extension systems will depend upon how these new extension resources are generated. Extension services that are carried out through farm organizations are an excellent example of a demand driven extension system. Now days, this system is being shifted to the general farmer organizations to various commodity based organizations through a small tax on each unit of production that is sold.

The commodity based organization; Interest based organizations *i.e.* Farmers Interest Group (FIG); Farmers federation, Farmers association, Farmers club, farmers Self Help Group, etc. are well known example of farmer led extension organized based on cluster approach. Some successful example are 'Grape Growers Association'; Mushroom growers association, Gulbarga crop based association, etc. Now a days, another concept is also arises 'Group farming' 'group marketing' (Kokate *et al.*, 2011) and 'Farmer Producer company'.

On the concept of co-operatives Indian Farmers Fertilizers Cooperative (IFFCO) and Krishak Bharti Co-operative (KRIBHCO) are the two major fertilizer producing companies actively involves in several extension activities.

8.2.1 Farm School

It is a innovative farmers operated school in village, spread advance agricultural related message, skills, method to the village farmers. A cader of grassroot farmer trainers can be built up through such farm schools.

- ☆ Farm schools can also be powerful instrument for participatory research and knowledge management.
- ☆ The host achiever farmer may be designated as a farmer scientist/farmer professor in the respective crops/enterprises.
- ☆ The farm school will help to impart a sense of grassroot realism to the capacity building programme.

8.2.2 Farmers Field Schools (FFS)

It consist of group of people with a common interest, who get together on a regular basis to study the 'how and why' of a particular topic.

- ☆ Farmer to farmer education through FFS will become an important vehicle for integrated crop management expansion.
- ☆ The purpose of FFS is to produce a healthy crops in a sustainable way.
- ☆ The ATMA model involves farmer's organizations in all stages of the extension delivery chain (Feder *et al.*, 2010).

8.2.3 Recognizing the Outstanding Farmers through Awards

Several state governments working in these line because this is the best way of promotion to farmers.

8.2.4 Need to Increase Farmers Participation

We have technological innovations but the definitely need a different mode of transfer among farmers. It is technology generation, technology assessment, technology refinement and then technology adoption by involving those stalkholders to whom we are working (R.S. Paroda, 2011) because farmers are great scientists and ultimate judge for value of the research.

- ✰ Increasing participation of farmers not only in research but also in extension as revised model of ATMA.
- ✰ To empowering farmers 'Primary Farmer producer company' should be established and make a 'brand' such like 'Grape Grower Association' in Maharashtra.
- ✰ Dr. Ayyappan, DG-ICAR, advocated a new initiative 'Farmer FIRST' may be the main focus of ICAR in XII five year plan. In 'Farmer FIRST', FIRST may mean 'F' Farm, 'I' Innovation, 'R' Resources, 'S' Science and 'T' Technology.

Thus, agricultural extension has to be more decentralized, participatory, pluralistic, demand driven and market led, with involvement of multiagency approach. Agricultural extension has to be focus on mobilization of community into Farmers' Organizations, Crop based associations, Self help group, etc. to empower farmers and to promote group or cluster based extension services. Such community based extension service (CBES) is perceived as an important strategy due to self dependent in nature as compare to depend system of public and private.

8.3 Non-Governmental Organisations (NGOs) driven Extension Approaches

8.3.1 Role of Non-Governmental Organizations (NGOs)

In India several NGOs plays significant role in agricultural development. ICAR also supported several NGOs, for establishing Krishi Vigyan Kendras (KVKs). Action for Food Production (AFPRO), Foundation for Ecological Security (FES), Professional Assistance for Development Action (PRADAN) and Bharti Agro-industries Federation (BAIF) are the some important NGOs working in more than one states. Several others like Civil Society Organizations are also transforming rural India such as *'Ralegaon Siddhi'* Village of Anna Hajare.

These organizations have a less bureaucratic and more participatory method of working. These are more competent at facilitating farmers to learn from their own experience and from each other.

8.3.2 Participatory Micro Credit

Micro credit is a viable and fast growing system to support and empower rural poor and farm women. Non Governmental organizations are major players in promoting microfinance through Self Help Group (SHG) (A.K. Singh, 2006).

9. Increases the Percolation of Knowledge and New Skills

Agricultural knowledge and skill not uniformly percolated to the all farmers even in few km. away from research and extension institute (P.K. Basu, 2011). These type poor effectiveness also seen in surrounding areas of SAUs/KVKs. A study of KVK Hamirpur (U.P.) also indicate that only 9.0 percent farmers know the name of KVK, Hamirpur district while only 4.0 percent farmer benefitted from KVK Hamirpur (S. Narain, 2012). That is the question that we have to address now. This agricultural lagness increases economic disparity among farmers. So, I think we have to give rethinking on the whole extension mechanism. Hence, reaching of advisory and diagnostic agricultural services to unreached area is a greater challenge for extension. Except these small, fragmented and scattered land holding have compounded with emerging problems like climate change, technology fatigue and inadequate market network, leading to distress sale of produce. There is nothing which can't be done. On this way a large gap between what we have and what we can have would be achieved.

10. Need Better Coordination and Management among all Extension Service Providers

Several extension players involves in technology transfer process for farmers but poor coordination and management among different extension players is a great challenge. To reducing several gaps of extension following future steps may be taken.

11. Need of Future Steps

- ✰ Dr.K.D. Kokate, DDG-Agril. Extension (2011), propose for 'Indian Agricultural e-Extension and Knowledge Management Institute, which can take care of extension research and subsequently.
- ✰ Need to initiate a National Coordinated Research Project on Extension for co-ordination of different extension researches, etc.
- ✰ Dr. M.S. Swaminathan, Former-DG, ICAR (2011) advocated establishment for Climate Risk Management Research and Training Center in each agro-climatic region.
- ✰ Developing National Institute of Agricultural Extension Management (*MANAGE*) as an International Center for Excellence in Agricultural Extension Management.

Conclusion

Indian agriculture has registered phenomenal growth during last four decades with manifold increase in production of major commodities like food grains, vegetables, fruits, milk, eggs and fish. As a result, the per capita availability of important food items has increased despite of increasing population. To achieving this goal public sector extension role is enormous. From Green Revolution to till date public extension has undergone several transformations. But changing agricultural development goals demands multifarious extension services, which can't only fulfill by public extension system. Therefore, it is necessary to adopt pluralistic extension approach as a supplementary or complementary to public extension system. On these lines several initiatives are already taken/proposed for strengthening or revitalizing extension services in India. Today's farmer is different from yesterday and their demands from extension are also changing. He knows more about advisory on climate change, cropping pattern, market outlook, new enterprizes, diversification, energy management, etc. So it is necessary to harvest pluralistic extension approaches and systems of extension for achieving changing goal of agriculture. There is a need to restructure government extension machinery with a new set of operational procedure with more flexible approach to meet the emerging needs of farmers at local level and improve the co-operation among different government departments media and e-connectivity system with other development agencies including corporates, private firms, NGOs, farmers organizations, etc. on participatory basis.

References

Alex, g., Zijp, W. and Byerke, D. 2002. Rural extension and advisory services – New direction. Rural Strategy Background Paper NO. 9, World Bank, Washington, DC.

Basu, P.K., 2011. Better Start Dreaming. Published in 'Future Agricultural Extension'. Westville Publishing House, New Delhi.

DAC, 2010. Guidelines for Modified Support to State Extension Programmes for Extension Reforms: Scheme, 2010. Available at http: //vistar.nic.in/project/revised_ATMA_Guidelines.pdf

Farrington, J., Sulaiman, V.R. and Pal Suresh, 1998. Improving the effectiveness of agricultural research and extension in India: An analysis of institutional and socio-economic issues in rainfed areas. Policy paper 8, National Center for Agricultural Economics and Policy Research, New Delhi.

Feder, G., Anderson, J.R., Birner, and Deininger, R.K. 2010. Promis and Realities of Community-based Agricultural Extension. IFPRI Discussion Paper 00959, Development strategy and Governance Division, International Food Policy Research Institute, Washington, DC.

Kokate, K.D., 2013. Technology Dissemination in Indian Agriculture : Lesson learnt and future Strategies. Division of Agricultural Extension, ICAR, New Delhi.

Kokate, K.D., Mehta, A.K. and Singh, A.K. 2011. Future Agricultural Extension. Westville Publishing House, New Delhi.

Narain, S., 2011. A Study on role of Krishi Vigyan Kendra in Hamirpur district (U.P.).

Neuchatel Group, 2002. Common framework on financing agricultural and rural extension. Swiss Center for Agricultural Extension and Rural Development, Lindau.

NSSO. 2005. Situation assessment survey of farmers: Access to modern technology for farming, National Sample Survey, 59 round (January to December, 2003). Report 499 (59/33/2). New Delhi: Government of India, Ministry of Statistics and Programme Implementation.

Paroda, R.S., 2011. Accomplishing Second Green Revolution. Published in 'Future Agricultural Extension', Westville Publishing House, New Delhi.

Prasad C., Choudhary, B.N. and Nayar, B.B. 1987. First line Transfer of Technology Projects, ICAR, New Delhi.

Rivera, W.M., Van Crowder, L. and Qumar, K. 2001. Agricultural and Rural Extension Worldwide: Option for Institutional Reform in the Developing Countries. Food and Agricultural Organization of United Nations, Rome.

Singh, A.K., Singh, L. and Burman, R.R. 2006. Dimensions of Agricultural Extension, Aman Pub. House, Meerut.

Thomas, K.V., 2011. Pluralistic Extension. Published in 'Future Agricultural Extension', Westville Publishing House, New Delhi.

Van den Ban, A.W. 1998. Supporting farmers' decision making process by agricultural extension. Journal of Extension System 14: 55-64.

2014, Sustainable Rural Development through Agriculture *Pages* ***247–261***
Editors: **Dr. Shobhana Gupta and Dr. S.S. Tomar**
Published by: **BIOTECH BOOKS, NEW DELHI**

Chapter 18

Farmer Field School: A Platform for Transformative Learning in Rural India

Sudhir Kumar Rawat, M.K. Awasthi, S.C. Singh, Sarju Narain and Shobhana Gupta

Introduction

Agriculture is now a day's entering in new era as production pressure is increasing due to population blast in India and other countries world vide. So there is huge requirement of strengthening a better and scientific approach to agriculture sector, we need to take care about multi cropping system, selection of improved seed, irrigation methods, harvesting and every other step of production from a new scientific approach. The farmer field approach is a *kadi* of this chain which enable us to meet challenges to meet millennium goals.

The Farmer Field School (FFS) is a participatory agricultural extension approach, based on 'learning by discovery'. The FFS approach was developed in the 1980s by an FAO project in South East Asia as a way for small-scale rice farmers to learn for themselves the skills required for, and benefits to be obtained from, adopting integrated pest management (IPM) practices in their paddy fields. Subsequently, the FFS approach was extended to several countries in Africa and Latin America. Simultaneously, the IPM emphasis shifted from rice-based systems towards other annual crops, vegetables, cotton and other crop management aspects were added to the curriculum. The FFS approach, although originally developed for IPM purposes, is an effective people-centered learning methodology. It allows farmers to learn about, and investigate for themselves, the costs and benefits of alternative crop and livestock management practices for improving farm productivity.

The FFS is cooperatively better approach from other participatory extension approaches. For example, under the old Training and Visit (T and V) system, farmers are simply passive recipients. Extension messages are developed elsewhere and then demonstrated to farmers by field extension agents. In contrast, the FFS is a learning process where farmers are gradually presented with new technologies, new ideas, new situations, and new ways of responding to problems and that in the real time. As in the old approaches, it takes almost a year to implement the knowledge gain in FFS farmers are using in in real time. The knowledge acquired during the learning process builds on existing knowledge, enabling farmers to adapt existing technologies to become more productive, more profitable, and more responsive to changing conditions, or to adopt new technologies. Over the past decade there is growing awareness that participatory approaches are required if extension is to respond usefully to farmers' needs technically, socially, environmentally and economically.

The History of the Farmer Field School

The Farmer Field School (FFS) approach was first developed in 1989 by the Food and Agriculture Organization of the United Nations (FAO). The term "Farmers' Field School" comes from the Indonesian Sekolah Lampangan meaning simply "field school". The first Field Schools were established in 1989 in Central Java (Indonesia) during the pilot phase of the FAO-assisted National IPM Programmed. This Programme was prompted by the devastating insecticide-induced outbreaks of brown plant hoppers (Nilaparvata lugens) that are estimated to have in 1986 destroyed 20,000 hectares of rice in Java alone. The Government of Indonesia's response was to launch an emergency training project aimed at providing 120,000 farmers with field training in IPM, focused mainly on recording on reducing the application of the pesticides that were destroying the natural insect predators of the brown plant hopper. The technicalities of rice IPM were refined in 1986 and 1987 and a core curriculum for training farmers was developed in 1988 when the National IPM Programmed was launched. It was based not on instructing farmers what to do but on empowering them through education to handle their own on-farm decisions, using experiential learning techniques developed for non-formal adult education purposes.

What is a Farmer Field School

FSS is simply an approach to gain the knowledge from own experience as well as from others. It is a chain process in which farmers regularly gain from doing and did better by the experience gained by doing. It is a capacity building method based on adult education principles using groups of farmers. Farmer field schools (FFS) is described as a platform and "School without walls" for improving decision making capacity of farming communities and stimulating local innovation for sustainable agriculture. It is a participatory approach to extension, whereby farmers are given opportunity to make a choice in the methods of production through discovery based approach. A Field School is a Group Extension Method based on adult education methods. It teaches basic agro-ecology and management skills that make farmers experts in their own farms.

It is composed of groups of farmers who meet regularly during the course of the growing seasons to experiment as a group with new production options. Typically FFS groups have 25-30 farmers. After the training period, farmers continue to meet and share information, with less contact with extensionists. FFS aims to increase the capacity of groups of farmers to test new technologies in theirown fields, assess results and their relevance to their particular circumstances, and interact on a more demand driven basis with the researchers and extensionists looking to those for help where they are unable to solve a specific problem amongst themselves. In summary, therefore a Farmer Field School (FFS) is a forum where farmers and trainers debate observations, apply their previous experiences and present new information from outside the community. The results of the meetings are management decisions on what action to take. Thus FFS as an extension methodology is a dynamic process that is practiced and controlled by the farmers to transform their observations to create a more scientific understanding of the crop- livestock agro-ecosystem. A field school therefore is a process and not a goal.

Why the Farmer Field School Approach

Extension work has traditionally been seen by research and extension institutions as a mechanism to transfer technologies to farmers. This approach, however, has proved inadequate in complex situations where farmers must frequently adjust their activities to changing conditions (crop protection, soil nutrient management, animal health and production). Technology packages, delivered in a 'top-down' approach, were often too complex, expensive or poorly adapted to farmers' needs. Extension workers realized that farmers were not sufficiently involved in identifying problems, selecting and testing options, and evaluating possible solutions. With declining government support for traditional extension work, it became clear that alternative methods were needed to identify the problems faced by farmers and to disseminate appropriate technologies. The FFS approach, in contrast, strengthens the capacity of farmers and the local communities to analyze their production systems identify their main constraints and test possible solutions. By adding their own knowledge to existing information, farmers eventually identify and adopt the most suitable practices and technologies to their farming system and needs to become more productive, profitable and responsive to changing conditions.

Objectives

The core objective of FSS is to bring people together train and educate them to gain the knowledge by doing and experience the result of the same and implement the better practices in the field in real time. FFS is not about technology but about people development. It brings farmers together for them to assess their problems and seek ways of addressing them.

Broad Objectives

To bring farmers together to carry out collective and collaborative inquiry with the purpose of initiating community action in solving community problems

Specific Objectives

1. To empower farmers with knowledge and skills to make them experts in their own fields.
2. To sharpen the farmer's ability to make critical and informed decisions that render their farming profitable and sustainable.
3. To sensitize farmers in new ways of thinking and problem solving
4. Help farmers learn how to organize themselves and their communities.

FFS also contribute to the following objective;

1. Shorten the time it takes to get research results from the stations to adoption in farmers' field by involving farmer's experimentation early in the technology development process.
2. Enhance the capacity of extension staff, working in collaboration with researchers, to serve as facilitators of farmers' experiential learning. Rather than prescribing blanket recommendation that cover a wide geographic area but may not be relevant to all farms within it, the methods train extensionists and researchers to work with farmers in testing, assessing and adapting a variety of options within their specific local conditions.
3. Increase the expertise of farmers to make informed decisions on what works best for them, based on their own observations of experimental plots in their Field schools and to explain their reasoning. No matter how good the researchers and extensions, recommendations must be tailored and adapted to local conditions, for which local expertise and involvement is required that only farmers themselves can supply.
4. Establish coherent farmer groups that facilitate the work of research and extension workers, providing the demand of a demand driven system.

Principles and Practice of FFS

Every FFS is guided by the following 10 principles:

1. Learning by doing

Adults do not change their behavior and practices just because someone tells them what to do or how to change. They learn better through experience than from passive listening at lectures or demonstrations. Discovery-based learning is an essential part of the FFS as it helps participants to develop a feeling of ownership and to gain the confidence that they are able to reproduce the activities and results on their own farm.

2. Farmer Led Learning Activities

Farmers, not the facilitator, decide what is relevant to them and what they want the FFS to address. This ensures that the information is relevant and tailored to their actual needs. The facilitator simply guides the farmers through their learning process by creating participatory exercises to provide farmers with new experiences.

3. Learning from Mistakes

Behavioral change requires time and patience. Learning is an evolutionary process characterized by free and open communication, confrontation, acceptance, respect and the right to make mistakes. This last point is key as more is often learned from mistakes than from successes. Each person's experience of reality is unique.

4. Learn How to Learn

Farmers are learning the necessary skills to improve their ability to observe and analyze their own problems and make conscious decisions. They also learn how they can educate and develop themselves further.

5. Problem Solving

Problems are presented as challenges, not constraints. Farmer groups learn different analytical methods to help them gain the ability to identify and solve any problem they may encounter in the field.

6. The Farmer's Field is the Learning Ground

The field (crop production system) is the main learning tool. All activities are organized around it. In the case of a livestock FFS, both the animal(s) and the field are the main learning tools. Farmers learn directly from what they observe, collect and experience in their fields instead of text books, pictures or other extension materials. Farmers also produce their own learning materials (drawings, etc.) based on what they observe. The advantages of these homemade materials are that they are consistent with local conditions, inexpensive to develop, and owned by the farmers.

7. Extension Workers are Facilitators, Not Teachers

Extension workers are called facilitators because their role is to guide the learning process and not to teach. The facilitator contributes to the discussions and aims to reach consensus on what actions need to be taken. Facilitators are trained in a formal Training of Facilitators (TOF) course developed by experienced FFS Master Trainers. Researchers and subject matter specialists are invited to provide technical and methodological backstopping support to an FFS and also learn to work in a participatory and consultative way with farmers.

8. Unity is Strength

Empowerment through collective action is essential. Farmers united in a group have more power than individuals. Also, when recognized as an active member within a group, the social role of individuals within a community is enhanced. The combination of two or more minds is often more successful than one mind on its own.

The FFS expresses this as 1 + 1 = 3; *i.e.* one mind + one mind creates a new, third mind.

9. Every FFS is Unique

Learning topics within the FFS should be chosen by the community. Training activities must be based on existing gaps in the community's knowledge and skills and should also take into consideration its level of understanding. Every group is different and has its own needs and realities. As participants develop their own content, each FFS is unique.

10. Systematic Training Process

All FFS follow the same systematic training process. The key steps are observation, group discussion, analysis, decision making and action planning. Past FFS experience has shown that the best results are achieved with weekly meetings. Longer gaps can slow down the learning process. The length of the FFS cycle depends on the focal activity. With livestock, a full year cycle is usually needed to allow for all seasonal variations to be studied. Crop- or poultry-based FFS usually base their length on the cycle of production; from land preparation to harvesting or egg to egg respectively. FFS increasingly include marketing and processing activities which may lengthen the FFS learning cycle.

Farmer Field School Activities

There are five activities that are repeated in each session to provide the framework for each FFS:

1. Agro-ecosystem analysis (AESA)
2. Field comparative experiments
3. Topic of the day (special topic)
4. Participatory monitoring and evaluation (PM and E)
5. Group dynamic exercises

1. Agro-ecosystem Analysis (AESA)

AESA is the cornerstone of the FFS approach and is based on the ecosystem concept, in which each element in the field has its own, unique role. It involves field observations, data collection and analysis, and recommendations. Through regular observation of a crop–livestock system, AESA exercises help establish the interaction between crops and livestock as well as other living and non-living factors. Data are collected based on key factors observed to help put a process in place for decision making. The analysis is performed in sub-groups of four to five members to enhance participatory learning. Each sub-group presents their observations and recommendations in plenary sessions for collective decision making on management actions.

AESA exercises improve decision-making skills by:

- ✰ Enhancing observational skills
- ✰ Developing record keeping skills by drawing simple forms
- ✰ Generating discussions and sharing of farmer-to-farmer experience
- ✰ Developing presentation skills to promote communal decisions.

2. Field Comparative Experiments

Field comparative experimentation, also known as participatory technology development (PTD), is a collective investigation process to solve local problems. Simple experiments are carried out to enhance farmers' observational and analytical skills to investigate the cause and effect of major production problems. They help farmers become experts and to design simple and practical experiments to test and select the best solution to their problems. Experiments also encourage the validation and adoption of new technologies or practices. In this case, the experiments compare farmer practices with a set of available solutions presented either by the facilitator, researchers or other farmers. By analyzing the results and developing recording skills, farmers are able to decide which solution (technology or practice) is best suited to theirsituation.

Each experiment should include a cost: benefit analysis using the data recorded during AESA exercises. Assessing the economics of each option improves decision-making skills for livestock health and production activities as farmers often do not know whether they operate at a profit or loss. Farmers can better understand the difference between production and productivity where cost per unit produced is calculated to determine the efficiency of their own systems. Besides recording and analyzing the financial costs and benefits of the options tested in the experiment, other indicators to validate the results of the experiment should be identified by FFS participants (*e.g.* labour needs, length and speed of growth, accessibility). Precise record keeping of indicators is required to monitor and evaluate the performance of a treatment or technology.

3. Topic of the Day (Special Topics)

Though adults learn best through a 'learning by doing' approach, where new knowledge is acquired from experience, basic technical information is usually needed before any handson activity can be implemented. Certain activities are also too risky to apply without proper expertise or information, as is often the case with animal health issues. The special topic or 'topic of the day' is used to introduce technical information. The objectives of special topics are to:

- ✰ Provide an opportunity for the facilitator, researcher or specialist to give theoretical inputs needed for a general understanding of the subject before any activities can be carried out.
- ✰ Enhance the farmers' technical knowledge and present the farmers with information they need at the time they need it.
- ✰ Ensure a demand-driven learning process
- ✰ Level knowledge among the participants.

Thirty minutes to 1.5 hours of each FFS session should be reserved to discuss a specific topic relevant to farmers' needs as the basic principal of FSS is to upgrade the current skill set of the farmers rather than developing the new one. The topic of the day is normally a livestock-related topic but could be any subject of concern. Participants may have other problems and feel a need to discuss issues such as crop production, horticulture, micro-finance, gender inequity etc. If the facilitator lacks the specific expertise, external scientists, specialists or other farmers can be invited to lead the discussion. The role of the facilitator is to target a specific topic at the most relevant time for FFS participants. This guide includes two participatory approaches to facilitate the special topics:

a) Focus group discussions where sub-groups of FFS participants are asked to answer questions followed by a plenary discussion; and
b) Participatory learning exercises of short and medium term duration (which can include simple demonstrations) to introduce technical topics and lead the group in discussing their experiences.

4. Participatory Monitoring and Evaluation (PM and E)

The PM and E plan is an extension of the participatory plan developed in the initial FFS stages. To implement the FFS approach, both the participants and facilitator need to be able to continuously assess whether they are making any positive changes and actually achieving the goals they set. Monitoring and evaluation (M and E) methods have been developed to help FFS practitioners (mainly project staff, facilitators and participants) actively observe and analyze situations and performances and help them understand what they are observing. Given the participatory nature of FFS, M and E should also embrace the established participatory principles. This Field Guide provides PM and E guidelines to:

- ☆ Monitor and evaluate the FFS performance and assess whether it is achieving its specific objectives
- ☆ Monitor and evaluate specific FFS sessions for self-evaluation purposes and monitor and evaluate a field comparative experiment.

5. Group Dynamic Exercises

Group dynamic exercises are used to create a pleasant learning environment, facilitate learning and create space to reflect and share. It promotes participant to enhance their ability to learning by doing. They also enhance capacity building in communication skills, problem solving and leadership skills.Specific examples that:

- ☆ Energies participants
- ☆ Enhance participation
- ☆ Strengthen learning topics
- ☆ Strengthen group work and cohesion
- ☆ Assist in solving conflicts

Steps in Conducting FFS (Organization and Management of Farmers Field School)

There are 8 key steps in conducting FFS

1. Ground working activities
2. Training of Facilitators
3. Establishment and running of FFS
4. Evaluating PTDs

5. Field days
6. Graduations
7. Farmer runs FFS
8. Follow up by facilitators

1. Conduct Ground Working Activities

- ☆ Identify focus enterprises
- ☆ Identify priority problems
- ☆ Identify solutions to identified problems
- ☆ Establish farmers' practices
- ☆ Identify field school participants
- ☆ Identify field school sites

2. Training of Facilitators on

- ☆ Crop/livestock production and protection technologies
- ☆ Field guides on how to effectively deliver crop/livestock production and protection topics using non-formal education methods (NFE)
- ☆ Participatory technology development (PTD) with emphasis on the approaches and developing guidelines on conducting PTD
- ☆ Non-formal education methods with emphasis on what, when and how to use NFE in FFS
- ☆ Group dynamics
- ☆ Special topics to be addressed at every stage of training.

3. Establishment and Running FFS

With the guidance of facilitators, the group meets regularly throughout the season, and

- ☆ Carries out experiments and field trials related to the selected enterprise.
- ☆ Implement PTDs (Test and Validate)
- ☆ Conduct AESA and Morphology and collect data
- ☆ Process and present the data
- ☆ Group dynamics
- ☆ Special topics

4. Evaluating PTDs

- ☆ Analyze collected data
- ☆ Interpret
- ☆ Economic analysis
- ☆ Presentation

5. Field Days

- During the period of running the FFS, field days are organized where the rest of the farming community is invited to share what the group has learned in the FFS.
- 1or 2 per season
- Farmers themselves facilitate during this day

6. Graduations

- This activity marks the end of the season long FFS. The farmers, facilitators and the coordinating office usually organize it.
- Farmers are awarded certificates

7. Farmer runs FFS

- FFS farmer graduates now have the knowledge and confidence to run their own FFS.

8. Follow up by Facilitators

Occasionally the core facilitators will follow-up on schools that have graduated preferably on monthly basis. The core facilitators also backstop on-going farmer run FFS.

Strengths and Weaknesses of the FFS Approach

Strengths

Conventional extension often fails due to incorrect recommendations being provided to farmers, causing a lack of trust between farmers and the extension worker. Rural extension staffs are generally not capable of dealing with the full spectrum of complex problems that farmers experience. Most sustainable agricultural practices are knowledge intensive (as compared to input intensive agriculture) and as such very few blanket recommendations exist and practices (especially among smallholders) need to be developed or adapted locally. Hands-on education is needed especially in order to improve farmer expertise in the management of site-specific agro-ecosystems for which there appears to be no shortcut alternative. Here FFS play an important role since the approach does not rely on highly trained external advisors but on farmers own discovery and reflection.

In the changing context for rural smallholders where no blanket recommendations exist in agriculture and collective action is required to access markets, farmers need to organize, be innovative and be able to adjust to changing situations. In this context FFS has an important role to fill in the development of locally based innovations, create knowledge for a framework of action and boost local management and leadership skills, aspects not normally catered for in regular training and extension based on technology transfer concepts. Human empowerment is often assumed as a precondition for the success of community-based interventions, services and project. However, often such interventions fail since the level of empowerment generally is low, particularly in Africa. Thereby FFS play an important role in serving as a platform for human capacity building and empowerment, which in turn can ensure the success of services provided for the community.FFS is best suited for problems and opportunities requiring site-specific decisions or management practices and for issues that entail articulation of changes in behavior within the farm enterprise, household, and community or among institutions at varying scales of interaction and situations that can be improved only through development of location dependent knowledge. Their comparative advantage relies on skilful incorporation of the following principles: (i) learner-cantered, field based, experiential learning; (ii) observation, analysis, assessment, and experimentation over a time period sufficient to understand the dynamics of key (agro-ecological, socio-ecological) relationships; (iii) peer-reviewed individual and joint decision making based on learning outcomes; and (vi) individual and group capacity building.

The FFS process builds self-confidence (particularly for women), encourages group control of the process, and builds group and management skills. Thereby the FFS is a means to enable vulnerable farmers to create their own cohesive economic empowerment groups that are capable to venture into collective, commercially oriented endeavors and ability to interact with service providers and market intermediaries. A major strength of the FFS is that it helps in strengthening civil society or social capital at village level. This happens when FFS mobilizes interest in a community, especially among those who do not belong to the "official" class of the community. Farmers gain voice and are taken more seriously as part of the decision making process. Related to the issue of lack of formal extension staff in many countries, particularly in drylands and pastoral areas, FFS provides an advantage in that it provides an opportunity for farmer-to-farmer extension. Farmer-led FFS have been a common strategy both for scaling-up up FFS interventions and for cost reduction in both Asia and Africa. FFS graduates are selected and appointed as FFS farmer facilitators that carry on the knowledge gained as participant in FFS as farmer facilitator for new FFS groups in the community. It is thus possible to scale up interventions even when there is very few extension staff. Since solutions are obtained jointly and through an experimentation process the FFS can function well even with facilitators of relatively low technical skills. This is a big advantage in the current situation of low investment in public extension systems and lack of extension staffing.

Due to the informal and participatory nature of FFS, with its inbuilt group dynamic and team building exercises, it provides an ideal entry point to deal also with broader livelihood issues such as nutrition, health and sanitation. The FFS approach can further act as a bridge between emergency and development by forming a platform for immediate input supply, agricultural training as well as building organizational capacities for future longer term interventions. Whereas inputs and emergency support are important among communities suffering from civil strife or returnees' efforts are also needed in terms of knowledge for efficient utilization of the inputs, food and income security and psycho-socio rehabilitation, something the FFS approach can fill. FFS provides a set of rules and processes that are fairly easily understood by most extension and community development facilitators. This helps in-experienced facilitators or staff that may have a somewhat top down attitude to still implement extension in a participatory manner. Even though FFS also depends on personal skills, the quality of participatory extension practice of more flexible nature without the "package" like structure of FFS is usually much more vulnerable to constraints in personal skills and attitude among the extension facilitators. The "package" like concept of FFS also makes it easier to scale up FFS in national extension systems, and has largely facilitated the institutionalization of FFS in many countries.

Weaknesses

FFSs are not a universal panacea for development, nor are they a substitute for more familiar technology centered, or profit driven approaches to rural development, such as extension, credit cooperatives, core-estates without growers, farmer training centers, or the use of mass media. The FFS supports an educational approach that emphasizes experiential learning, action research and critical thinking, to enable farmers to take the lead in local adaptation of practices. Clearly, the FFS is not the best instrument for achieving quick and wide application of standardized recommendations. There are instances in which "technology transfer" is useful and for such issues, non FFS methods, such as radio and community meetings are often more appropriate. Extension campaigns and the FFS were thus implemented side by side and could be considered complementary. Often FFS is specified as costly, particularly under the current situation of structural adjustment and declining agricultural budgets. Efforts have been made to compare FFS costs vs. other methods of extension but the comparisons falling short due to the difficulties in comparing outcomes of the investments, particularly in relation to aspects of empowerment, which are very difficult to cost. The debate shifts when FFS are regarded as a form of public investment in farmer education to tackle rural poverty and hence as a tool for achieving the Millennium Development Goals from issues of diffusion and absolute costs, to issues of quality and comparative effectiveness of different forms of rural adult education. The cost issue is currently being addressed in ongoing FFS programmers, where various models of revolving FFS funds, self-financing and FFS loan and repayment schemes have been explored. The use of farmer facilitators also provides drastic reductions in costs.

FFS are vulnerable to loss of quality (and thus impact) particularly in terms of poor or inappropriate curriculum design and inadequate attention to the quality of the learning process. With the current popularity of the approach practitioners and policymakers sometimes "pick and choose" sub-aspects of the approach, not paying attention to the necessary adult education and experiential learning principles woven into FFS. Field experiences show that the approach often loses its effectiveness when the fundamental principles and components are overlooked and that FFS needs to be implemented as a complete package to achieve desired results. An aspect that often is overlooked is the need to train FFS facilitators thoroughly in facilitation skills. Often priority is given to technical training of facilitators rather than provide opportunities for personal development and mentality change among facilitators which requires time to enable staff to make the shift in thinking feasible. Further, to implement FFS well it is imperative that the management and supervisory levels have a participatory mindset and are well versed with the approach, something often lacking in FFS development projects.

Despite FFS attracting mostly women farmers, there are concerns that vulnerable individuals may find it hard to participate in the relatively intensive FFS learning processes. Due to poverty, short term needs is a priority for many poor families, particularly single parent headed households, and many poor need to spend considerable time in search for casual work. Participation of the most vulnerable though is a general problem in development work and being addressed where available through "food for training" arrangements, which allow the poorest to join in development activities including FFS.

Challenges Faced by Farmer Field Schools

Sustainability

Many FFS activities are started up with assistance from donor projects, which may provide funds for training facilitators, initial running costs of newly established schools, external inputs for demonstrations and experiments. Once external funding stops, schools may find it difficult to sustain themselves. It is therefore important that every FFS finds its own source of income and that the FFS approach is institutionalized and fully supported by national extension services.

Monitoring and Evaluation; Quality Control

Once schools are on their ownand new ones are formed spontaneously, 'quality control' may become problem. It is important to ensure that the learning process remains intact, demonstrations and experiments are carried out correctly, and lessons are of high quality. For this to happen, strategies to institutionalize are important. Information exchange between schools, peer learning and continuing training of facilitators are also important.

Relevant Learning Materials

If schools are to remain effective under ever changing physical, social and economic farming conditions, then learning materials and messages should be continuously updated. This is a challenge for FFS facilitators, which must develop and provide relevant, accessible materials.

Other Participatory Extension Approaches

FFS need to recognize and engage with other participatory approaches, such as Participatory Extension Approach (PEA), Participatory Technology Development (PTD) and People Centered Development with a Livelihoods Perspective (PCDL). All these approaches aim to promote farmer learning and development. There is a needfor greater openness, interaction and learning among organizations using different approaches. For example, FFS can benefit from collaborating with other approaches in several areas: social mobilization, fostering and spreading local innovations, and development of farmer groups and organizations.

References

Braun, Arnoud R., J. Jiggins, N. Röling, H. van den Berg and P. Snijders, (2005). A Global Survey and Review of Farmer Field School Experiences. International Livestock Research Institute, Nairobi

Braun.A and Duveskog.D. (2010). The farmar field school approach-history, global assessment and success story.background paper for the IFAD Rural poverty report.

CIP-UPWARD, (2003). Farmer Field Schools: From IPMN to Platforms for Learning and Empowerment. International Potato Center – Users' Perspectives with Agricultural Research and Development, Los Banos, Laguna, Phillippines. 87 pp.

Duveskog, D. and Friis-Hansen, E. (2008). Farmer Field Schools: a platform for transformative learning in rural Africa. In "Transformative Learning in Action: Handbook of Practice", edited by Mezirow, J. and Taylor, E., Jossey-Bass Press

Fakih, M. (2002). Gender mainstreaming in IPM. Paper presented at International Learning Workshop on Farmer Field Schools (FFS): Emerging Issues and Challenges, 21-25 October 2002, Yogyakarta, Indonesia.

FAO, (2005c). Kenya Farmer Field School Networking and Coordination Workshop (2005). Blue Post Hotel, Thika, Kenya. 4-5 May 2005. FAO, Nairobi, Kenya. 23 p.

FAO/IIRR, (2008). Discovery-based Learning on Land and Water Management – A Practical Guide for Farmer Field Schools. FAO, Rome, Italy.

FAO/WFP, (2007). Getting started! Running a junior farmer field and life school. FAO/WFP, Rome, Italy. 146 p.

FARMESA, (2003). Soil and water conservation, with a focus on waterharvesting and soil moisture retention. A study guide for Farmer Field Schools and community-based study groups. Version 1.1, March 2003. FAO Project GCP/RAF/334/SWE. Harare, Zimbabwe.

Godrick khisa, (2004). Farmer field school methodology.training of trainarsmanuals,FAO Kenya.

Groeneweg K.et.al, (2006). Livestock farmer field school. ILRC 2006.

Hughes, O. and J.H. Venema (eds), (2005). Integrated soil, water and nutrient management in semi arid Zimbabwe. Farmer Field Schools Facilitators' Manual, Vol 1. FAO, Harare, Zimbabwe.

Hughes, O. and J.H. Venema (eds), (2005). Integrated soil, water and nutrient management in

Khisa, G. and E. Heineman, (2004). Farmer Empowerment through Farmer Field Schools: A Case Study of IFAD/FAO IPPM FFS Programme in Kenya. Paper presented at the Nepad-IGAD Conference on Agricultural Successes in the Greater Horn of Africa, Nairobi, Kenya, 22-25 November 2004. 13 pp.

Minjauw, B. (2001) Training of Trainers manual for Livestock Farmer Field Schools. Based on a participatory workshop held on 17–19 September 2001, Mabanga Farmer Training Center, Bungoma, Kenya.

Minjauw, B., H.G. Muriuki and D. Romney, (2002). Development of the Farmer Field School Methodology for Smallholder Dairy Farmers in Kenya. Paper presented at International Learning Workshop on Farmer Field Schools (FFS): Emerging Issues and Challenges, 21-25 October 2002, Yogyakarta, Indonesia.

Mureithi, J.G., F.M. Murithi, C.C. Asiabaka, J.W. Wamuongo, L. Mose and B.A.M. Mweri, (2002). Methodology development for participatory monitoring and evaluation of farmer field school approach for scaling up the adoption of agricultural technologies. Paper presented at International Learning Workshop on Farmer Field Schools (FFS): Emerging Issues and Challenges, 21-25 October 2002, Yogyakarta, Indonesia.

Okoth, J., Arnoud R. Braun, R. Delve, H. Khamaala, G. Khisa and J. Thomas, (2006). The emergence of Farmer Field Schools Networks in Eastern Africa, paper presented at the Research Workshop on Collective Action and Market Access for Smallholders, 2-6 October 2006, Cali, Colombia.

Pontius, J., R. Dilts and A. Bartlett (Eds.), (2002). Ten Years of IPM Training in Asia - From Farmer Field School to Community IPM. FAO, Bangkok. 106 pp.

Praneetvatakul, S. and H. Waibel, (2003). A socio-economic analysis of farmer field schools (FFS) implemented by the National Program on Integrated Pest Management of Thailand. Paper presented at the CYMMIT impact assessment conference, 4-7 February 2002, San Jose, Costa Rica.

Rahadi and H. Widagdo, (2002). Applying the Farmer Field School approach to farmer-based advocacy in Indonesia. International Learning Workshop on Farmer Field Schools: Emerging issues and challenges, 21-25 October 2002, Yogyakarta, Indonesia.

Sones, K.R., Duveskog, D. and Minjauw, B. (eds), (2003). Farmer Field Schools: The Kenyan experience. Report of the Farmer Field School stakeholders' forum held on the 27th March 2003 at the International Livestock Research Institute (ILRI), Nairobi, Kenya.

Van den Berg, H. and B.G.J. Knols, (2006). The farmer field school: A method for enhancing the role of rural communities in malaria control. Malaria Journal 5-3: 33

Van den Berg, H. (2004). IPM Farmer Schools: A synthesis of 25 impact evaluations. Wageningen. University. 53 pp.

Vuthang, Y. (2002). Farmer Empowerment through Farmer Life School, adapted from Farmer Field School Approach. International Learning Workshop on Farmer Field Schools: Emerging issues and challenges, 21-25 October 2002, Yogyakarta, Indonesia.

2014, Sustainable Rural Development through Agriculture *Pages* ***262–275***
Editors: **Dr. Shobhana Gupta and Dr. S.S. Tomar**
Published by: **BIOTECH BOOKS, NEW DELHI**

Chapter 19

Rural Development through Agro-based Industries in India

G.L. Meena, S.S. Burark, V. P. Chahal and Prem Chand

Introduction

Nearly 2/3rd of the Indian population depends on agriculture and agro-based industries. Agro-based industries have been identified as one of the most important activities for gainful employment in the rural non farm sector and for overall growth of the economy. The agricultural sector in India is undergoing a rapid transformation from a traditional to modern agricultural systems, driven by emerging supermarkets and by a modernizing agro-industry sector. The agro-based industries in India have given a major boost to agriculture by raising its productivity. These depend on the agriculture for raw materials and sell their products to the farmers. Cotton mills, sugar mills, jute mills were promoted during pre-independence period for development of agro-based industries in India. During post-independence period with a view to providing more employment and using local resources, small scale and village industries were encouraged. In the present day world of globalization, our industry needs to be more efficient and competitive.

Definition and Classification of Agro-based Industry

Agro-based industry refers to any activity involved in cultivation of agricultural and horticultural crops including floriculture, fruits and vegetables under controlled conditions.

Agro-based industries are those industries which have either direct or indirect links with agriculture.

Agro industries not only those industries which are concerned with the processing of agricultural products but also such industries which are involved in the production of farm inputs and farm implements.

The agro industry mainly based on capital and capacities are;

(i) Village industries: These industries are owned and run by rural households with very little capital investment and high level of manual labour. For example, pickles, papad, basket-making.

(ii) Small scale industries: These industries are characterized by medium capital investment and semi automation. For example, edible oil and rice mills.

(iii) Large scale industries: These industries involving large capital investment and high level of automation. For example, sugar, jute and cotton mills.

According to nature, the industries are broadly classified into following types.

(i) Extractive industries: These industries are concerned with the extraction and utilization of natural resources. For example – fishing, fruit gathering, forestation.

(ii) Genetic Industries: These industries include breeding of plants, seeds, cattle, breeding farm, fish hatcheries, and poultry farms.

(iii) Manufacturing Industries: These industries are engaged in the conversion of raw material into semi finished or finished goods. Some important examples are cotton textile industry, spinning and weaving mills etc. Manufacturing industries can further be classified into five types: (i) Analytical industry, (ii) Processing industry, (iii) Synthetic industry, (iv) Service industry, and (v) Assembly industry.

Agro industries can also be categorized into another two classes *i.e.* processing industries and supply industries. The processing industries are those which process agricultural produce for further use while supply industries are those which produce inputs for agriculture. These are further grouped into four categories.

i. Agro produce processing units. For example, rice mill
ii. Agro produce manufacturing units. For example, sugar factories
iii. Agro inputs manufacturing units. For example, fertilizers factories
iv. Agro service centre. For example, repair and service centre for farm implements and equipments.

Importance of Agro-based Industry in Indian Economy

Agro-based industries are considered an extended arm of agriculture and their development has assumed crucial importance in the economic planning and growth of the national economy because agro-based industries;

- ☆ Facilitates in modernizing agriculture.
- ☆ Expand more employment opportunities and helps in eradication of unemployment and poverty.
- ☆ Aimed at bringing down regional disparities by establishing industries in tribal and backward areas.
- ☆ Enlarge trade and commerce and brings foreign exchange.
- ☆ Assists in strengthening the farm sector and stabilising rural incomes.
- ☆ Aid in processing agricultural products such as; field crops, tree crops, livestock and fisheries and converting them to edible and other usable forms. Thus, it also attracts the value addition.
- ☆ Helps in improving the social, human, financial and physical infrastructure.
- ☆ Would cause diversification and commercialization of agriculture; thus it ultimately enhances the income of farmers and creates food surpluses for Indian population.

Major Agro-based Industries in India

Cotton Industry

Cotton is one of the principal crops of India and plays a vital role in the country's economic growth by providing substantial employment and making significant contributions to export earnings. Cotton is the major raw material for domestic textiles industry. India is first in cotton cultivated area and second in production (accounting for 16 per cent of the global production) among all cotton producing countries in the world next to China. India is the second largest exporter of cotton behind the US. Recent facts about cotton industry are given as below:

- ☆ In ancient, cotton textiles were produced with hand spinning and hand loom weaving techniques. After the 18th century, power-looms came into use.
- ☆ The first Indian modernized cotton cloth mill was established in 1818 at Fort Gloaster near Kolkata but this mill was not successful.
- ☆ The first successful cotton textile mill named 'Bombay Spinning and Weaving Co.' was established in 1854 at Mumbai by KGN Daber.
- ☆ Various movements like Anti Bengal Partition, Non-Cooperation Movement, Quit India Movement etc., created a wave of boycotting foreign clothes and accepting swadeshi clothes which helped a lot in developing cotton industries in India.
- ☆ The partition of India in 1947 had adversely affected the cotton industry of India. At the time of partition, 40 per cent of cotton producing are became the part of Pakistan and only 60 per cent area was transferred to India. That is the reason why India was compelled to import raw cotton to meet the input requirement of cotton mills.
- ☆ Cotton industry is the largest organized agro-processing industry in India.

- There were 1,947 cotton mills in the country (1763 spinning mills and 184 composite mills). There were 552 cotton mills closed by the end of Nov. 2010.
- Cotton industry is most important industry in terms of employment and production of export goods. In Maharashtra (Mumbai, Sholapur, Pune, Kolhapur, Satara, Wardha, Hajipur), Gujarat (Ahmedabad, Vododara, Rajkot, Surat, Bhavnagar), Tamil Nadu (Coimbatore-Manchester of South India). Tamil Nadu has the largest number of cotton textile mills in India.
- The cotton industry provides employment to nearly one millions workers which include farmers, cotton boll pluckers, skilled and semi-skilled workers engaged in ginning, spinning, weaving, dyeing, designing, packaging, tailoring and sewing of cotton.
- The major cotton industries in India are Arun Textiles (P) Ltd., Arun Kumar Spinning Mill Private Ltd., Asha Cotton Industries, Atmaram Maneklal Industries Ltd., Balaji Awnings Industries, Century Textiles and Industries Ltd., Cottonto Yarn, Durairaj Mills Ltd., Dwarkadas Cotton.com, Farook Thread Machinery, Gangotri Textiles, Gargi Weaving Mills, and Ginni Filaments Ltd.

Jute and Jute Textiles Industry

Jute sector plays very important role in Indian Textile Industry. Jute is called Golden fiber and after cotton it is the cheapest fiber available. Indian jute industry is an old industry and predominant in West-Bengal and eastern part of India. India is the largest producer of raw jute and jute products in the world and second largest exporter of jute goods in world. Current facts about jute and jute textiles industry are described here:

- The first jute mill was set up near Kolkata in 1859 at Rishra. After partition in 1947, more than 70 per cent of area under jute cultivation was transferred to Pakistan (now in Bangladesh) which substantially reduced jute production.
- The total Jute Mills in India are 77. Out of these 60 in West Bengal, 3 in Bihar, 3 in Uttar Pradesh, 7 in Andhra Pradesh, 1 in Assam, Tripura, Chattisgarh, and Orissa. There are 6 mills under Government of India, 1 mill (Tripura) under State Government, 2 mills (Assam and New central) under co-operative sector and 68 under private sector.
- The jute sector occupies an important place in the Indian economy. The jute industry provides direct employment to about 0.26 million workers, and supports the livelihood of around 4.0 million farm families and around 0.14 million people are engaged in the tertiary sector and allied activities, supporting the jute economy.
- The jute industry is traditionally export oriented. India is the largest producer of raw jute and jute goods and stand at second place in export of jute goods after Bangladesh.

- Department of Agricultural Research and Education, Ministry of Agriculture, launched the Mini Mission I, II, III of the Jute Technology Mission on 2nd June 2006.

Wool and Woolen Textile Industry

Wool is only natural fibre. The Indian wool industry is an important and prominent source of livelihood for the rural India. The India wool industry is smaller in size in comparison to other textile industries in India. The bulk of the wool produced in India is of coarse quality and used mainly in manufacture of hand-knitted carpets. The rest are being used for the manufacture of apparel, blankets, finished textiles, garments, knitwear, etc. Further, the Indian wool industry also caters to civil and defense requirements for warm clothing for the Indian Army. Current facts about wool and woolen textiles industry are as under:

- The Indian wool industry is the 7th largest in the world and it accounts for about 1.80 per cent of total world production of wool.
- **The woolen industry** is basically located in the states of Punjab, Haryana, Rajasthan, U.P., Maharashtra and Gujarat. 40 per cent, of the woolen units are located in Punjab, 27 per cent in Haryana, 10 per cent in Rajasthan, while the rest of the states account for the remaining 23 per cent of the units. A. few of the larger units are located in **Maharashtra, Punjab, U.P., Gujarat** and **West Bengal.**
- Currently, there are 718 woolen units in the country.
- Out of the total production of raw wool, only 5 per cent is of apparel grade, 10 per cent is coarse grade and 85 per cent is the carpet grade.
- This industry is rural based and employment oriented industry, providing employment to 27 lakh workers.
- Domestic production of wool is not adequate. Hence, this industry is heavily dependent on imported raw material.

Sugar Industry

Sugar industry occupies an important place among agro-based industries. This industry took a shape of a large industry in beginning of the 20th century. Sugar industry is the second largest agro industry after cotton textile industry among agro-based industry of the country. Various facts about sugar industry are given as below:

- India is the largest consumer of sugar and second largest producer of sugar in the world after Brazil.
- India occupies the first place in the production of gur and khandsari with a share of over 15 per cent of world sugar production.
- Presently there are 453 operating sugar mills in the country.
- The Indian sugar industry uses sugarcane in the production of sugar. Hence, maximum number of the companies found in the sugarcane growing states of India such as: Uttar Pradesh, Bihar, Maharashtra, Karnataka, Tamil

Nadu, Andhra Pradesh and Gujarat along with Punjab, Haryana, and Madhya Pradesh. Sixty per cent mills are located in Uttar Pradesh and Bihar. Uttar Pradesh alone accounts for 24 per cent of the overall sugar production in country.

- ☆ The sugar industry generates employment for about 50 million sugarcane growers and over 7.50 per cent of the rural population of India is directly and indirectly dependent on the sugar industry. Besides this, the industry also provides employment to about 2 million skilled / semi skilled workers and others mostly from the rural areas.
- ☆ The major sugar companies in India are Bannari Amman Group, Dhampur Specialty Sugars Ltd., Dwarikesh Sugars, M. B. Sugars, Rajshree Sugars, Rana Sugars, Sakthi Sugars, Shree Renuka Sugars, Srisivan and Co., and Triveni.

Food Processing Industry

India comes first worldwide in the production of milk, pulses, sugarcane and tea and the second largest producer of wheat, rice, fruits and vegetables. Though India is one of the major producers of food globally, but it accounts for less than 1.50 per cent of world trade in this sector. Thus, India is the world's second largest producer of food in the world after China. The food processing industry is one of the largest industries operating in India and it is ranked fifth in terms of production, consumption, export and expected growth.

Food processing as such is a large sector that covers various sub sectors like agriculture, horticulture, plantation, animal husbandry and fisheries. It also includes other industries that use agricultural inputs for manufacturing of edible products. The Ministry of Food Processing, Government of India, indicates that following segments within food processing industry like;

- ☆ Dairy, fruits and vegetable processing
- ☆ Grain processing
- ☆ Meat and poultry processing
- ☆ Fisheries
- ☆ Consumer foods include packaged foods, beverages and packaged drinking water.

Food processing levels in various sub sectors: Processing in organized as well as un-organized sector is reported to be around 1.70 per cent in fruits and vegetables, 37 per cent in milk, 21 per cent in meat, 6 per cent in poultry and 11 per cent in marine fish. The highest share of the processed food is in the dairy sector, where 37 per cent of the total produce is processed, of this only 15 per cent is processed by the organized sector. Segmentation of different sub sectors in food processing industry is given as under:

Dairy	Whole milk powder, Skimmed milk powder, Condensed milk,Ice cream, Cheese, Butter and Ghee
Fruits and Vegetables	Beverages, Juices, Concentrates, Pulps, Slices, Frozen andDehydrated products, Potato wafers/chips, etc.
Grains and Cereals	Flour, Bakeries, Starch glucose, Cornflakes, Malted foods,Vermicelli, Beer and Malt extracts, Grain based Alcohol
Fisheries	Frozen and Canned products mainly in fresh form
Meat and Poultry	Frozen and packed mainly in fresh form, Egg powder
Consumer Foods	Snack food, Namkeens, Biscuits, Ready to eat food, Alcoholicand Non-alcoholic beverages

The major Indian player in food processing are Dabur India Ltd., Gitz Food Products Pvt. Ltd., Godrej Industries Ltd., Haldiram Marketing Pvt. Ltd., MTR Foods Ltd., Parle Agro Private Ltd., Hindustan Lever Ltd., Britannia Industries Ltd., Agro Tech Foods., ITC Ltd., Nestle India Pvt. Ltd., PepsiCo India Pvt. Ltd., and Cadbury India Ltd.

The major Indian player in Milling products, Bread, Biscuits and Breakfast are Bambingo Agro Industries, Britannia Industries, General Mills, Godrej Foods Ltd., Hindustan Vegetable Oil Corporation, ITC Ltd., Kellogg's, Modern Food Industries, Parle Agro Private Ltd., Shakti Bhog Foods, and Surya Foods and Agro.

The major Indian player in fruit and vegetable processing are Capital Foods, Dabur India Ltd., Green Valley, Hindustan Unilever Ltd., Mother Dairy (Safal), Priya Foods, Temptation Foods, Goderaj F and B, and Mother Dairy.

Dairy Industry

The development of Indian dairy industry took its shape after the white revolution. Dr. Verghese Kurien engineered the white revolution in India. The dairy industry in India has been witnessing rapid growth. **India's milk product mix:** Dairy industry account for the major share of processed fluid milk (46.00 per cent) followed by ghee (27.50 per cent), curd (7.00 per cent), **butter** (6.50 per cent), khoa (6.50 per cent), milk powder (3.50 per cent), panner and chhana (2.00 per cent) and others (1.00 per cent) including cream, ice cream marketed in the country. **Some facts about the Indian dairy sector are given as follows:**

- India is the largest producer of milk in the world with a share of 16 per cent of the total world's milk production.
- India's milk production in 2009-10 was estimated at 112.50 million tonnes.
- Per capita availability of milk is 258 grams per day in 2009-10
- The annual value of India's milk production was about Rs. 1745 billion in 2009-10.
- Milk is processed and marketed by 177 milk producer cooperative unions, operating in 346 districts, which federate into 22 state cooperative milk marketing federations covering nearly 140227 village level societies.

- The packaged milk segment is dominated by the dairy cooperatives. Gujarat Co-operative Milk Marketing Federation (GCMMF) is the largest player.
- Dairy cooperatives generate employment opportunities for around 14 million farm families, of which 4 million are the women.
- Over the years, brands created by cooperatives have become synonymous with quality and value. Brands like Amul (GCMMF), Vijaya (AP), Verka (Punjab), Saras (Rajasthan), Nandini (Karnataka), Milma (Kerala) and Gokul (Kolhapur) are among those that have earned customer confidence.
- Some of the key domestic private players manufacturing a variety of dairy products are J K Dairy, Heritage Foods, Indiana Dairy, Dairy Specialties, Cadbury, H. J. Heinz Ltd., Indodan Industries Ltd., Milkfood Ltd., Mother Dairy, Nestle India Ltd., BNZF, Ved Ram and Sons, Hatsun Agro, Uniliver and SmithKline Beecham Ltd.

Seed Industry

Today, the Indian seed programme boasts one of the biggest seed markets in the world. India has the potential to become the foremost player in the seed export business in the developing world with prospective markets in Asia, Africa and South America. Indian seed market is the 6th largest in the world. Both public and private sector companies/corporations are involved with the production of seed.

Public Sector

The giant public sector players include the National Seed Corporation (NSC) and State Farms Corporation of India (SFCI) and 13 State Seed Corporations (SSCS). NSC was the first public sector organization, established in 1963. NSC undertakes production, processing and marketing of agricultural seeds. It's product range includes cereals, pulses, oilseeds, fodder, fibre and vegetable crops. NSC pioneered the development of Indian Seed Industry on scientific lines with its involvement in the formulation of seed certification standards. It's seed testing laboratories located at Delhi, Secunderabad, Pune, Bhopal and Kolkata. NSC is the single largest producer of certified seeds in the country in terms of quantity and third largest company in terms of turnover amongst all the seed producing companies including private sector. The public sector seed companies, however, lag behind in research; they are mostly dependent on public research institutions, under the aegis of Indian Council of Agricultural Research (ICAR) and State Agricultural Universities (SAUs) for their breeder seed requirements.

Private Sector

The easing of government regulations and the implementation of a new seed policy in 1988, the private sector seed companies have started playing a major role in seed development and marketing. The private sector comprises around 200 seed companies, which include national and multinational companies and other seed producing/selling companies and these tend to focus on low volume, high value crops with the principal effort being placed on creating hybrids for oilseeds, maize, cotton and vegetable crops. Private sector accounts for 80 per cent turnover in seed.

Private companies are spending 10-12 per cent of the turnover in R&D. Almost 1/3rd companies have global technology. Most large multinational seed companies now have their presence in India with their main focus on biotechnology. These include Monsanto, Bayer Crop Science, Syngenta, Advanta, Hicks-Muse-Tate, Emergent Genetics, Dow Agro, Bioseed Genetics International Inc., Tokita Seed Co; and Nunhems Zaden BV.

Public-Private Sector Cooperation

Cooperation between private sector seed companies and public research institutes under ICAR, SAUs, and the International Crop Research Institute for Semi-Arid Tropics (ICRISAT), supported by the Consultative Group on International Agricultural Research (CGIAR), is growing. Public sector breeder seeds are available free of charge to private seed companies with no strings attached.

Other seed industries/companies are Ankur Seeds, Aryan Mushrooms and Bio Fuels, Gujarat Growell Agro Forestry Pvt. Ltd., HLL, Indo-American Hybrid Seeds (India) Pvt. Ltd., J.K. Seeds, JK Agri Genetics Ltd., Maharashtra Hybrid Seeds Company Ltd., Mahyco Seeds, Namdhari Seeds Pvt. Ltd., Nuziveedu Seeds, Pioneer Seeds, Pro-Agro Seeds, Sunagro, Vibha Agrotech Ltd., A.G.Sunseeds (I) Pvt.Ltd., Advanta India Ltd. (formerly ITC zeneca Ltd.), Agri Genetic Research Organisation Pvt. Ltd., Agro Biotech, Ajeet Seed Pvt. Ltd., Ajinkya Seeds etc.

Leather Industry

The leather and leather products industry was started from the middle of the 19th century. The leather industry occupies prominent place in the Indian economy in view of its substantial export earnings, employment potential and growth. It is spread over organized and unorganized sectors. Some facts about leather industry are given as follows:

- ✰ After China and Italy, India ranks at the third place amongst the leather production and export countries.
- ✰ Today the leather industry ranks 8th in the export trade in terms of foreign exchange earnings of the country.
- ✰ About 10 per cent of the world's supply of leather is processed in India.
- ✰ The small, cottage and artisan sector account for over 75 per cent of the leather production. India has traditionally a rich advantage in this industry both in terms of raw material and skilled manpower. People employed in this sector are predominantly from the minorities and disadvantaged sections of the society. The leather industry is spread in different segments, namely, tanning and finishing, footwear and footwear components, leather garments, leather goods including saddlery and harness, etc.
- ✰ The major production centres for leather and leather products are located at Chennai, Ambur, Ranipet, Vaniyambadi, Trichi, Dindigul, Kolkata, Kanpur, Jalandhar, Bangalore, Delhi and Hyderabad.
- ✰ The leather industry employed about 2.5 million people, of which 50 per cent are the skilled and semi-skilled workers. Further, more than 30 per cent of the work force employed in this sector constitutes women.

- ☆ **Top ten Indian leather exporters are** Tata International Ltd., Florind Shoes Ltd., Punihani International, Farida Shoes Ltd., Mirza Tanners Ltd., T. Abdul Wahid and Company, Hindustan Lever Ltd., Super House Leather Ltd., RSL Industries Ltd., Presidency Kid Leather Ltd.
- ☆ The leather sector in India has a comparative advantage due to natural endowment of raw hides and skins, level of technology of preparedness, vast human resources and a skill based industrial sector.

Meat Industry

India has largest population of livestock in the world. India has 16 per cent of cattle, 57 per cent of buffaloes, 17 per cent of goats and 5 per cent of sheep population of the world. India is one of the meat producers in the world. Poultry meat is fastest growing meat sector in India. The total meat production in the country is 4 million tones. Animals, which are generally used for production of meat, are cattle, buffaloes, sheep and goat, pigs and poultry. Methum and rabbit are also slaughtered for meat in some parts of India. In meat processing sector, India has 3600 slaughter houses, 10 modern abattoirs and 171 meat processing units. Major players involved in meat industry are A. P. Meat and Poultry Corporation, Al-Kabeer Exports Ltd., Allanasons, Arabian Exports, Deejay, Fair Exports, Frigo Refica Allana Ltd., Godrej Agrovet, Hind Industries Ltd., P.M.L Industries, Punjab Agro, Rajasthan Meat and Wool Marketing Federation, Ranchi Bacon Factory, U. P. Pashudhan Udyog Nigam Ltd., and Venkateshwara Hatcheries.

Floriculture Industry

Floriculture as an industry began in the late 1880's in England. The present day floral industry is a dynamic, global, fast-growing industry. India has a long tradition of floriculture. In Rig Veda, Ramayana, Mahabharata and Kalidasa, there are many references to flowers and gardens. Floriculture industry comprises of florist trade, nursery and potted plants, seed and bulb production, micro propagation, extraction of essential oil from flowers, dehydration of flowers and flower crafts, cut and loose flowers, dry flowers, protected cultivation, processing and value addition. India is the second largest flower grower after china. The major flower growing states are Karnataka, Tamil Nadu and Andhra Pradesh, West Bengal, Maharashtra, Rajasthan, Delhi and Haryana. Tamil nadu is leading producer of loose flower closely followed by Karnataka both in terms of area and production. Flower cultivation is concentrated in the surroundings of big cities like; Mumbai, Pune, Banglore, Mysore, Chennai, Kolkata, Delhi, etc. Biggest wholesale market of flower in India is located at Connaught place. APEDA and NHB are major organizations involved in R&D of floriculture. Reliance, Tata Tea, ITC, Bharti Group's Field Fresh, Thapar Groups and Century Textiles are some of the corporate organizations planning to invest and set up world size floriculture units.

Fisheries Industry

Till Independence Indian production was just sufficient to meet its domestic requirements and no much surplus was available for export. After 1950-51, India could make beginning in the exports of marine products. Fisheries have been a traditional occupation in India and both marine and inland fishery has been practiced. The vast coastal line (8118 KM), reservoirs (3 million hectares) and brackish water (1.4 million hectares) provides vast opportunity for production of inland, marine and ornamental fish culture in India. Annual fish production of India is 76.08 lakh tones that provides livelihood to 14.49 million people in the country. India occupies the third position in fish production and second in fresh water fish production in the world. There are 2.4 lakh fishing crafts operating in the Indian coast apart from seven major fishing harbors, 62 minor fishing harbors and 194 landing centers which are functioning presently. The three major carps, namely *Catla, Mrigal* and *Rohu* contribute the bulk of production followed by silver carps, grass carp and common carps.

The **fisheries and aquaculture sector** is recognized as the sunshine sector in Indian agriculture. It provides employment to 30 lakh people contributes 1 per cent to Indian GDP and 4.5 per cent to agriculture and allied products. Major products are shrimps, frozen fish, cuttlefish, squid and dried items. Major marine products exporting states are AP, Kerala, Tamil Nadu and West Bengal. The players in fisheries are ASF Sea Foods, Bell Foods, and Deep Sea Products.

Spice Industry

A spice is a dried seed, fruit, root, bark or vegetative substance used in nutritionally insignificant quantities as a food additive for the purpose of flavoring, and sometimes as a preservative for killing or preventing the growth of harmful bacteria. No Indian meal is considered complete without the tangy and delectable flavor of Indian spices, locally known as *'masala'*. Some facts about Indian spice industry are as under:

- ✰ No other country in the world produces as many spices as India does. The list vary from the hot spices like; chilli, pepper, ginger to mild pungent and spicy items like; cardamom, coriander, cumin and herbal spices like; thyme, rosemary, mint and vanilla.
- ✰ India is one of the largest spice producer and consumer in the world.
- ✰ The Indian spice industry is booming with a substantial increase in exports over the past few years. India occupies a share of 44 per cent in quantity and 36 per cent in value in the world spice trade.
- ✰ Annual production of spices in country is more than four million tons. Only 9 to 10 per cent of the total production is exported and the rest is consumed domestically.
- ✰ Spices have a share of 6 to 7 per cent of India's total agricultural exports.
- ✰ Major players in Indian spice industry are Asian Spices, KGN Enterprises Ltd., Marwari Snacks Foods Pvt. Ltd., Nani Ka Masala, Rajarana Impex, Reho Natural Ingredients, Swastiks Masala Foods Pvt. Ltd., Vaspal Salt and Chemicals Pvt. Ltd.

Agro-forestry Industry

Agro-forestry refers to growing trees with crops and/or animals. In agro-forestry systems, trees or shrubs are internationally used within agricultural systems. In

India, main industries of agro-forestry are match, pulp and paper and plywood briefly described as under:

Match Industry

The origin of the safety match industry in India goes back to the beginning of 19th century. Around 1910 immigrant Japanese families who settled in Kolkata began making matches with simple hand- and power-operated machines. Local people soon learned the necessary skills and a number of small match factories sprang up in and around Kolkata. After the First World War, handmade match production shifted from Kolkata to southern India, especially in the Ramanathapuram and Tirunelveli districts of Tamil Nadu State. Mechanization came to the Indian match industry in 1924 when WIMCO Ltd., started operations in 1924. There are now 12,000 units in the small-scale, non-mechanized sector, of which 75 per cent to 90 per cent are situated in Southern India. About 70 per cent of the match box production is contributed by the small scale industry. A large number of Indian tree species such as; *semul, aspen, white mutty, poplar, linden, rubber tree* etc. have been found suitable for use in the match industry. The demand for upstream market of paper products, like, tissue paper, tea bags, filter paper, light weight online coated paper, medical grade coated paper, etc., is growing up. The major Indian players in the paper industry are WIMCO Ltd, Arasan Match Industries, Rajasekaran Match Industries, Galaxy Match Company, Himalaya Match Industries, Thangavel Match Industry, Vivekananda Match Company, South Indian Lucifier Match Works, Venkateshwara Match Industries etc.

Pulp and Paper Industry

The paper industry has played a vital role in socio-economic development of the country. The first mechanized paper mill was set up in 1812 at Serampur (West Bengal). At present, there are around 700 pulp and paper mills producing nearly 6.30 million tones of paper and paper board and about 1.04 million tones of newsprint. The Indian pulp and paper industry is 15th largest industry in the world. It provides employment to nearly 3 lakh people directly and 10 lakh people indirectly. Nearly 65 per cent of the production is based on non-conventional raw materials, such as; agro-residues and recycled fibre and only 35 per cent paper production comes from good quality raw material, *i.e.* forest resource. The major Indian players in the pulp and paper industry are Ballarpur Industries, J. K. Industries, Tamilnadu Newsprint, Andhra Pradesh Paper, West Coast Paper and Century Pulp and Paper Industry.

Plywood Industry

First time two plywood factories were set up in Assam during 1923-24. Thereafter, some well equipped factories were set up in different places *viz.*, Sitapur, Banglore, Baliapatan, Dandeli and Coochbehar. In India, the organized players hold only 10 per cent market share and rest with the unorganized sector. Diversified products of this industry are block boards, flush doors, commercial plywood, decorative plywood, marine and aircraft plywood etc. The major Indian players in the plywood industry are Century Plyboards India Ltd., Greenply Industries Ltd., Kitply Industries Ltd., Sarda plywood Industries Ltd., Sharon veneers Pvt. Ltd., MB Plywood Pvt. Ltd.

Sericulture and Silk Industry

Sericulture or silk production is the breeding and management of silk worms for the commercial production of silk. Silk is one of the oldest fibers known to man. Sericulture is an important industry in Japan, China, India, Italy, France and Spain. Sericulture has been growing in India as an agro-based industry and playing a vital role in the improvement of rural economy. Silk was first introduced in China by Hoshomin, the Queen of China. Sericulture was introduced into India about 400 years back. Current facts about sericulture industry are given as follows:

- ✰ Sericulture is the most important cottage industry and is practiced in nearly 54 thousand villages all over India.
- ✰ India producing all the five known commercial varieties of silk *viz.*, Mulberry, Trophical Tasar, Oak Tasar, Eri and Muga.
- ✰ Sericulture is cultivated in Karnataka, Bengal, Tamil Nadu, Andhra Pradesh, Jammu and Kashmir, Gujarat, Kerala, Maharastra, Uttar Pradesh, Rajasthan, Bihar, Orissa etc.
- ✰ Sericulture is a labour intensive industry and provides employment to approximately 61 lakh people of rural and urban areas.
- ✰ India is the second largest producer of raw silk in the World after China, with a production of about 19,690 tonnes in 2009-10. This accounts for 15.50 per cent of global production.
- ✰ Central Silk Board was set up in 1949 to act as a facilitator for planning, development and monitoring of sericulture industry between the States and Central Govt. Two Central Silk Research Training Institutes were also established at Mysore and Barhampur.
- ✰ Central Silk Board has initiated National Sericulture Project (1989-90 to 1995-96) with the help of World Bank, Switzerland, April, 1989. The government has launched the Silk Mark Scheme for the brand promotion of silk. Central Silk Board (Amendment) Act, 2006 was enacted to regulate the quality of silk-worm seeds and came into force *w.e.f.* September 14, 2006.
- ✰ Silk is an export oriented product and is exported to more than 50 countries like USA, U.K., Italy, UAE and Saudi Arabia. Some European and Asian countries are main buyers of Indian silk.

Apiculture Industry

Apiculture is also known as bee-keeping. The art and science of rearing bees is called apiculture. Honey bees are used for production of honey and other valued products such as; bees wax, bee venom, propolis, and royal jelly. Beekeeping is one of the subsidiary professions and is recognized as *a khadi and village industry activity* for rural upliftment in India. China has captured 40 per cent of the world honey market. India is the six largest producer of honey in the world. The main honey producing states are Punjab, Haryana, Uttar Pradesh, Bihar and West Bengal. Punjab accounts for almost half of the India's total honey production in the organized sector. Honey is also produced in the states like; Rajasthan, Madhya Pradesh, Maharasthra and

Eastern Ghats in Orissa and Andhra Pradesh. Kashmir Apiaries Exports in Doraha (Ludhiana) is the best entrepreneurial drive for Punjab farmer. This company exports to 35 different nations. It is also supplies to Nestle, Kellog's, Carrefour, Metro Super Store Chain (Germany) and Dollar Stores (US). This industry provides employment opportunities to over 6 million rural and tribal families.

Lac Culture Industry

Lac is a resinous substance secreted by a tiny insect called *Laccifer lacca* (popular name "lac insect"). Shellac is the purified lac usually prepared in the orange or yellow flakes. Lac culture is an economically important vocation practiced by millions of farmers. It plays an important role in life of tribal communities of various states like; Jharkand, Orissa, Chatishgarh, West Bengal, Madhya Pradesh, U. P. and Maharasthra. The product of lac insect namely, lac, dye and wax have many useful industrial applications such as; paint, varnishes, perfumery, pharmaceuticals, dying of textile yarn, etc. The many cottage and small scale lac based rural industries such as; bangle making, sealing wax manufacturing, and preparation of French polish, etc., are means of livelihood for millions people. About 55-60 per cent of the world requirement of lac is met by India. More than 80 per cent of lac produced in the country is exported mainly in refined / semi-refined form. Lac Research Institute in Ranchi (Jharkhand) conducts research on the various aspects of the lac insect. Lac based cottage industries are Surface Coating Industries, Adhesive Industries, Electrical Industries, Food Industries, Leather Industries, Pharmaceutical Industries and Cosmetic Industries.

2014, Sustainable Rural Development through Agriculture *Pages* ***276–284***
Editors: **Dr. Shobhana Gupta and Dr. S.S. Tomar**
Published by: **BIOTECH BOOKS, NEW DELHI**

Chapter 20
Agri-Business in India: An Overview

G.L. Meena, D.C. Pant, V.P. Chahal and Prem Chand

Introduction

Agriculture is considered as the backbone of our economy. Indian agriculture contributes approximately 14 per cent of the country's GDP, provides large scale employment and fulfills the food and nutritional requirements of the nation. It provides important raw materials some major industries. Indian agriculture faces numerous challenges with a rapidly changing business environment, pace of technological change, globalisation, competitive environment and changing role of government. Today, agriculture has achieved commercial importance and has changed from subsistence farming to commercial farming, import oriented to export oriented, supply driven technology to demand driven technology etc. New inputs and new technologies are hitting market every day. The market for processed and packaged food products is increasing day by day and therefore there is a vital need of trained manpower in this business. Only 14 per cent of total food products are being processed in India, whereas 35 per cent food is wasted during packaging and transportation. Many businesses started building up in and around agriculture. This resulted in growth of agri-business. Agri-business management is becoming a popular career choice for agricultural students and there is great need to develop the professional agri-business managers who can, not only fill the management requirements of the changing agriculture scenario but also prove to be a great support to the farmers. Agri-business education is one of the promising qualifications helps to mould the personnel into good managers having managerial expertise.

Concept of Agri-business

The term agri-business is combination of two words namely; agriculture and business. But, both agriculture and business are different from each other. The word agriculture indicates ploughing a field, planting seed, harvesting a crop, and cattle feeding, breeding, milking. But today's agriculture is radically different. Thus, agriculture is an economic activity refers to all those activities aiming to fulfill the basic needs of life, primary food, fodder, fiber and other associated goods. Literally speaking business means bushes. In simple words "business means the state of being busy". Broadly, business involves activities connected with the production of wealth. Business concerns with buying and selling goods, manufacturing goods or providing services in order to earn profit. The oxford dictionary defines the word "business" as buying and selling or trade or commercial work.

Agri-business includes all business enterprises that buy from or sale to farmers. The transaction may involve either a product, a commodity or a service and encompasses items such as (a) productive resources *e.g.* feed, seed, fertilizers equipment, energy, machinery etc. (b) agricultural commodities *e.g.* food and fibre etc. (c) facilitative services *e.g.* credit, insurance marketing, storage, processing, transportation, packing, distribution etc.

Agri-business can also be defined as science and practice of activities with backward and forward linkage related to production, processing, marketing and trade, distribution of raw and processed food, feed, fiber including supply of inputs and service for these activities.

Agri-business can be defined as science and practice of activities with backward and forward linkage related to production, processing, marketing and trade, and distribution of raw and processed food, feed, fiber including supply of inputs and service for these activities.

Sectors of Agri-business System (ABS)

Agri-business system with forward and backward linkages consists of following four sectors in India shown in Figure 20.1

1. Agricultural Input Sector

This sector deals with manufacturing and supplying the farm inputs such as feed, seed, fertilizers, agricultural chemicals, farm equipments, credit, insurance, farm machinery, pharmaceuticals, veterinary services, repair services, technological services, agri-clinic services, etc. used by the production sector of agri-business. Many traditional agri-business firms have been functioning in this sector such as Monsanto, John Deere, Jain irrigation etc.

2. Agricultural Production Sector

This sector deals with the growing of crops and rearing of animals. Farmers are buying yield increasing farm inputs from the market, agriculture is growing fast and agriculture production is getting increasingly transferred to manufacturing-processing sector. This sector includes the actual production of various agricultural

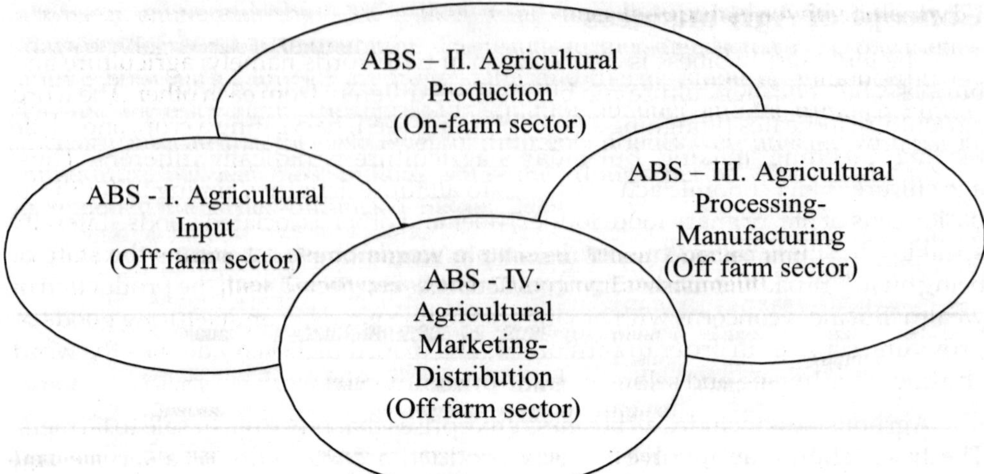

Figure 20.1: On Farm and Off-Farm Sectors of Agri-business System.

commodities of food, feed and fibres like cereals, pulses, oilseeds, vegetables, fruits, spices, condiments, milk, meat, egg, wool, and fish so on. It must focus on natural resources management and imprcvement in productivity.

3. Agricultural Processing-Manufacturing Sector (Agro processing) Sector

This sector includes the entire individual and firms that process raw agricultural commodities (*e.g.* convert wheat into *flour, maida, suji*; raw milk into *milk products* etc.) and manufacture food products (*e.g. bread, ice-cream, breakfast* etc.) for making these available to the final consumer. This sector is attracting many Indian and overseas companies.

4. Agricultural Marketing-Distribution Sector

This sector includes the millions of people and thousand of firms that handle agricultural products from farm to final consumer. On one side the sector made available farm inputs to the production sector, on the other side, the sector made available processed products to the ultimate consumers. It includes different activities like; packaging, transport, storage, warehouses, advertising, insurance, wholesale firms and retailer outlets services etc. Some of the well-known firms are engaged in this sector.

Need for Agri-business Education

Commercialization of agriculture calls for specialized production, post-harvest management, expansion of processing, transportation, packaging activities and positioning of products both in the domestic as well as international markets. Every year about 20-30 per cent of crop is wasted at farm level in India due to improper handling, spoilage, floods, droughts, pests and diseases attack and lack of knowledge of post-harvest technologies. In the case of fruits and vegetables, this wastage is around 30 per cent. The policy of liberalization and the establishment of WTO have created more opportunities for globalizing agriculture. Certain sectors in India such as floriculture, aquaculture, poultry, processing of fruits and vegetables are reaping the benefits of advanced technology. The agri-business sector encompasses the many activities of agricultural sector under one umbrella like integration of agricultural inputs, agricultural productions, agro-processing and agricultural marketing and trade which add value to the agriculture produce. Nowadays, agri-business has become one of the most important fields in the developing county like India to boost up its economy and taking care of globalization opportunities and challenges.

In the case of fruits and vegetables, this wastage is around 30 per cent. Certain sectors in India such as; floriculture, aquaculture, poultry, processing of fruits and vegetables are reaping the benefits of advanced technology. The agri-business sector encompasses the many activities of agricultural sector under one umbrella like; integration of agricultural inputs, agricultural productions, agro-processing and agricultural marketing and trade which add value to the agriculture produce. Nowadays, agri-business has become one of the most important fields in the developing county like, India to boost up their economies and meeting globalization opportunities and challenges.

The organizations are looking for professionally competent and trained agri-business managers. The State Agricultural Universities (SAUs) can take a lead in this direction to provide them the desired manpower. There is greater need of agri-business education, research and capacity buildings. This is the right time for SAUs to start post--graduate programmes in agri-business management, so that managerial skills of meritorious agricultural graduates can be developed and can prove as effective agri-business managers. Now, most of the State Agricultural Universities (SAUs) are offering post-graduate programmes in agri-business management.

Currently, some of the premier institutes providing post graduate courses in agri-business management programs are:

1. National Academy of Agricultural Research Management (NAARM), Rajendranagar (Hyderabad) - www.naarm.ernet.in
2. National Institute of Agricultural Extension Management (MANAGE), Rajendranagar (Hyderabad) - www.manage.gov.in
3. National Institute of Agricultural Marketing (NIAM), Jaipur (Rajasthan) - www.niam.gov.in
4. Institute of Rural Management Anand (IRMA) (Gujarat) - www.irma.ac.in
5. Indian Institute of Management (IIM), Vastrapur (Ahmedabad) - www.iimahd.ernet.in
6. Indian Institute of Management (IIM), Prabandh Nagar (Lucknow) - www.iiml.ac.in

Other Institutions/Universities Offering Courses in Agri-business Management

7. Acharya N.G. Ranga Agricultural University (College of Agriculture) Rajendranagar (Hyderabad) - www.angrau.net
8. Allahabad Agricultural Institute-Deemed University, Allahabad (Uttar Pradesh) - www.aaidu.org
9. Amity University Rajasthan (Amity school of Rural and Agricultural Management) Jaipur – www.amity.edu
10. Anand Agricultural University (International Agri-business Management Institute) Anand (Gujarat) – www.aau.in
11. Banaras Hindu University, Varanasi (UP) - www.bhu.ac.in
12. C. S. Azad University of Agriculture and Technology, Kanpur (Uttar Pradesh) - www.csauk.ac.in
13. CCS Haryana Agriculture University (College of Agriculture) Hissar (Haryana) - www. hau.ernet.in
14. Dr. Panjabrao Deshmukh Krish Vidyapeeth, Akola (School of Agri-Business Management), Nagpur (Maharashtra) - www.pdkv.ac.in
15. Dr. Y. S. Parmar University of Horticulture and Forestry (College of Horticulture), Nauni, Solan (Himachal Pradesh) - www.ysuniversity.ac.in
16. G B Pant University of Agriculture and Technology (College of Agri-business Management) Pantnagar (Uttarakhand) - www.cabm.ac.in
17. Institute of Agri-business Management, Noida - www.iabm.in
18. Jawaharlal Nehru Krishi Vishwavidyalaya (College of Agriculture) Jabalpur - www.jnkvv.nic.in
19. Jhunagarh Agricultural Uviversity (Post Graduate Institute of Agri-Business Management), Jhunagarh (Gujarat) - www.jau.in
20. Kerala Agricultural University (College of Cooperation, Banking and Management) Thrissur, (Kerala) - www.kaumba.com, www.kau.edu.
21. Maharana Pratap University of Agriculture and Technology (Rajasthan College of Agriculture) Udaipur (Rajasthan) - www.mpuat.ac.in
22. Mahatma Phule Krishi Vidyapeeth, Rahuri (Maharashtra) - www.mpkv.mah.nic.in
23. National Institute of Rural Development, Rajendranagar (Hyderabad) - www.nird.org.in
24. Navsari Agricultural University (Institute of Agri-Business Management Navsari (Gujarat) – www.nau.in
25. Punjab Agricultural University (College of Agri-business Management) Ludhiana (Punjab) - www.pau.edu
26. Rajasthan Agricultural University (Institute of Agri-Business Management, IABM) Bikaner (Rajashtan) - www.iabmbikaner.org

27. Rajendra Agricultural University, Pusa (Bihar) - www.pusavarsity.org.in
28. Regional Institute of Cooperative Management, Chandigarh E-Mail-ricmchd@hotmail.com
29. Symbiosis Institute of International Business, Pune (Maharashtra) - www.siib.ac.in
30. Tamil Nadu Agricultural University, Coimbatore (Tamil Nadu) – www.tnau.ac.in
31. University of Agricultural Science, Bangalore (Karnataka) - www.uasbangalore.edu.in
32. Vaikunth Mehta National Institute of Cooperative Management, Pune (Maharashtra) - www.mah.nic.in

Distinguish Features of Agri-business in India

1. The various stakeholders involved in the agri-business sector includes such as farmesr-producers, agri-input dealers and companies, certification agencies, extension agencies, financing institutions, agricultural and food R and D institutions, cold storage and warehouse operators, storage firms, restaurants, food chains, graders, packagers, porters, weighmen, transporters, village merchants, pre-harvest contractors, traders, wholesalers, brokers or commission agents, shipping agents, processors, cooperative societies, manufacturers, food retailers, etc.
2. Most of Indian agriculture is in the hand of small and marginal farmers. Thus, agri-business is built around several millions of farmer producers; these farmers produce hundreds of different food and fibre products. Most agri-businesses deal with farmers both directly and indirectly. Other industry is not built principally around the basic producer of the raw product.
3. Agri-business of several million farm units and several thousand business units. Agri-business is the sum total of hundreds of trade associations, commodity organizations, farm organizations, quasi-research bodies, conference bodies and committee. Government is also part of agri-business. Agricultural Universities with their teaching, experiments and extension function form another sector of agri-business.
4. Agri-businesses are small and compete in a relatively free market where there are many sellers and buyers. The numbers and size of agri-business do not allow monopoly like in other enterprises.
5. Husband and wives are often involved heavily in both the operational and decision making phases of agri-business. Agri-businesses tend to be community oriented that too in small town's rural areas.
6. Agri-businesses are highly seasonal in nature. Agricultural products can not produce throughout year; their productions are limited by a particular season (*Kharif*, *Rabi* and *Zaid*). Seasonality in production affects market price to a greater extent, if no storage facility.

7. Agri-business deals with vagaries of biological nature of production such as erratic rainfall, temperature, sunshine, drought, storm, floods, humidity, insect-pest and diseases etc. Therefore, farmers have no control over the output of production. Agricultural output comes from many small units. The production is to a largely extent dependent on weather and biological pattern of production. It is not possible to quickly shut off or switch on agricultural production.
8. Agri-business is characterized by the small production from comparatively small land holdings and thus, provides little scope for division of labour. Assembling of agro-products from such a large number of small production units is a huge task.
9. Agri-business is distinguished by the perishability of products. Most agricultural products are extremely perishable in nature. The storage period of farm products range from few hours to few days in the case of fruits and vegetables and few years for cereals, pulses and oilseeds. The Perishability influences the marketing of produce. It is also results in a wide price variation. Perishable product requires speedy handling and cold chain infrastructure.
10. Agri-business is also distinguished by bulkiness of products. Bulkiness of agricultural products makes difficulties for transportation and storage. The bulkiness product requires large storage capacities.
11. The producers-farmers and consumers of agricultural products are far and wide scattered throughout the length and breadth of the India. Most of the producers are kept small size of land holdings. This creates difficulty in marketing and estimation of demand and supply.
12. Processing has got special significance in the marketing of farm products. Therefore, processing of agricultural products in agri-business is necessary; because very few farm products are consumed directly in the form as such they obtained by producers-farmers. Thus, most of the agricultural products have to be converted into consumable form before their consumption by the ultimate consumers. Processing is important both for the agri-business producers and consumers. For example, wheat converted into flour, *maida, suji*; sugarcane converted into *gur, khandsari*, sugar etc.
13. There is a large variation in the agricultural products regarding size, shape, colour, flavor, appearance and quality due to different varieties of crops. This variation in quality of products occurs from year to year and from season to season. This causes problems in grading and standardization in different quality products. Therefore, grading and standardization of agricultural products become difficult and costly.
14. Financing in the agricultural sector is a problem because high risk and uncertainty involved in agricultural production, productivity and prices.
15. Agriculture is mainly concerned with the production of food and raw material. As income increase, the demand for products will increase less

rapidly. Agricultural products generally being necessaries of life, the demand (especially food) is relatively inelastic.

16. Farmers possess low holding-back capacity and generally sell their produce immediately after the harvest at low prices to meet their pressing demands. About 60-80 per cent of the food grains are still sold in the first quarter of the harvest season.

Scope for Agri-business in India

1. India has a vast geographic spread, varied agro-climatic conditions, soils, which facilitates and promoting the production of variety of food and non-food crops. India is the seventh-largest country in the world, with the total land area of 3,287,263 KM^2 and also blessed with varied agro-climatic zones. There are 20 agro-climatic zones (ACZ) and nearly 46 out of 60 soil types in the country.
2. The domestic market for processed foods is not only huge but is growing at a fast pace. The domestic demand for processed food products is very high and ever increasing, the scope for export earning is also a major factor to push up agricultural sector for enhanced income and increased employment options.
3. At present, processing is done at primary level only and the growing standard of living expands opportunities for secondary and tertiary processing of agricultural commodities. Most of the food processing units are in the unorganized sector and Indian food market continues to be dominated by the fresh food segment. According to one estimate, Indian food market comprises 10 per cent processed segment, 15 per cent semi-processed segment and 75 per cent as fresh food segment. Processing in organized as well as un-organized sector is reported to be around 2 per cent in fruits and vegetables, 37 per cent in milk, 21 per cent in meat, 6 per cent in poultry and 11 per cent in marine fish. The highest share of the processed food is in the dairy sector, where 37 per cent of the total produce is processed, of this only 15 per cent is processed by the organized sector. The overall value addition in food products, which is currently 8 per cent, is likely to increase to 35 per cent by 2025. Thus, there is huge scope for processing in the food sector for increasing the income, exports and employment.
4. There is growing demand for agricultural inputs like seed, feed and fodder, inorganic fertilizers, bio-fertilizers etc.
5. Applications of Biotechnology in agriculture have an immense scope in production of quality seed, bio-control agents, microbes for bakery products.
6. The vast coastal line (7517 KM) and internal water courses provides vast opportunity for production of inland, marine and ornamental fish culture. With an annual fish production of 6 million metric tones. India occupies the third position in fish production and second in fresh water fish production.

7. India has largest popuiation of livestock in the world. In livestock sector, India has 16 per cent of cattle, 57 per cent of buffaloes, 17 per cent of goats and 5 per cent of sheep population of the world. The livestock wealth gives enormous scope for production of meat, milk and milk products etc.
8. The forest area (21 per cent) and resources can be utilized for production of by products of forestry.
9. The worldwide market is enormous for sugar, coffee, tea and processed foods such as; sauce, jelly and honey. The market for processed meat, spices and fruits is equally large. Only with mass production aided by modern technology and intensive marketing can the domestic market as well as the export market be exploited to the fullest extent.
10. Mushroom production can be taken up on large scale in country for domestic consumption and its export can be enhanced with upgrading in the state of art of their production.
11. Organic farming has excellent prospective in India, because the application of pesticide and inorganic fertilizer is less in country, as compared to industrial countries. Therefore, the farmers can be encouraged and educated to switch over for organic farming.
12. There is a ample scope for production and promotion of bio-pesticides and bio-control agents for protection of crops.
13. Hybrid seeds and genetically modified crops have good potential in India in the future.
14. Micro-irrigation systems *viz.*, drip, sprinkler, pipe and labour saving farm equipments have good potential in the coming years due to declining groundwater level and labour scarcity for agricultural operations, especially at peak harvesting time like weeding, transplanting and harvesting.
15. India account for about 8 per cent of the world's fruits and 15 per cent of the world's production of vegetables. Production of vegetables and flowers under poly house and green house conditions can be taken up to harness the export market.
16. The improved agricultural production provides open opportunities for employment in marketing, transport, cold storage and warehousing, credit and insurance services.
17. India has vast potential to improve its present position in the world trade of agricultural commodities both in raw and processed forms. Export can be harnessed as a source of economic growth.
18. Agri-business management has wide scope in developing the trained manpower in different area of operations. Careers in agri-business are varied from farming to commodity broker, food broker, loan officer, marketing researcher/specialist, product analyst, purchase agent, statistician, and wholesaler.
19. There is a good scope of agri-business in medicinal and aromatic plants cultivation *viz.*, Cinchona, Isabgol, Opium, Poppy, Lemongrass, Japanese Mint, Henna etc.

2014, Sustainable Rural Development through Agriculture *Pages* ***285–306***
Editors: **Dr. Shobhana Gupta and Dr. S.S. Tomar**
Published by: **BIOTECH BOOKS, NEW DELHI**

Chapter 21

Sustaining Vegetables Production through Farmers' Own Seed: An Integrated Approach

Krishan Pal Singh, Beena Nair, Prem Chand, S.K. Sharma and A.S. Bhati

Introduction

Vegetables are protective food rich in vitamins and minerals which are essential for maintaining good health. Vegetable crops assume great importance in view of widespread malnutrition that exists in India. Among food crops vegetables are the most easily affordable food. Increased production and consumption of vegetables could alleviate the malnutrition and improve nutritional standard of our people. But unfortunately the vegetable production in India is far below the requirement leading to low consumption. Further, yields of vegetable crops are below their potential yield. Short supply of quality seed of improved vegetable varieties has been identified as one of the major constraints for this low yield.

Vegetables occupied an area of 8.9MH during 2011-12 with a total production of 155.9 million tonnes having average productivity of 17.4 tonnes/ha. The overall growth rate of 2.08 per cent in area, 3.72 per cent in total production and 1.64 per cent in productivity has been achieved during the last five years. Vegetable production registered a quantum jump of 77 per cent from 2001-02 to 2011-12. Globally, India ranks second in vegetable production and contributing 16.7 per cent to global vegetable area and 15.4 per cent to production. Vegetables reside in about 2.8 per cent of cropped area in the country.

More than 40 kinds of vegetables belonging to different groups are grown in India in tropical, sub tropical and temperate regions. Important vegetable crops grown in the country are potato, tomato, onion, brinjal, cabbage, cauliflower, peas, okra, chillies, beans, melons, etc. The leading vegetables growing states are West Bengal which accounts for 15 per cent of production followed by Uttar Pradesh (12 per cent), Bihar (10 per cent), Andhra Pradesh (8 per cent), Madhya Pradesh (6.5 per cent), Gujarat (6.4 per cent), Tamil Nadu (5.8 per cent), Maharashtra (5.7 per cent), Karnataka (5.0 per cent) and Haryana (3 per cent) which altogether contributes about 83.4 per cent of the total vegetable production in the country.

Among vegetables, potato is the major vegetable accounting for (27.0 per cent) followed by tomato (12 per cent), onion (11 per cent), brinjal (8 per cent), cabbage (5.4 per cent), cauliflower (4.7 per cent), okra (4 per cent), peas (2.5 per cent) and others (25.4 per cent) in the country. India is the second largest producer of vegetables after China and is a leader in production of vegetables like peas and okra. Besides, India occupies the second position in production of brinjal, cabbage, cauliflower and onion and third in potato and tomato in the world. Vegetables such as potato, tomato, okra and cucurbits are produced abundantly in the country.

While production of horticulture crops is growing, the 'demand' side is also witnessing a marked growth. As incomes rise and consciousness about 'healthy foods' increases, there is a significant change in the consumption basket of consumers. Households are spending significantly higher amounts of their expenditure on food to the fruits and vegetables category. However, it is important to note that the availability of vegetables has kept pace with the growing demand. The per capita availability of vegetables has also increased from 230g in 2010-11 to 245g in 2011-12. Our demand of vegetables will be 225 million tonnes by 2020 and 350 million tonnes by 2030. However, this spectacular growth of vegetable production in the country has accompanied with several challenges like low and uneven productivity across the country; perishability and high cost of cultivation; inadequacy of multipurpose varieties and eco-friendly agro-techniques for improving total factor productivity; poor management of dwindling natural resources and poor quality of the produce including food safety issues.

The major challenge, which lies ahead, is to develop technologies that enhance quality and productivity of vegetables under reducing land, declining natural resources and increasing biotic and abiotic stresses.

India has taken a bold step towards self sufficiency in food. However, self sufficiency in the true sense can be achieved only when each individual in the country is assured of balanced diet. This created an urgent need for providing health security to our population by supplying nutrition through balanced diet. Indians are predominantly vegetarians and depend on vegetables for bulk of their nutrients and minerals. More recently, the role and usefulness of anti-oxidants present in vegetables in human health has been demonstrated, adding value to this set of crops. Vegetables cultivation has been known to stimulate development because it is labour intensive, earns higher returns and involves extra skills. Development in vegetable production will therefore contribute not only to food and nutritional security but also to poverty alleviation and income generation.

We can grow variety of vegetables all the year round. Varied agro-climatic conditions in India make it possible to grow a wide variety of vegetable crops all the year round in one part of the country or another. India can claim to grow the largest number of vegetable crops compared to any other country of the world and as many as 61 annual and 4 perennial vegetable crops are commercially cultivated.

Although India's economy is booming, agriculture has lagged behind and growth has been sluggish. The national and state governments have recognized the need to stimulate the sector to increase productivity, enhance farmer incomes, and promote agribusiness. To earn more income, increasing numbers of farmers are producing vegetables on a portion of their land, and new marketing arrangements are emerging, along with some export opportunities. With its vast population 25 per cent of which lives below the poverty line and a larger proportion are malnourished. Key opportunities are: to intensify production for urban markets, to promote more production in rain-fed areas, and to encourage more garden and village production in remote regions to improve nutrition and diversify village economies.

The annual requirement of quality vegetable seeds in India is estimated to be approximately 10,000 tonnes, out of which only 3,500 tonnes of quality seed are produced in organized way. The organized production of vegetable seed has just been started in the country.

Table 21.1: Vegetable Seed Production

Seeds	*2009-10*		*2010-11*	
	Purchase	*Sale*	*Purchase*	*Sale*
Breeder's seed (q)	56.00	63.00	270.00	456.00
Foundation seed (q)	143.00	82.00	624.00	393.00
Certified seed (q)	4973.00	4303.00	24035.00	22155.00
Sale of certified quality seeds (Rs. lakhs)	994.68	2893.28.00		

Source: Annual Report, 2010-11, National Seeds Corporation, GOI, http://iivr.org.in/Publications/Vegetable per cent 20Statistics.pdf.

Seed Production Technology of Major Vegetable Crops

Tomato

Tomato (*Solanum lycopersicon*) grows best in the dry season under day temperatures of 21–25°C and night temperatures of 15–20°C. Vines will struggle to set fruit if temperatures exceed 30°C. Humidity levels higher than 60 per cent at the time of fruit maturity will increase disease problems and reduce seed yields. Seed production during the rainy season leads to poor seed quality. Avoid fields where the previous crop was tomato; this prevents the new seed crop from being contaminated with seeds from volunteer tomato plants. Growing tomato after paddy rice reduces the incidence of diseases and nematodes. Training of tomato plants generally results in early ripening, fewer diseases, higher yields and better seed quality.

Isolation

The minimum isolation distance between different cultivars of tomato for seed production is relatively short. This is because of the crop's high level of self-pollination. However, the minimum recommended isolation distance between different varieties varies from 30 to 200 m in different countries. Tomato produces perfect, self-pollinating flowers. Anthers are fused together into a little cone that rarely opens until pollen has been shed and the stigma pollinated. Alternatively, bagging the whole flower cluster can prevent cross-pollination.

Selection of the Plant

Look for early maturing and attractive plants. Selected plants should be marked, staked, and inspected during the growing season for resistance to diseases.

Roguing

Plants showing different characters to the type must be removed. Roguing is done at different stages of crop growth.

- Before flowering, plants showing different growth habit and foliage characteristics than the particular variety should be rogued out.
- Early flowering and fruit setting stage-Off-types are rogued out judging the size and shape of immature fruits.
- Fruiting stage, the off-types are identified examining the fruit characteristics like shape, size, colour etc.

Harvesting

Seed fruits are allowed to ripen to maturity on the plant. Only completely colored and matured seed fruits are harvested. The mark of the two sepals (calyx) cut off should be checked carefully to ensure that only pollinated fruits are harvested.

Seed Extraction

The following methods have been suggested by many workers for tomato seed extraction.

Fermentation Method

In this method the selected ripe fruits are harvested and kept in wooden or plastic containers for 2 to 3 days until the fruits become soft. They are crushed by hand and no fruit juice is allowed to drain out. Entire mass is kept for 24 to 72 hours depending upon temperature. Flesh will float at the top and seed will settle down at the bottom. The fermented mass is removed and the seeds are sieved and cleaned with fresh clean water and dried. Longer fermentation may damage the seed.

Separation with Sodium Carbonate

This method is relatively safe and can be used for small quantities of seed in cooler temperate areas where the fermentation method is not used. The pulp containing the extracted seeds is mixed with equal volume of a 10 per cent solution of sodium carbonate (washing soda). The mixture is left up to 2 days at room temperature after which time the seed is washed out in a sieve and subsequently dried. The sodium carbonate method of extraction tends to darken the testa of the seed and is, therefore, not normally used for commercial seed.

Separation with Hydrochloric Acid

This method is often favoured by large commercial producers as it produces a very bright clean seed sample. The hydrochloric acid treatment is often combined with later stages of fermentation. George (1985) reported that 567 ml of concentrated hydrochloric acid stirred into 10 liters of seed and pulp mixture and left for half an hour is successful. After the extraction seeds must be dried as quickly as possible. Drying of tomato seed is done up to the moisture content of 8 per cent.

Brinjal

Brinjal (*Solanum melongena*) is a warm season crop. It requires a long and warm growing season for successful production. It is more susceptible to lower temperatures than tomato and pepper. A day temperature of 25–32°C and a night temperature of 21–27°C are ideal for seed production.

Isolation

Brinjal, although self-pollinated, can outcross to a considerable extent. It is, therefore, essential to isolate the seed crop from other varieties to avoid contamination and to produce pure seeds. Agrawal (1980) recommended an isolation distance of 400, 200 and 100 m for breeders, foundation and certified seed production, respectively. The breeder's seed plot should also be isolated from commercial seed crops.

Roguing

Seed growers should be well acquainted with the characterization of the variety so that they may effectively rogue out the off-types and undesirable plants at different stages of crop growth. The following three roguing stages:

- ☆ Before flowering by examining plant colour, growth habit and foliage characteristics such as shape, size and posture.
- ☆ At early flowering and fruit development-by observing general plant habit, vigour, degree of spinyness.
- ☆ At fruiting off-types can be identified on the basis of fruit characteristics like shape, size, colour etc.

Selection of the Plants

Select the most vigorous and healthy plants, mark fruits on the second branch, and leave them until they are fully mature. Keep one or two fruits from one plant and several fruits from different plants of the same variety to maintain crop vigour.

Harvesting

Harvesting is done when fruits are fully ripe (the skin of fruit turns brownish-yellow in green varieties or brownish in purple varieties). Harvest and store the fruits in a shed for a week until the fruits get soft.

Seed Extraction

There are two basic methods used for the extraction of brinjal seeds: wet extraction and dry extraction. The wet extraction is favoured for large-scale seed production while the dry extraction is employed for small-scale seed production.

- ☆ **In wet extraction**, the harvested fruits are stored for 5-7 days at room temperature until they become soft. This allows the seeds to mature fully. The fruits are crushed or cut into thin slices. These are then softened by soaking till the seeds are separated from the pulp. Since the brinjal fruit pulp is relatively dry, it requires extra water during and after crushing and would be allowed to stand overnight to facilitate seed separation from the flesh.
- ☆ **In dry extraction**, the ripened fruits are harvested and dried in the sun until they shrivel. During drying of purple and purple black fruits the skin colour turns to coppery brown. The fruits are then hand beaten to extract the seed. This method is used for small- scale seed extraction.

Cleaning and Drying of Seeds

After extracting and washing, the seeds are cleaned and dried. Drying is done by spreading the seeds in the partial sun light for few hours for one to two days up to moisture content of 8 per cent or below.

Chilli

Chilli (*Capsicum annuum*) grows best in the dry season with temperatures in the range of 21–33°C. The night temperature is especially critical; generally, plants will not set fruits if night temperatures remain above 30°C.

Isolation

Chilli is considered as self-pollinated crop, but significant cross-pollination does occur if plants are placed together. A minimum distance of about 400 m between two varieties is recommended.

Selection of the Plants

The earliest maturing and more attractive plants should be marked and inspected during growth. Select healthy, attractive fruits for seed saving. Seeds from off-type plants or fruits should not be saved.

Roguing

Plants should be rogued based on the plant and fruit characters as a whole rather than the individual character. Off-types should be removed as soon as they are observed. When the fruits begin to show their final colour of red or yellow, occasional plants with off-colour fruits have to be removed. In addition to off-types, diseased plants are also to be removed.

Harvesting and Threshing

Red-ripe fruit is picked, and macerated mechanically to separate the seeds. Early harvest of immature fruits will affect germination. Seeds are cleaned to free pulp and skins and dried in the partial sun to below 8 per cent moisture content before storage.

Spinach (*Palak*)

Spinach (*Spinacia oleracea*) belonging the family Chenopodeaceae. Plants are grown as an annual for edible leaves and as a biennial for seeds. The crop prefers temperate weather but is widely adaptable. Seed production can be done in cooler regions of the tropics and subtropics.

Method of Seed Production

Two methods are in general used:

1. **Seed to seed**: Sowing of seeds in late summer. Mulch in late fall to ensure winter survival. The following spring, select the finest young plants and transplant using 45-cm spacing. The optimum transplant diameter is 2.5 cm. The tops may be trimmed, but not the root.
2. **Root-to-seed**: Harvest first-year roots in fall. Select desirable roots and trim tops 2–5 cm above root. Store at 4°C with 75-80 per cent relative humidity. Re-plant in early spring at 45-cm spacing with tops just showing above soil. In either method, only transplant the most desirable plants. Stalks may become tall (over 1 m) and are susceptible to lodging; support with staking if needed.

Isolation

The flowers are perfect and borne in groups of 2–3 in axes of leaves. Flowers produce pollen that is carried long distances by wind; thus, it is highly cross-pollinated. Isolate different varieties 500–1000 m apart.

Selection

Rogue out off-types taking into consideration the shape and color of leaves and roots. Save seed from at least six plants to prevent inbreeding.

Harvesting, Threshing and Cleaning

Seeds are ready for harvest when they are fully matured. Seed does not ripen uniformly on the plant and sheds easily when mature. Cut stalks when most flowering clusters have turned brown and stalks have turned yellow and dried, the first seeds are shed at this time. The seeds are cleaned and separated from plant parts by winnowing and are dried well in the gentle sun light and stored.

Pea (*Pisum sativum*)

Pea (*Pisum sativum*) is a short lived, herbaceous annual and climber. The cultivars may be dwarf, semi-dwarf or tall. The flowers are solitary, axillary or up to 3 flowers per raceme, bracts very small, calyx oblique, corolla white, pink or purple; stamens diadelphous, filaments broad, anthers uniform, style falcate, flattened, stigma minute terminal. Pods swollen or compressed, straight or curved on short stalk. Seeds are angular, smooth or wrinkled. Pea is a self-pollinated crop.

Roguing and Isolation

The off-type and diseased plants affected by pea mosaic, foot-rot and blight should be rogued out from the seed field. Pea flowers are almost self - pollinated. Hence, the isolation distance for peas is relatively short and aims mainly to avoid mechanical mixtures. However, the isolation distance may be at least 20 meter from one variety to another.

Harvesting, Threshing and Drying

Maturity of seed crops take about 130-140 days after sowing. To test the maturity a common practice is to squeeze the seed between fingers. If the cotyledons break away from each other and free moisture is not visible, the crop may be considered mature enough for harvest. Plant along with the pods are harvested from the field and dried in the threshing floor under sunshine. Threshing is done by beating with stick when sufficiently dry. Care should be taken during threshing so that the seed coats are not injured. Threshed seeds are cleaned by winnowing, dried to reduce seed moisture content to 12 per cent for temporary storage. For longer storage pea seed should be stored in sealed containers at 10 per cent moisture content and in air cooled rooms.

Okra

Okra (*Abelmoschus esculentus*) can be grown on a wide range of soils with good drainage, but sandy loam soils are preferred. Temperatures between 26–30°C promote rapid and healthy seedling development. Seeds will not germinate below soil temperatures of 18°C. Seeds should be soaked in water for 24 hours before sowing. Plants grow well in raised beds (20–30 cm high).

Isolation

A minimum isolation of 500 m is desirable.

Selection

Plants for seed multiplication can be selected before flowering, taking into consideration the vigor and habit of the plants. Once flowering begins, plants with off-type flowers should be removed. After the first pods are developed, remove plants with off-type pods. Plants with viral symptoms should be removed as soon as symptoms appear.

Harvesting, Threshing, Cleaning and Drying

Pods of okra are angular in shape and have the tendency to open along structures when dry. This will cause shattering loss and the seeds may be damaged due to entry of rain water inside if there is rain. Pods, therefore, must be harvested as soon as they have become mature and before shattering. Pods are best harvested by hand individually and there after dried. Threshing is also done by hand after the pods are sufficiently dry. Light seed are then removed by cleaning and winnowing. Collected seeds are sun dried to moisture level not exceeding 12 per cent for open storage and up to 9 per cent for sealed container storage.

Onion

Onion (*Allium cepa*) is one of the most widely cultivated vegetables in the world. It is a biennial crop and takes two seasons to produce seeds. In the first year bulbs are formed and in the second year stalks develop and seed are produced. It requires cool weather during its early development and during the early growth of the seed stalk varieties bolt readily between 10 to 15°C. A moderately high temperature and a dry atmosphere favour the bulb maturity as well as seed production. Temperatures around 18–23°C favor vegetative growth while temperatures around 12-13°C favor seed stalk formation. Onion flowering is sensitive to day length for most varieties grown in the tropics; short days are conducive to seed production. One bulb can produce 20 or more stalks and may be in bloom for over 30 days. Onion produces perfect flowers, most of which are cross-pollinated.

Two methods of producing seed are used: Most commonly used method of seed production is the bulb-to-seed method. Another method is seed-to-seed.

Bulb-to-Seed Method

Production of mother bulbs

Nursery raising: For raising a crop for bulb production onion seeds are sown in nursery beds to raise seedlings for transplanting in the field. Seeds sowing on well prepared beds in lines at a spacing of 5-6 cm and are covered with soil. Before sowing seeds are treated with Thiram 2-3 g per kg of seeds. Thiram is also applied in seed bed soil for drenching to protect the seedlings against damping off. Seed sowing is done in October-November. Since the medium sized onion bulbs have been found to be satisfactory, many seed growers use higher seed rates, about 8-10 kg/ha then commonly used for the production of market crop (2.5-4.5 kg/ha).

Seed-to-seed

Plant seeds in summer. Immature onions are more winter hardy than larger, full-grown bulbs. Mulch in late fall to ensure winter survival. Thin to 30-cm spacing in the spring.

Roguing

In order to detect and eliminate different plant types, roguing should be started before the bulbs are harvested. It is easier to remove late maturing bulbs at this stage. After the bulbs are harvested they may carefully be rogued for colour and off-types.

Isolation

The minimum isolation distance recommended between different varieties is 1000 m. Some authorities stipulate shorter distances than this for cultivars with the same bulb colour.

Selection of the Plants

In using the bulb-to-seed method, replant only disease-free bulbs that are true-to-type. Remove doubles or long and thick-necked bulbs. For bulbs preparing to flower, remove any types of undesirable shape or color; do this before flowering begins. Save seeds from several plants to maintain crop vigor.

Harvesting and Threshing

The seed is harvested when the fruit opens and exposes the black seed. According to Hawthorn and Pollard (1954), a field is considered ready to harvest when about 10 per cent of the heads have black seeds exposed. At this stage practically all the seed is well matured to give a good germination. Two to three pickings may be necessary to harvest the heads. The seed heads with a small portion of the stalk attached are cut with sharp knife. When cutting the umbel are supported in the palm of the hand and held between fingers to avoid seed loss. Seed heads after harvest are thoroughly dried on canvas. Heads can be threshed when seeds separate easily from them. Much of seeds fall from capsules during drying. The remaining seed is removed by flailing. Under the humid conditions, seeds may be dried in sheds with air circulation. Frequent stirring may be needed when the seed is dried in shed. Since natural seed drying often requires 2 to 3 weeks, some growers prefer to dry seeds quickly using artificial dryers or dehydrators. Before storage, the seed must be dried to six to eight percent moisture.

Carrot

Carrot (*Daucus carota*) seed production requires two years. Roots form during the first year and then require a cold period to stimulate flowering and seed production. Generally, two methods of seed production are used:

Seed-to-seed

Pencil-sized or larger roots are left in the ground over winter. In late fall, plants are thinned to 5 cm apart, tops are cut back to 5 cm high and mulched. Once temperatures rise in spring, the mulch is removed and leaves will re-grow. After several weeks, a seed stalk will appear. Superior plants are thinned to stand up to 75 cm apart.

Root-to-seed

Harvest eating-sized roots for replanting in early spring. Cut back tops to 5 cm and store at 4°C in a humid location or layered in sawdust or sand. Replant of the roots at 75 cm apart with soil just covering shoulders. This method is most reliable and allows for inspection of roots for seed production.

Isolation

Carrot plant produce perfect flower and honeybee is mainly responsible for pollination. Seed fields must be isolated from other variety of carrot, and the same variety not conforming to varietal purity requirements by at least 1000 m for foundation seed and 800 m for certified seed production.

Selection of the Plants

Rogue off-types taking into consideration the root color and shape, plant habit, and plant vigor. Plants that bolt and go to seed early should be removed. Save seed from many carrots to maintain crop vigor.

Harvesting

The seed turns brown 6 weeks after pollination. Before the seed shatters, cut and place umbels into paper bags to dry completely. Late-season rains will reduce seed quality. For small amounts, handpick each umbel as it dries brown. Large amounts of seed can be harvested by cutting the entire stalk as umbels begin to dry.

Processing

Allow seed to mature in a cool, dry location for an additional 2–3 weeks. Seeds can be removed by hand-beating or rubbing umbels between hands. Winnow to clean. Remove spines from dry seed by rubbing.

Radish

Radish (*Raphanus sativus*) is a member of the cabbage family and grown worldwide. The harvested roots, prized for their pungent taste, come in a wide variety of colors, shapes and sizes. There are two major types. First, there are the biennials of temperate origin that require a cold period for flowering. These include the Japanese, American and European radishes. Second, there are radishes of tropical origin that do not require a cold period for flowering.

Method of Seed Production

Both seed-to-seed and root-to-seed methods are employed for radish seed production.

Seed-to-Seed

Seed-to-seed method is preferred for raising certified seed. In this method the roots are uprooted and planted again before the onset of reproduction phase. In Bangladesh, radish seed is largely produced by root-to-seed method.

Root-to-Seed

In root-to-seed method, fully matured roots (before pith development) are harvested, true-to-type roots are selected and after giving proper root and shoot cuts they are transplanted in a well prepared field.

Isolation

Radish is a cross-pollinated crop and honeybee is mainly responsible for pollination. Seed fields must be isolated from other variety of radish, and the same variety not conforming to varietal purity requirements by at least 1600 m for foundation seed and 1000 m for certified seed production.

Selection of the Plants

In the root-to-seed method, foliar characteristics, root shape, size, skin and flesh colors, pungency, pithiness and bolting behavior are considered. Hairy or forked roots and early bolting plants are removed. Selection is more difficult in the seed-to-seed method because one cannot see the entire root. Nevertheless, growers using this method should aggressively rogue out undesirable plants taking into consideration the same factors as with the root-to-seed method.

Harvesting, Threshing and Storage

The seed plants are allowed to mature fully before harvesting, since there is no natural dehiscence, and there is often considerable difficulty in threshing the seed from the pod. Plants are cut when most of the pods are brown. The crop is cut by sickle and brought to the threshing floor for thoroughly drying. The drier the pods, the more easily will they break open during threshing process. Threshing can be done by beating with sticks. The seed after sifting should be dried to 6-8 percent moisture content before storage.

Cucurbits

The family Cucurbitaceae, commonly referred to as 'cucurbits', includes cucumbers, melons, squashes, pumpkins, and gourds. They are all warm season crops and very susceptible to frost. Many cucurbits are susceptible to foliar diseases that attack plants during periods of high humidity and rainfall. Therefore, regions having high temperatures and low humidity are ideal for the production of cucurbit seeds.

Sex Expression

Cucurbits exhibit a wide range of sex form *viz.*

(i) Monoecious
(ii) Dioecious
(iii) Hermaphrodite
(iv) Gynomonoecious
(v) Andromonoecious
(vi) Trimonoecious

In sex forms, although a species character, a wide range of exceptions have been reported in cucurbits. Sex of *Lagenaria vulgaris* has been reported to be typically monoecious (Whitaker, 1931). Kalia and Dhillon (1964) however, observed trimonoecious forms in *Lagenaria siceraria* grown during the extreme cool season. Apart from it sex ratio of each species, varied from the characteristic type (Whitaker, 1931).

Pollination and Fruit Set

Anthesis, pollen dehiscence and fruit set in cucurbits are influenced by environmental factors. Usually fruit set takes place early in the morning between 6:00 A. M. to 8:00 A. M. in crops like cucumber, pumpkin and watermelon etc. Optimum temperature during this period would range between 12.8°C to 18.3°C (Bose *et al.*, 1986). There are other cucurbits which flower later in the day and fruits set at higher temperature of mid day as in bottle gourd etc.

Monoecious condition in cucurbits imposes a situation conducive to cross-pollination; however, a limited percentage (20-40 per cent) of natural self-pollination takes place within the same plant. The andromonoecious condition favours a higher degree of natural self-pollination than in the monoecious condition.

Isolation

Most cucurbit plants produce separate male and female flowers on the same plant. Female flowers can be identified by locating the ovary (a small looking cucumber, melon, gourd, etc.) at the base of the flower (Figure 21.1). The flowers are insect-pollinated, and easily cross within species. However, seed savers can grow more than one variety at a time in a single location by using hand pollinating techniques:

Roguing

Roguing is done at 4 stages as follows:

Early Vegetative Stage

The plants, whose vegetative characters (*e.g.* bush or trailing type), foliage and vigour and resistance to specific pathogens are not in accordance with the cultivar description, should be removed.

Figure 21.1: Female Flower of Squash.

Figure 21.2: Male Flower of Squash.

Before First Flower Open

Plants having under developed fruit or female flower buds, whose characters are not true to type, should be removed.

First Fruit Setting

Developing fruits of such a plant which are not typical of the cultivar should be removed along with the whole plant itself.

Hand Pollination

Bag female and male flower buds on the same plant or nearby plant of the same variety. Then select male flowers when they bloom, turn over their petals to expose their anthers, and gently roll the anthers over the stigma of the just bloomed female flowers; you can see a layer of pollen has been transferred on the stigma. After pollination, bag the female flower again to exclude insects. Mark the pollinated female flower by wrapping a string to the pedicel.

Selection of the Plants

Select early flowering type and vigorous plants. Hand-pollinate the female flowers located 10–20 nodes from the base of the plant. Remove any deformed fruits.

Harvesting of Fruit for Seed Extraction

For seed extraction fruits should be harvested at full maturity. The fruits should be left to fully ripen and turn color. The fruits of luffa and bottle gourd should be left on the plant until they dry. For cucumbers, fruits will turn brownish color. Bitter gourd fruits will turn orange. Some wax gourds will be covered with a pale-white powdery wax on the surface of the fruit. Several factors are taken into consideration to judge the maturity in cucurbits which are: (i) When fruit colour changes to yellow or yellow orange or straw colour, (ii) when the peduncle becomes straw coloured and (iii) when the vines start drying. Any deformed fruit should be removed earlier.

Seed Extraction

Before seed extraction fruits should be stored in room temperature for 4 to 7 weeks spreading in one single layer with a space between the fruits preferably in a cooler dry place. Afterward, the fruits are cut into half and scoop out the seed by hand. Some placenta may remain with the seeds which are to be separated by rolling and raking simultaneously. Then the seeds are to be washed with water in troughs.

Processing

For 'wet seeds' such as cucumber, wax gourd, bitter gourd and melons, cut the fruit lengthwise and scrape seeds out with a spoon. Allow seeds and the jelly like surrounding liquid to sit in a container at room temperature for 1–2 days. Fungus may start to form on top. Stir daily. The jelly will dissolve and good seeds will sink to bottom while remaining debris and immature seeds can be rinsed away. Spread seeds on a paper towel or screen until dry. For 'dry seeds' such as luffa and bottle gourd, keep the seeds in the fruit until they naturally separate from the flesh. This can be identified when you shake the fruit; the sound of seeds moving inside is heard. Cut off the bottom of the fruit and shake the seeds out, winnow to clean the remaining chaff, then place the seed on a screen for further drying before storage.

Seed Drying

The washed seeds should be dried quickly. For this, trays with screen wire or burlap bottoms may be used. The seeds are spread on trays and placed in the shade and gradually to sun to dry and continued up to a moisture level of 7 percent. Frequent turning of seeds will ensure uniform drying. Seeds may be dried more rapidly in a drier or dehydrator employing artificial heat and forced air circulation for large quantity. Seed should be dried carefully at a temperature not exceeding 38-42°C.

Pumpkin

Pumpkin (*Cucurbita maxima)* is one of popular vegetables among the India. It is native to Central America. Pumpkin is a nutritious vegetable vine plant which date back many centuries. It can be prepared various ways. There are about 26 species of cucurbita.

Isolation

To maintain genetic purity of the seed, the recommended isolation distance between seed crops of different cultivars of the same species is 800 m for foundation seed and 400 m for certified seed. Crops for basic seed production should be isolated by at least 1500 m (George, 1985). If adequate isolation distance could not be provided, hand pollination must be followed.

Roguing

Roguing is done at 4 stages in pumpkin as follows:

Early Vegetative Stage

The plants, whose vegetative characters (*e.g.* bush or trailing type), foliage and vigour and resistance to specific pathogens are not in accordance with the cultivar description, should be removed.

Before First Flower Open

Plants having under developed fruit or female flower buds, whose characters are not true to type, should be removed.

First Fruit Setting

Developing fruits of such a plant which are not typical of the cultivar should be removed along with the whole plant itself.

Harvesting of Fruit for Seed Extraction

For seed extraction fruits should be harvested at full maturity. Several factors are taken into consideration to judge the maturity in pumpkin which are: (i) When fruit colour changes to yellow or yellow orange or straw colour, (ii) when the peduncle becomes straw coloured and (iii) when the vines start drying. Any deformed fruit should be removed earlier.

Seed Extraction and Washing

Before seed extraction fruits should be stored in room temperature for 4 to 7 weeks spreading in one single layer with a space between the fruits preferably in a cooler dry place. Afterwards, the fruits are cut into half and scoop out the seed by hand. Some placenta may remain with the seeds which are to be separated by rolling and raking simultaneously. Then the seeds are to be washed with water in troughs.

Seed Drying

The washed seeds should be dried quickly. For this, trays with screen wire or burlap bottoms may be used. The seeds are spread on trays and placed in the shade and gradually to sun to dry and continued up to a moisture level of 7 per cent. Frequent turning of seeds will ensure uniform drying. Seeds may be dried more rapidly in a drier or dehydrator employing artificial heat and forced air circulation for large quantity. Seed should be dried carefully at a temperature not exceeding 38-42°C.

Seed Storage

For safe storage, moisture content of the seeds should be 7 per cent (Harrington and Douglas, 1970). Moisture determinations should be made on properly drawn samples of seed at the temperatures prevailing in the seed storage facility. The well dried seeds are placed in containers and stored in a cool, well-ventilated room, preferably provided with some means of dehumidification, and with protection from rats and other pests.

Cucumber

Cucumber (*Cucumis sativus*) is a widely cultivated plant in the gourd family Cucurbitaceae. It is a creeping vine that bears cylindrical fruit that are edible when ripe. There are three main varieties of cucumber: slicing and pickling.

Isolation

As described in pumpkin

Roguing

Roguing is done at 4 stages in cucumber as follows:

Before 1st Flowers Open

At this stage roguing should be done considering growth habit and vigour of the cultivar.

Early Flowering

Roguing is done on the basis of observable characters of immature fruit, especially colour of spines and whether any specific seed-borne diseases are present.

Fruit Setting

Off-types are rogued out considering the following factors such as (a) satisfactory level of productivity, (b) fruit characters, including size, shape and colour.

Ripe Fruit

Off-types are rogued out considering the colour of ripe fruits in accordance with cultivar description, *e.g.* fruits either green, yellow, white and orange.

Fruit Harvest for Seeds

Any malformed or deformed fruits should be removed earlier and only healthy fruits are selected for seeds. The fruits are allowed to ripen fully. The factors are taken into consideration to judge full maturity in cucumber as follows : (i) Yellow or brown or brownish-yellow skin colour of fruit, (ii) Carpel separation in transverse section of fruit, (iii) fruit stalk adjacent to the fruit withers, (iv) mature seeds separate easily from the interior flesh. After full maturity fruits are harvested and kept 5-7 days for post-harvest maturity spreading in one single layer with a space between the fruits in a shade dry place under ordinary condition.

Seed Drying and Seed Storage

As described in pumpkin

Bottle Gourd

Isolation

As described in pumpkin.

Roguing

As described in pumpkin.

Fruit Harvest for Seed

Any deformed or malformed fruits should be removed earlier. The rest of the fruits are allowed in the plant to ripen fully to dry. When fruits become dry and seeds rattle inside the shell, the fruits are harvested for seed extraction.

Seed Extraction

Seed extraction done by breaking the dry and mature fruits.

Seed Drying and Cleaning

Seeds are washed in clean running water and dried up to 7 per cent moisture level.

Seed Storage

As described for pumpkin.

Watermelon

Watermelon (*Citrullus lanatus* Thunb.), family Cucurbitaceae is a vine-like plant originally from southern Africa. Its fruit, which is also called watermelon, is a special kind referred to by botanists as a pepo, a berry which has a thick rind (exocarp) and fleshy center (mesocarp and endocarp). Pepos are derived from an inferior ovary, and are characteristic of the Cucurbitaceae.

Pollination

Watermelon is monoecious in nature and its fertilization is entomophillous. Sometimes, problem of fruit setting occurs in some part of our country due to the non-availability of insects in early morning during winter season. In this case, hand pollination showed some good result in fruit setting. Hand pollination is accomplished by transferring pollen of a new staminate flower to the stigma of a pistillate flower which blossoms on the same day.

Roguing

Early Vegetative Stage

The plants, whose vine growth, leaf shape and colour and resistance to specific pathogens are not in accordance with the cultivar description, should be removed.

At Early Flowering

Plants, having underdeveloped fruit on female flower buds whose characters are not true to type, should be removed earlier to prevent the out crossing of surrounding plants by the off-type pollen.

Fruit Developing

Developing fruits of such plants which are not typical of the cultivar, should be removed earlier along with the whole plant.

Marketable Fruit

Fruits, whose skin, colour, size, shape and quality are not in accordance with cultivar description, should be removed.

Harvesting

For seed extraction fruits should be harvested at full maturity.

Seed Extraction

Seed extraction done by breaking the dry and mature fruits.

Seed Drying and Cleaning

Seeds are washed in clean running water and dried up to 7 per cent moisture level.

Seed storage

As described in pumpkin.

Cruciferous

Cole crops requires cool climate especially during a definite stage of their growth for thriving seed production. Throughout this period, these crops must have the vernalization requirement a pre-condition essential for breaking the dormancy of the plant thus stimulating the conversion of the vegetative phase into the reproductive phase.

After leafy development ceases, as for illustration the completed growth of the head of the cabbage, the flowering stem elongates. It is characterized by numerous branches (mostly from a main stem), small leaves, and numerous bright yellow or occasionally white flowers. The flowers of all cruciferous have four petals half to one inch long, that appear to form a cross, hence the name cruciferous (cross bearing). All Cole crops are hardy and thrive best under condition of cool climate. They are grown in the plain during winter season and can be cultivated throughout the year in the hill regions of the country.

Table 21.2: Name of the Cole Crops

Crops	*Botanical Name*	*2n*	*Plant Parts Used*
Cabbage	*Brassica oleracea* (L.) var. capitata	9	Heads
Cauliflower	*Brassica oleracea* (L.) var. botrytis	9	Curd
Knol-Khol	*Brassica oleracea* (L.) var. gongylodes	9	Swollen stem
Brussels sprouts	*Brassica oleracea* (L.) var. gemmifera	9	Sprouts (Swollen axillary buds)
Broccoli	*Brassica oleracea* (L.) var. italica	9	Flower heads
Kale	*Brassica oleracea* (L.) var. acephala	9	Top leaves

Constraints for Seed Production

- ☆ Problems of satisfactory isolation due to cross-pollination by insects.
- ☆ Crops have to be carried over in to the second season
- ☆ Plant attains morphological shape and size during the additional growing period, which is not known to majority of the seed growers.

Seed Production Areas in India

- ☆ Srinagar valley (J and K)
- ☆ Upper Kullu valley (Himachal Pradesh)
- ☆ Lahaus valley (Himachal Pradesh)
- ☆ Kalpa valley, Kinnaur (Himachal Pradesh)
- ☆ Saproon valley, Solan (Himachal Pradesh)
- ☆ Kumaon hills (Uttar Pradesh)
- ☆ Kalimpong – Darjeeling hills (West Bengal)
- ☆ Nilgiris (South India)

Table 21.3: Per cent of Cross Pollination in Cole Crops

Crop	Cross Percentage	Crop	Cross Percentage
Cabbage	73 per cent	Cauliflower	70 per cent
Knol-Khol	91 per cent	Broccoli	95 per cent
Brussels sprouts	72 per cent	Kale	83 per cent

Table 21.4: Flowering Season and Vernalization Temperature

Crop	Flowering Season	Vernalization
Cabbage	March - May	4.4-10ºC
Cauliflower	February- April	10-16ºC
Knol-Khol	March - May	4-10ºC
Broccoli	February- April	10-15ºC
Brussels sprouts	April-May	2.2-10ºC
Kale	March- April	4-8ºC

Vernalization

Vernalization of *Brassica oleracea* is performed on mature vegetative plants, which are uprooted in autumn from the fields, potted and over wintered in a greenhouse at temperatures between 5-10°C. The cabbages are decapitated (±3 cm of head remains). Other brassicas are defoliated, leaving ±6 leaves, to prevent rotting. Early cabbages are decapitated in the field, and the new growing shoots are cut and rooted. The cuttings are placed in a cooling compartment for one day, to desiccate the cutting surface. No leaves are removed, to avoid the creation of fresh tissue wounds. The cuttings are planted in trays with soil and covered with cheesecloth. After roots are formed, the cuttings are potted. The plants from these rooted cuttings over winter.

Cabbage

Cabbage (*Brassica oleracea* (L.) var. capitata) is one of the important vegetables grown on a large scale in India, which is formed by the development of densely overlapped leaves around the growing point. Cauliflower is mostly grown in North India whereas cabbage is popular in the south and South Eastern parts on India. Cabbage is commonly used fresh as salad, Cole slaw, boiled vegetable cooked in curries and processed.

Methods of Seed Production

Being a biennial, the cabbage requires two seasons to produce seed. In the first season the heads are produce, and in the following season seed production follows. The seed crop can be left in situ or transplanted during autumn. In situ method is usually followed for certified seed production. Three methods have been devised to produce seed of cabbage.

Stump Method

In this method, when the crop in the first season is fully mature, the heads are examined for trueness to type. Then heads are cut just below the base by means of a sharp knife, keeping the stem with outer whorl of leaves intact. The beheaded portion of the plant is called "stump". The heads are marketed and the stumps either are left in situ, or replanted in the second season *i.e.*, during autumn. The following spring, after the dormancy is broken, the buds sprout from the axils of all the leaves and leaf scars.

Stump with Central Core Intact Method

In this method, when the crop is fully mature in the first season. The heads are examined for trueness to type. Then the heads are chopped on all sides with downward perpendicular cuts in such a way that the central core is not damaged. This is an improvement over stump method in that the shoots arising from the main stem are not decumbent. During the last week of February and until mid March when the heads start bursting, two vertical cross-cuts are given to the head. The operation is completed by going around the field twice or thrice during this period.

Head Intact Method

The head is kept intact and only a cross-cut is given to facilitate the emergence of a stalk.

Roguing

The loose-leaved poorly heading plants, and those having a long stem and heavy frame, must be rogued out at this state. It is highly undesirable to keep such poor plants in the seed plots.

The following two roguing stages:

1. The first rouging is done at the time of handling the mature heads.
2. The Second roguing is done before the heads start bursting.

Harvesting and Threshing

To avoid shattering of seeds, the whole crop is harvested in two or three lots with sickles. Generally, the early plants are harvested first and when the pod colour in about 60-70 per cent of the rest of the crop changes to yellowish-brown, it is harvested completely and piled up for curing. After 4-5 days, it is turned upside down and allowed to cure for another 4-5 days, in the same way. It is then threshed with sticks and sifted with hand sifters. After thoroughly drying the seeds they are cleaned and stored.

Seed Storage

The serious problem faced by stored cole crop seeds is a rapid deterioration which occurs at an increasing rate in uncontrolled storage environments. Seed deterioration cannot be fully arrested, but may be slowed down by storing the seeds in suitable containers such as laminated bags and polythene bags.

Cauliflower

India is the biggest producer of cauliflower (*Brassica oleracea* (L.) var. botrytis in the world. With the development of tropical types in cauliflower in addition to the temperate type, it has now become possible to grow this vegetable almost throughout the year particularly in the northern and central part of India. Cauliflower is thought to have been domesticated in the Mediterranean region since the greatest range of variability in type of *Brassica oleracea*.

Climate

It requires cool, moist climate for seed production. The optimum monthly average temperature is 15° to 20° C. the early varieties, however, require high temperature and longer day lengths. It is less tolerant to extreme high or low temperatures, or strong winds. It is also susceptible to cold injury after the curds have appeared. Extreme rains and snowfall, after curd formation, causes rotting in curds. Periods of low temperature are not essential, but cool conditions are required. Therefore, these conditions must be given due consideration in selecting suitable areas for seed production. In India, the seed production of early and mid-season varieties can be done in the plains. However, the seed of late varieties can only be produced in temperate regions of the country. Lately, Himachal Pradesh has emerged as the major producer of quality cauliflower seed of late varieties.

Isolation

Cauliflower is mainly cross-pollinated. Pollination is mainly done by honey bees. The seed fields must be separated from fields of other varieties, at least by 1600 meters for foundation seed class, and by 1000 meters for certified seed class.

Method of Seed Production

- *In situ* method (seed to seed method)
- Transplanting method (head to seed method)

For seed production, seed to seed method is recommended since the head to seed method in India has not been very successful. In seed to seed method (in situ method) the crop is allowed to over-winter and produce seed in the original position, where they are first planted in the seedlings stage. Main season and late varieties (seed production in hills)

The seed production is dependent upon curd formation and its bolting. Bolting further depends on the curd type *i.e.*, degree of compactness of the curd. The loose curd bolts easily while compact curds takes more time to bolt. To facilitate bolting, different curd- cutting methods like scooping, half curd cutting and curd pruning are recommended. These practices have impact on branching, seed yield and seed quality.

- **Scooping** (Single Scoop): Approximately half of the curd was removed from the center with the help of a sharp knife.
- **Curd pruning:** The outermost curdles were pruned 5 cm, away from the center.

☆ **Half- curd removal**: The curd was cut vertically into two parts from the middle and one of them was removed leaving half portion of the curd intact for seed production.

Roguing

Selection of curds is done when the curds are well developed. Off-type plants, and those forming poor curds, should be removed at this stage. Subsequent roguing for off-types, and diseased plants affected by blackleg, black rot and leaf spot should be done from time to time as required.

Harvesting and Threshing

Harvesting is done when pods brown. Too ripe pods dehisce. Harvesting is done, when about 60 to 70 per cent of the pods turns brown and the rest of the crop changes to yellowish-brown. After harvesting it is piled up for curing. After four to five days it is turned upside down and allowed to cure for another four to five days in the same way. It is then threshed with sticks and sifted with hand sifters.

References

Agrawal, R. L. (1980). Seed Technology. Oxford and IBH Publishing Co. PVT. Ltd.

Bose, T. K., Som, M. G. and Kabir, J. (1986). Vegetable Crops. Nayaprakash, Calcutta.

George, R. A. T. (1985). Vegetable Seed Production. Longman, London and New York.

Harrington, J. F. and Douglas, J. E. (1970). Seed Storage and Packaging: Application for India. National Seeds Corporation Ltd and The Rockefeller Foundation, New Delhi.

Hawthorn, L. R. and Pollard, L. H. (1954). Vegetable and Flower Seed Production, Blackiston Co. NY.

Kalia, H. R. and Dhillon, H. S. (1964). Alternations in the Genetic Pattern of Sex Ratio in *Lagenaria siceraria* Standl. by Asafoetida - A New Sex Regulant. *P. Agr. Univ. J. Res.* 1 (i): 30-49.

Rashid, M. A. and Singh, D. P. (2000). A Manual on Vegetable Seed Production in Bangladesh. Karshaf Printers (Pvt.) Ltd. Dhanmondi, Dhaka.

Sukprakarn, S., Juntakool, S. and Huang, R. (2005). Saving Your Own Vegetable Seeds - A Guide for Farmers. AVRDC publication number 05-647. AVRDC - The World Vegetable Center, Shanhua, Taiwan. 25 pp.

Whitaker, T. W. (1931). Sex Ratio and Sex Expression in Cultivated Cucurbits. Bot. Gaz., 18: 359-66.

2014, Sustainable Rural Development through Agriculture *Pages* **307–316**
Editors: **Dr. Shobhana Gupta and Dr. S.S. Tomar**
Published by: **BIOTECH BOOKS, NEW DELHI**

Chapter 22

Scope of Utilization of Under Utilized Feeds for Enhancing Animal Production

Pankaj Lavania and Shobhana Gupta

Introduction

According to FAO's Economic and Social Development,2004 Report, India's Milk Production has increased from a mere 17 million tones in 1951 to 100 million tones in 2007, with a share of close to 15 per cent of world milk Production. India ranks first in bovine population holding of about 51 percent of Asia and about 19 per cent of world bovine population but the productivity of these animals is very low. In india, inadequate supply of nutrients is always a problem primarily due to shortage of grains, fodders and over dependence on crop residues for the feeding of dairy animals. The gap between availability and demands of feeds is very wide. Currently, total DM requirement for Indian livestock is 623 million tones MT (62 per cent for cattle and 38 per cent for buffaloes),which will be about 670MT (53 per cent for cattle and 47 per cent for buffaloes) in the year 2020. DM availability through all the feed resources in the country was 509.5. MT in the year 2003 which is likely to be 633.1 MT in the year 2020.CP requirement in year 2003 was 59 MT, which is likely to grow to 66 MT by the year 2020. The TDN requirement in corresponding years is estimated at 354 and 387 MT against the availability of 238.5 and 371.0 MT, respectively. The critical assessment indicated that straw and stoves supplied 69 per cent of the available DM while it is envisaged that these will contribute 76 per cent in the year 2020. The CP availability during 2004 was about 32 MT and likely to be 45 MT by the year 2020. This analysis showed that presently there is a shortage of nutrients and it will continue to persist in coming years. Therefore, concerted efforts have to be made to bridge the gap between supply and demand of nutrients for improving the animal productivity as animals maintained on the poor plane of nutrition can not exhibit their production potential.

To bridge the gap between supply and demand of nutrients we have to explore new feed resources in the country, which are not being utilized due to one reason or other. For example abundant availability of monsoon grasses, and some industrial by products such as cottonseed hulls, sugarcane biomass. Monsoon grasses are utilized to some extent by allowing the animals for grazing but large chunk remained unutilized because of lack of management and proper method of their preservation at right stage of maturity. Similarly, industrial by products are not being utilized of their poor palatability and low nutritional value these are not being utilized for animal feeding. Similarly, by products of food processing industry such as fruit and vegetable wastes are available in abundance in many pockets in the country but due to their high moisture content there is problem for their handling and transportation therefore, these are not being utilized optimally to make use of their nutrients. Traditional crop residues such as straws of wheat and paddy and stoves of sorghum and maize are also not being properly utilized as these are not harvested completely due to the mechanized harvesting and the residual crop are being burnt for the early preparation of land for next sowing, which not only reduce the availability of straws but also create environmental pollution.

Monsoon Grasses

After the scorching heat of may and June, monsoon season starts and the empty field fill with natural grasses such as dub grass (*Cynodon dactylon*). During the summer season, animals lose their productivity and also the body weight due to the non availability of adequate nutrients as they usually maintained on straws and stoves. Fresh early vegetative dub grass, also known as Bermuda grass, contains 21.9 per cent CP, which decreased up to 10 per cent on maturity (Ranjhan, 1998). Other grass such as Baru (*Sorghum helepens*), wild cholai,kamakki, Gadbad (*Trianthema protulascastrum)* etc are also available in plenty. Grazing on grasses provide the nutrients and resulted in recovering the loss of body weight and productivity. In addition to this grass various other vegetation also provides the nutrients and wide choice for the animals to select most nutritious part of the available vegetation. In dry areas including the desserts the monsoon grasses are the sole source of nutrients for animals to maintain them. In addition to the grasses, tree leaves also from the source of nutrients during monsoon season, especially for the goats and sheep. In certain areas, where stall feeding is practiced, 'cut and carry' grasses in are the source of nutrients in monsoon season. However, keeping in view the abundance, the grasses, harvested at the right stage of maturity may be preserved in the form hay, to feed the animals in scarcity period. Through this approach appears simple but it involves labour and hence this is the big impediment. In certain pockets tree leaves are preserved, known as *pala*, for the supplementation of green biomass to the animals.

Industrial Byproducts in Animal Ration

(i) Sugarcane Byproducts (SB)

India is one of the major sugarcane producing countries and many a times it is difficult to manage the bumper crop for crushing. Sugarcane tops are being utilized for animal feeding in many areas in India, in spite of its Poor nutritive value and palatability. Chopped sugarcane tops are mixed with wheat straw and water soaked concentrate mixture/oilcake and this mixed ration is offered to animals to overcome the palatability problem. Sugarcane bagasse (SB) is another crop residue, which is abundantly available (about 35.55 MT) from sugar mills. Though the nutritive value of SB is poor than the straws due to its higher lighter lignin content, however, efforts have been made to utilize it as a source of roughage with varying success. Recently, experiments at National Dairy Research Institute, Karnal have shown that 40 per cent of WS in complete feed (roughage: concentrate 60:40) can be replaced with SB without any adverse affect on the animal performance (Table 22.1).

Table 22.1: Average DM Intake, Growth, Feed Efficiency and Cost of Feeding of Crossbred Calves on Two Types of Feed Blocks.

Attributes	*Wheat Straw Based*	*Wheat Straw and Sugarcane Bagasse Based*
Intital b.WT. (Kg)	203.24±8.07	200.54±7.69
Final b. wt. (Kg)	261.50±9.40	262.07±10.99
Total gain (Kg),120 d	58.26±3.07	61.53±5.80
Average b.wt. gain (g/d)	485.46±43.51	512.76±35.81
DMI (feed block, kg/d)	5.82±0.30	5.90±0.31
Total DMI (kg/d)[1]	6.21±0.30	6.29±0.31
Feed efficiency (kg DMI/kg gain)	12.79±0.83	12.27±0.80
Cost of feed block (Rs/kg)	3.85	3.54
Total cost in 120 days (Rs.)	2785.94±120.88	2602.73±109.79
Feed cost/day/animal (Rs.)	23.22±1.01	21.69±0.91
Feed cost/kg gain (Rs.)*	47.82[a]±1.28	42.29[ab]±2.09

1: Additionally 0.39 kg DM was offered to each calf of all the groups daily through green maize fodder.

A,b values in a row bearing different superscripts differ significantly ($p<0.05$).

SB have higher bulk density than WS and its incorporation in complete feed followed by its densification using hydraulic pressure (about 3000 psi for 30 seconds) turn the feed in block form, which increased the bulk density by 4.5 times thereby the transportation, storage and handling cost of feed can be rescued besides reducing the feed cost (Hazhabri,2005).

Addition of banana tops to basal ration of either chopped whole sugarcane or of chopped cane stalk the intake of dry matter increased by approximately 20 per cent without altering the apparent digestibility of the ration. Higher intake was due to increased efficiency of microbial fermentation of dietary dry matter in rumen which in turn was brought about through improved availability of substrates (mainly protein and amino acids) to these microorganisms.

Cottonseed Hulls (CSH)

Cottonseed hulls can be utilized for the feeding of ruminants as a source of roughage as it is having low protein and high fiber contents. Though the availability of CSH in the country is limited due to the lack of cotton decortications facilities, however, it can be utilized for animal feeding. Complete feed system for utilizing the CSH opens a new vista to balance the nutrients in the diet of animals. CSH does not need any processing to reduce its particle size and it can be mixed with other feed ingredients with the help of normal mixers. Moreover, molasses, a cheap and easily available energy source, can be absorbed in higher quantity over CSH as it is having higher molasses absorption capacity than other feed ingredients.in an experiment at National Dairy Research Institute, Karnal, complete feed CSH at 60 per cent level was evaluated. Growth rate of the calves was found better than those fed on wheat straw based complete feed (Table 22.2). Voluntary feed intake and growth rate of CSH based complete feed were significantly higher than straw based feed, however physical form of CSH based feed did not make any significant difference on the feed intake as well as growth rate of calves.

Table 22.2: Performance of Calves Fed on Cottonseed Hulls based Complete Feed

Particular	*Wheat Straw Based Control*	*Cottonseed Hulls based Mash*	*Cottonseed Hulls based Flakes*
B.wt.gain	245.4	585.8	533.03
DMI (kg/day)	3.99	5.20	5.19
DMI (per cent b.wt.)	2.78	3.55	3.55
Feed efficiency per cent	15.61	8.49	9.29
DM digestibility per cent	58.7	54.8	59.1
CP digestibility per cent	69.3	54.2	56.3
NDF digestibility per cent	46.4	43.9	51.3
DCP per cent	8.8	6.9	7.4
TDN per cent	54.3	53.6	58.9

Other Hulls

Availability of groundnut shell, the outer covering of groundnuts, is abundant as India is leading producer of groundnut. Groundnut shells are fibrous and poor in protein and mineral contents. Unlike cottonseed hulls, these are very brittle, hence can be turned in to powder form and can be utilized for animal feeding at lower level in ration, when fiber is not a limiting factor.

Horticultural Residues in Animal Feeding

India is the second largest producer of vegetable in the world and its share is over 13 percent of the produce globally. It is estimated that the annual yield of fruit and vegetables in India is about 64.6 and 86.0 million tones, respectively. Horticultural byproducts as feed can be classified in the following categories

1. Byproducts of fruits processing industry
2. Byproducts of vegetable processing
3. Byproducts of other horticultural crops.

All the categories of horticultural products are being processed to some extent and during the processing varying type of residues are obtained. These residues contain sizable nutritional value (Table 22.3) besides good quality of fiber, however, these contain high moisture content, which make them perishable. Therefore they must be utilized after quickly processing to make best use in animal rations. Keeping in view their low protein and high carbohydrate contents, they must be enriched with protein supplements. Since the availability of these residues is seasonal, therefore, these may be preserved by adopting technologies.

Table 22.3: Chemical Composition of Various Industrial Byproducts (per cent DM basis)

Particular	CP	CF	ASH	EE	NFE	CA	P
Brewer's yeast	49.9	1.5	8.5	1.3	38.8	0.13	1.56
Spent barley	27.8	12.6	4.9	8.0	46.7	0.16	0.65
Cottonseed hulls	3.9	30.0	3.6	8.8	43.7	0.13	0.06
Groundnut shell	4.9	68.4	7.4	0.6	18.7	0.25	0.06
Groundnut haulm	14.7	30.9	11.8	3.6	38.9	2.27	0.39
Rich hulls	3.8	43.9	21.6	1.7	29.0	0.08	0.08
Cassava leaves	29.0	16.7	8.6	6.2	39.5	1.17	0.62
Potato haulm	10.9	27.0	13.5	4.3	44.3	0.05	0.15
Orange pomace	6.9	13.1	7.1	2.8	70.1	–	–
Orange peels	6.8	6.2	3.7	1.9	81.4	1.30	0.12
Cabbage waste	20.0	10.3	27.3	3.5	38.9	5.95	0.46
Tomato waste	27.8	27.6	6.6	22.0	19.0	–	–
Apple pomace	5.5	16.4	–	–	–	0.03	0.12
Grape pomace	13.4	33.3	6.6	7.0	–	–	–
Sunflower head (seed free)	7.2	15.8	11.4	2.9	62.6	–	–
Coconut fiber	1.8	19.3	5.4	2.9	70.6	0.3	0.1
Pea pods	15.4	23.4	4.6	2.2	54.4	–	–
Carrots waste	16.7	8.7	8.1	1.9	64.6	–	–
Okra waste	20.0	34.5	7.6	9.8	28.1	–	–

To give information about the scope of horticultural residues in the animal ration, following examples are cited.

1. Apple Pomace

Apples is a main fruit crop in the country and sizable part of this fruit is processed for manufacturing the juice, Jam and other products. Apple pomade, the residual material from pressing the apples for juice extraction, contains pulp, peels and cores. Its moisture content is about 85 per cent. Apple yields about 25 to 35 per cent of fresh pomace, which is highly palatable for animals and can be fed as fresh or after preservation or processing.its nutrients content in terms of DCP and TDN is 1.62, and 64.84 per cent respectively. Apple pomace can be preserved by sum drying and takes 52 h for complete drying with occasional stirring (Sharma *et al.*, 1989).

Urea treatment of apple pomace under air tight condition for 2 days following its drying increased its nitrogen content, which was found similar to the maize grain. When the urea treated apple pomace replaced the maize grain in the concentrate mixture of growing and lactating cows, the growth and milk production performance remained uninfluenced and cost of feeding reduced significantly. (Table 22.4).

Table 22.4: Performance of Growing and Lactating Cows on Feeding Urea Treated Apple pomace in Place of Maize Grain in Concentrate Mixture

Parameter	*Control*	*Urea treated AP*
Growth		
DM intake (kg)	483.6	4834
Live wt gain (kg)	53.0	51.9
Daily wt gain (g)	541.6	529.8
Cost/kg gain (Rs)	13.8	10.75
Lactation		
DM Intake (kg)	6154	6268
Milk production (kg)	5137	5248
Total fat production (kg)	233.7	232.0
DM intake/kg milk yield (kg)	1.2	1.2
Feed cost/kg milk (Rs.)	1.25	1.04

Sharma *et al.* (1989).

Apple pomace can be preserved following ensiling process. Apple pomace and straw can be mixed in equal quantity and mixed with 4 kg urea. The mixture can be kept under air tight conditions after covering it with polythene sheet for 3 to 6 months (Singhal *et al.*, 1991). This technology not only preserves the apple pomace but also improves nutritive value of straw. Thus, the utilization of apple pomace not only improves the financial viability of apple processing industry, but enhances the feed availability in the country. Similar technology may be applied for the residues obtained from the pineapple, orange and other similar residues having high moisture content and low protein value for the feeding of animals.

Other Residues

Maize bran, residue obtained from the wet milling of maize grain for starch production, is also a moist residue. This can also be fed as such and in case of abundant availability, can be preserved by ammonization along with straws (Singhal and Sharma., 1991, Singhal and Grant,2000).

Spent grain, a residue from brewery industry, having sizable nutritional value can also be preserved by the ammonization process (Singhal and Sharma,1991).

Sea Weeds for Animal Feeding

India is having a 2000 km long coastal line and there is abundant availability of sea weeds, which have sizable nutritional value, however, instead of boon these are considered problem as they block the water ways. Sea weed is a novel phyto -nutrient resource with great potentialities. The kind of sea weed used as feed includes *Ascophyllum nodosum, Alaria esculanta, Fucus vansecan Laminaria and Sargassum.* These algae contain abundant and balanced nutrition which can enhance animal performance (Table 22.5), immune status and productivity of animals.

Table 5. Chemical Composition of Seaweeds (DM basis)

	Asco-phyllum nodosum	*Laminaria digitata*	*Alaria esculenta*	*Palmarira palmate*	*Porphyra sp.*	*Porphyra Yezoensis*	*Ulva sp.*
Type	*Brown*	*Brown*	*Brown*	*Red*	*Red*	*Red*	*Green*
Water(per cent)	70-85	73-90	73.86	79.88	86	nd	78
Ash per cent	15-25	73-90	73-86	15-30	8-16	7.8	13-22
Protein(per cent)	5-10	8-15	9-18	8-25	33-47	43.6	15-25
Fat (per cent)	2-7	1-2	1-2	0.3-0.8	0.7	2.1	0.6-0.7
Tannins	2-10	0.1	0.5-6.0	ND	ND	ND	ND
Potassium	2-3	1.3-3.8	ND	7-9	3.3	2.4	0.7
Sodium	3-4	0.9-2.2	ND	2.0-2.5	ND	0.6	3.3
Magnesium	0.5-0.9	0.5-0.8	ND	0.4-0.5	2.0	ND	ND
Lodine	0.01-0.1	0.3-1.1	0.05	0.01-0.1	0.0005	ND	ND

ND: Not detected.

Sea weed meal of good quality has constituted up to 10 per cent of cattle feeds with good results, and 35 g a day have been fed experimentally to sheep with increased gains as compared to control animals. Sea weeds imbibe many rare minerals from the sea. These rare minerals are transformed to cheated organic matter which can be absorbed by the animal.

Supplementation of livestock ration with sea weeds improve feed utilization and productivity, provide iodine and other trace elements, builds immunity and reduce incidences of mastitis (due to its active biological ingredients) besides improving the growth rate in heifers by boosting cobalt and selenium level and conception rate. In a recent study at National Dairy Research Institute, Karnal it was demonstrated that supplementation of sea weed (*Sargassum wettai*) @10 per cent of total ration on DM basis can supplement adequate minerals and improved the milk production in Sahiwal cows. Sea weeds have great potential for animals feeding, however, their harvesting and processing are the major constraints in their utilization.

Non Conventional Oilcakes for Animal Feeding

Oil cakes of mustard, groundnut, cottonseed, sesmum etc are the conventional protein supplements in animal ration, However, some non-conventional oil cakes are also available in sizable quantity in the country as a result of oil extraction from oilseeds of mainly tree origin such as neem, mahuwa, karanj etc. The availability of these oil cakes and areas where these are available in plenty are presented in Table 22.6. These oilcakes are non-conventional as these are containing one or other incriminating factor, which restricts their utility as feed resource, however, research work carried out to develop the techniques to make them edible in animal ration by removing their incriminating factor completely or partially.

Table 22.6: Availability of Non-conventional Oilseeds in the Country*

Oilseed	*Availability 000 ton*	*Oil per cent*	*States*	*Incriminating Factor*
Mahwa	2176	35	AP, UP, MP, Bihar, Guj Orrisa, TN, Karnataka	Sapoglucoside (Mowrin) Sapogenin
Karanj	111	27	AP, Karnataka, TN, Bihar	Karanjin
Kosum	90	33	Bihar, UP, MP, Orissa	Prussic acid
Neem	418	20	UP, AP, MP, TN, Karnataka, Guj, Raj.	Nimbin and its derivatives
Sal	5504	13	MP, UP, WB, Bihar, Orissa	Tannius
Mesta (Thumba)	153	30	WB, Guj, Bihar	Fiber
Piludi	46	33	Guj, Raj, UP	Not identified
Nahar	6	40	Assam, WB, Orissa, Kar, Kerala, Maha	Not identified
Kokam	2	40	Maha, Karnataka	Not identified
Undi	4	30	Kerala, Kar., M.H., AP, WB	Not identified
Dhupa	10	25	Karnataka, Kerala, TN	Not identified
Mango kernel	15	7	Whole India	
Rubber seed	3360	20	Kerala, TN	Prussic acid
Castor	407	37	Guj, AP, Bihar, Orissa	Ricin, Ricinin

Mudgal and Singhal (1993).

The elaborate techniques for the removal of incriminating factors in non-conventional oilcakes have been described by Mudgal and Singhal (1993) besides presenting the data on the performance of animals fed on treated oilseed cakes based diets. In addition of these oilcakes, information has been provided about other protein rich non-conventional feeds such as oak kernel, tobacco seed cake, Plash seeds, Coffee seeds residue, babool seed, cassia tore seeds, taramica seeds etc.

The non-conventional feed resources are important for livestock feeding and these must be preserved and processed to make their best, wherever it is possible and feasible.

Future Projections

CP availability in cattle and buffalo rations would continue to be deficient in future, however, magnitude of deficiency is like to narrow down in future.the projections revealed that the overall deficiency of CP was 46.5 per cent in the year 2016 and likely to be narrow down to the magnitude of 31.3 per cent by the year 2020. Similarly, the TDN deficiency was about 32.7 per cent of requirement in th year 2016 and likely to be about 4.2 per cent by the year 2020. Keeping in view these deficiencies, we have to mobilize protein resources for the livestock feeding to bridge the gap between requirement and availability for enhancing the livestock productivity. Since importing of the animal feed ingredients have remote possibility, therefore, we have to augment our feed resources and utilize them judiciously to improve the deficiency of nutrients. Sea weeds are having great potential as these are not being utilized currently. Besides this, we have to develop the systems for the preservation and storage of our feed resources and educate the farmers for the benefits of balanced feeding to avoid the feeding excess or deficient nutrients.

References

FAO. 2004. Production Yearbook. Food and Agricultural organization of United Nations, Rome, Italy

Hozhabri, F. (2005) Effect of roughage sources and protected protein in complete feed on fiber digestion kinetics, nutrient utilization and growth performance of crossbred cavils. Ph.D. Thesis submitted to NDRI, Karnal, India.

Mudgal, V.D. and Singhal, K.K. (1993) Agro-industrial byproducts in animals Nutrition (Hindi) Rajasthan Hindi Granth Academy, Jaipur

Ranjhan, S.K.(1998) Nutrient Requirements of Livestock and Poultry. Indian Council of Agricultural Research, New Delhi.

Sharma, D.D., Bakalkar, R.K. and Singhal, K.K. (1989) Apple pomace: a new feed resource for animals, Publ No. 149, NDRI Karnal

Singhal, K.K., Thakur,S.S. and Sharma, D.D. (1991) Nutritive value of dried and stored apple pomace and various methods of its utilization for ruminant feeding. Indian j. Anim. Nutri. 8(3): 213-216

Singhal, K.K. Sharma,D.D. (1991) Effect of ammonization of brewer's Spent grain along with crop residues for preservation and utilization of the product for ruminant feeding. Indian J.Anim. Sci 61(6): 620-623

Singhal, K.K. and Sharma, D.D. (1991) Effect of urea added or ammoniated wet maize bran and paddy straw on nutrient utilization in buffaloes. Indian.J. Anim. Nutri. 8(3): 209-214

Singhal, K.K. and Grant, R. (2000) Effect of ammonization of wheat straw alone or in combination of wet corn gluten feed on fiber digestion kinetics. Indian j.Anim.Sci.70(5): 513-517

2014, Sustainable Rural Development through Agriculture *Pages* ***317–337***
Editors: **Dr. Shobhana Gupta and Dr. S.S. Tomar**
Published by: **BIOTECH BOOKS, NEW DELHI**

Chapter 23

Enhancing Marketing Efficiency in Domestic Trade of Milk and Milk Products

B.S. Chandel, Prem Chand and J.P. Dhaka

'Actually the public may prefer to keep some known inefficiencies rather than to adopt new methods- especially if prospective improvements in efficiency might reduce employment, decrease price competition, or lead to greater concentration of economic power'

Waugh, F.V. (1954)

Introduction

Milk is highly perishable commodity. Self-life of raw milk is few hours depending on the season. Milk produced in a day need to be either consumed or processed or converted into milk products in few hours. Thus, time (short shelf life) is one of the major considerations in marketing of milk. Even after processing the self life for marketing of liquid milk can only be enhanced to 12 hours at ambient temperature while that of milk products a week. Ultra High Temperature (UHT) milk has the highest six months self-life. The short self life has both marketing cost and efficiency implications besides health dimensions. There need to be an expensive cold chain in vertical integration of the supply of milk and milk products. Shortage of power supply adds extra cost to the marketing of perishable commodities like milk. A milk product is required to be recollected from the market after its expiry date which further increases the marketing cost. The point has been further elaborated later in the chapter.

Second major issue in marketing of milk and milk products (MMPs) is the dominance of unorganized sector of marketing and processing. In unorganized marketing system, the milk vendors/suppliers market 69 per cent of the total milk surplus from the adjoining villages of an urban consumption centre carrying 10 litres to 100 litres of milk at a time in unrefrigerated cans. The total duration of his marketing is approximately six hours comprising of three hours collection and three hours selling, keeping in view the temperature. Only 18 per cent of the marketed surplus of milk is marketed by the organized sector like cooperatives, private milk plants, government schemes, etc. Due to small quantity of production and marketed surplus at individual level, mostly organized sector has to incur high procurement cost, which inflates the marketing cost. However, the scale of milk production per household is very low; 63 per cent households producing less than or equal to 2.75 litres of milk per day (Birthal, 2008).

Thirdly, market infrastructure especially for MMPs is very poor and adds to marketing inefficiencies. The market infrastructure lacks on various fronts like all weather roads, interlinked cold chain, specially designed vehicles for transportation and mechanical handling. It results into spoilage and milk losses adding to the marketing costs and margins. Down the process, producer gets smaller share in consumer's rupee and consumer have to pay higher prices in turn.

Early considerations of the efficiency of marketing, in general, focused on efficiency from the standpoint of the producer or marketer. There is qualitative dimension to marketing output from the consumer's perspective leading to the concept of consumer market efficiency. It refers to a concept of market efficiency from the consumer's perspective, *i.e.*, value received from consumption of a product category relative to the cost of consumption (Gaski, 1987). Considering the recognized importance of such issues as consumerism and the social responsibility of business, development of a method for evaluating the net contribution of consumer marketing to human welfare or satisfaction would be useful and significant. In agriculture in general and dairy in particular, delivering product of a desired/fare quality is very important because of its health implications which has social cost in the sense that lot of tax money is to be spent in providing medical treatment besides loss of man hours of working. As the demand is looming large over production, the quality has taken a back seat in marketing of milk and milk products. Not too long the concept of consumer market efficiency in delivering quality products can be ignored.

The chapter elaborates on the issues raised above regarding market efficiency of milk and milk products in the context of the India. The first section is focused on scale of marketing in milk, marketing system and marketing costs. The second section defines and measures marketing efficiency and finally, innovative efforts towards improving the marketing efficiency of milk in the country.

Scale of Milk Marketing

Dairying has been a tradition in rural India rather than a business. It can not be seen in isolation from agriculture. It is the dairying to which the four major roles and functions of livestock assigned by Rangnekar (2006) could be accredited namely (1) output function: related to producing food and non-food products, (2) input function:

related to providing inputs for crop production, transport, etc. (3) risk coverage or asset function: related to raising moneys in times of need, and (4) Socio-cultural functions: related to social status, culture, etc. For a long time, the only objective of the keeping dairy animal had been to meet dietary requirement of the family and at the same time making best use of crop residues. Selling milk was not regarded well. Any marketing of milk started with the urbanization and commercialization came too late. Farmers living near the cities took advantage of their proximity to the cities and began supplying milk to the urban population; this gave rise to the fluid milk-sheds we see today in every city of our country and the on set of modern milk marketing. Semi-commercial dairying started with the establishment of military dairy farms and co-operative milk unions throughout the country towards the end of the nineteenth century. Due to lack of suitable means of transportation and refrigeration, most of the milk produced within a short distance of the place of consumption. Evidence on marketed/marketable surpluses of livestock products is anecdotal. Marketed surplus of milk is estimated 54 percent of the total production (Dairy India, 2007). Though the scale of milk production per household is very low; 63 per cent households producing less than or equal to 2.75 litres of milk per day, but these may be called commercial from the definition when more than fifty per cent of the milk is sold.

In milk, there is every likelihood of having marketed surplus more than the marketable surplus because of dairy playing function of risk cover and meeting the immediate cash requirement. Therefore, there is higher probability of selling the milk

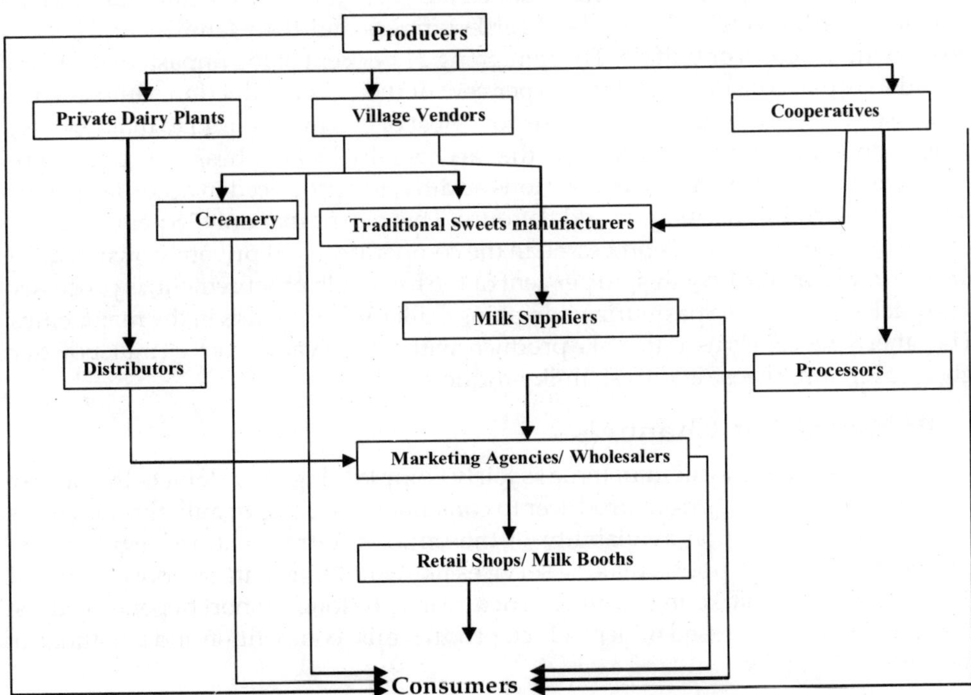

Figure 23.1: Major Milk Marketing Channels in India.

by undernourished households than the houses with surplus milk. Other factors determine marketed surplus of milk are the production, family size and income.

The Indian Dairy Industry made a rapid progress since Independence. A large number of modern milk plants and product factories have since been established. These organized dairies have been successfully engaged in the routine commercial production of pasteurized bottled milk and various Western and Indian dairy products. With modern knowledge of the protection of milk during transportation, it became possible to locate dairies where land was less expensive and crops could be grown more economically.

In India, milk market technology may be considered to have commenced in 1950, with the functioning of the Central Dairy of Aarey Milk Colony, and milk product technology in 1956 with the establishment of AMUL Dairy, Anand. Indian dairy sector is still mainly an unorganized sector as barely 18 per cent of our total milk production undergoes organized handling. due to close proximity of producing and consuming centers of milk and absence of processing units. However, the market structure for milk is constantly changing. Taking up of an ambitious project of 'Operation Flood' in 1970-71 in order to link the producing and consuming centre, strengthened the establishment of dairy co-operative societies and Federations. This resulted into that, cooperative sector accounts for nearly 50 per cent of milk marketed through organized sector.

Of India's total milk production, no less than 65 per cent is consumed unpasteurized. Of this percentage, 44 per cent is consumed in the rural area in which it is produced, meeting the needs of cattle farmers and their families and sold to others with no cows or buffalo. The remaining 21 per cent of the unpasteurized milk is sold to urban consumers. Of the 35 per cent of the milk production, almost half is processed by the unorganized dairy sector. Most of this milk is used to make sweets. This is done in 'halwai', workshops that are usually located beside a shop. Here, sweetened condensed milk, with various additives is produced in accordance with the region's traditions, religion and flavours. This means that only 18 per cent of the Indian milk procurement is processed in the co-operative and privately owned dairy industry. The majority of this, 10 per cent of the total milk procurement, is processed into packaged or loose pasteurized drinking milk for consumers in the major cities. The other 8 per cent is used to make products with added value, such as milk powder, ghee, ice cream, cheese and fresh milk products.

Milk Marketing Channels

Milk marketing system in India is fairly complex. Figure 1 depicts the flow of milk and milk products from producer to consumer. Disposal of milk through these channels depends upon availability of the options, infrastructure, regularity of payments and also rapports making services like extending loan, supply purchased items. Further the marketing channel structure may be long or short depending upon the nature of the processed milk product, product mix, competition and situations of individual producer.

In general about 65 percent of milk is marketed through the highly fragmented un-organized sector, which includes local milk vendors, wholesalers, creameries, retailers, and producers themselves. On the other hand, the organized dairy industry, accounts for about 35 percent of total milk production. Even though co-operatives provide a remunerative price to the producer, the unorganized sector plays a major role in milk marketing because of three factors. The first factor is the pricing policy of the co-operatives: their purchase price is based on the fat content of the milk, whereas the private sec-tor pays a flat rate per liter of milk. The second fac-tor, which motivates the milk producers to sell milk to private vendors, involves the type of milk animals reared by the producer. Crossbred cows yield more milk with a lower fat than do buffalo. The crossbred cow population has increased over years because management practices. The third factor is payment policy. The private sector can pay their producers everyday, whereas the co-operatives pay weekly or fortnightly. Producers sometimes have to fight with the co-operatives to get their payments.

Within the organized sector, the co-operative sector is by far the largest in terms of volumes of milk handled, installed processing capacities, and marketing infrastructure. Milk is processed and marketed by 170 Milk Producers' Cooperative Unions, which federate into 15 State Cooperative Milk Marketing Federations. Over the years, several brand names have been created by cooperatives like Amul (GCMMF), Vijaya (AP), Verka (Punjab), Saras (Rajasthan). Nandini (Karnataka), Milma (Kerala), Gokul (Kolhapur) to name a few.

There are however large regional differences in the milk marketing channels. In areas with high production like Gujrat, Rajasthan low consumer concentration or few alternative market outlets, marketing through cooperative still dominates, with over 50 per cent of marketed milk passing through their factories.

Cost Components in Milk Marketing

An efficient marketing system is one capable of moving goods from producer to customer at the lowest cost consistent with the provision of the services that customers demand. Science of market management has identified number of ways to cut costs both fixed and variable. Per unit costs can be reduced by increasing the volume, mechanical handling, improve technologies in storage and processing, using locational advantage, and follow horizontal and vertical integration.

Marketing costs are incurred when commodities move from the farm to the final market, whether they are moved by farmers, intermediaries, cooperatives, marketing boards, wholesalers, retailers or exporters. Larger the number of marketing agents, higher is the cost of marketing.

Milk marketing costs can be broadly categorized procurement cost, processing costs and distribution costs and, spoilage and miscellaneous co. ts. Cost of milk procurement included costs on collection, transportation, chilling and delivery of milk at the reception dock. The processing cost of dairy products comprises expenditures on electricity, water, steam, refrigeration, maintenance and repairs, stationery and stores, labour, packing materials, detergents, besides quality control expenditure, salaries and administrative expenses, depreciation on buildings, equipments and machinery, interest on investment in buildings, plant equipment and machinery costs. Distribution cost includes expenses on advertisement, sales promotion, rent of booths and parlours, salary of marketing and sales personnel, stationery, telephone, conveyance, sales commission to wholesalers, retailers and commission agents, transportation of milk and milk products to booths, parlours and sales outlets, storage of milk and milk products under refrigerated condition, loading and unloading of milk and milk products in the dairy plant, sales tax, and depreciation, interest and other miscellaneous costs.

Milk Procurement Pattern and Cost

Efficiency of the marketing depends on the way a producer sells milk or a processing plant procures milk from producers. As has been discussed earlier, the larger share of milk is sold through unorganized traditional marketing systems. This partly due to lack of infrastructure development at village level for collection and transportation of milk and partly due to existence of alternate system with attached convenience. The milk procured by organized sector was about 4 per cent in 70s, which could increase to 18 per cent by the turn of century. The dismal share of processed milk is also because of tastes and preferences of the consumer in the country. There is a preference of raw liquid milk and only the milk, which is not consumed liquid, is converted into milk products that too traditional milk products. Processed milk and milk products are considered substandard for direct consumption and preferred occasionally for consumption. Milk procurement by cooperative sector increased very fast (22.5 per cent ACGR) during Operation Flood II (1978-85). Now, it is growing little lower than the increase in production.

Among all, the procurement cost of milk is the highest especially among cooperatives due to widespread milk shed area on account of meager marketed surplus with individual farmer. Table 23.1 shows the quantum of milk collected from each farmer for the three major state cooperative milk federations.

Table 23.1: Quantum of Milk Procured by Agencies of Major Cooperative Federations in the Country

	Gujarat Cooperative Milk Marketing Federation		*Rajasthan Cooperative Dairy Federation*		*Punjab State Cooperative Milk Producers Federation*	
Year	*2008-09*		*2006-07*		*2006-07*	
	Total	*Milk Collection per Unit (Litres)*	*Total*	*Milk Collection per Unit (Litres)*	*Total*	*Milk Collection per Unit (Litres)*
Cooperative Milk producer Union (No.)	12	460083	19	81579	–	–
Village societies (No.)	11132	495	9050	171	5989	130
Producer Members (No.)	22,80,000	2.42	6,15,000	2.52	3,60,000	2.16
Milk procurement per day (LLPD)	55.21		15.50		7.78	

The information on milk collection per unit in the table exhibits that milk collection is less than two and a half liters per producer in all the three federations.

There is not much variation in average amount of milk collection per producer in the three federations. However, the milk collection per village level milk society was as low as 130 liters in Punjab. The average milk collection per society recorded in Gujarat and Rajasthan was 495 and 171 liters, respectively. If it is assumed that a society has area spread from five to ten kilometers and there are two to five milk pick up points, the cost of milk collection per liter comes around one rupee at the society level and forty five minutes of time. The quantity of milk procured during lean season (March to August) reduces 20 to 40 per cent. Houde *et al.* (2007) observed flush lean ratio 1.37, 1.40 and 1.18 during 2000-03 in case of cooperatives in Karnataka. The situation was not much different in case of private milk processing plants.

Milk collection system followed by the plant has bearing on the cost of the milk product. Generally, private plant follows more then one milk procurement systems. The existing milk procurement systems are as following:

1. Direct Collection

Milk is collected directly by the company from farmers through their collection centres. Company provides different incentives to the farmers to remain loyal to the company and sustain milk supply chain. These incentives are generally non-pecuniary for example higher prices, timely payment, loan for purchase of animal, etc.

2. Cooperative System

The farmers are enrolled as member of the milk producer's society which follows cooperative by-laws and have democratic set up. All these members supply milk to the society and get the benefits of services provided by the society. The procurement cost in cooperative system is high per litre of milk procured but less per dairy farm since it serves large number of farmer producing one litre to 10 litres of milk.

3. Venders/Milk Contractor

Milk venders and milk contractors also supply milk to the plants.

4. Contract Farming

It is an agreement between processing and marketing/producing firms to supply milk at predetermined prices. This stipulates a commitment on the part of the farmers to provide a specific commodity in terms of quality and quantity as determined by the purchaser and commitment on the part of company to support the farmer for production through inputs and other technical support.

Contract farming in dairy being practice by the corporate sector in Punjab and Tamil Nadu. Here the plants in corporate sector are directly contracting the dairy farmers to supply them the milk. This practice improves efficiency in two ways- one by pushing out the middlemen and another by ensuring the quality milk production. The added benefits are that farmers are sure of price and market for their produce. Through contracts, the processors get an assured supply of raw material and thus can utilize optimally their installed capacity, infrastructure and manpower. Producers also benefit from assured off-take of produce, reduced price uncertainty, lower marketing and transaction costs, and an easy access to inputs, technology, credit and services. However, producers will join contract farming if the expected net benefits are higher from it compared to alternative marketing options. Several studies have confirmed that producers do benefit from contract farming. In case of milk, net revenue to contract producers was double than that for non-contract producers. It was largely because of reduction in marketing and transaction costs, which were less by 93 percent for contract producers. Lower marketing and transaction costs reduced the total cost of production by one-fifth, which otherwise was not significantly different from that for non-contract producers. However, there is a need to encourage contract farming by framing institutional structure which protects interest of the farmer and processor.

About 80 per cent of the small plants procure milk from cooperative societies. On the other hand, majority of the large plants have their own procurement system collecting milk directly from the farmers through their collection centres. While medium plants follow more than one method of procurement like milk vendor/milk contractors and cooperative societies. Large plants also take the help of milk contractors to supplement their supply.

Low amount of milk procurement by the organized sector could also be explained through the mode of payment followed. The fortnightly and weekly payments were found very common. One third of the plants make fortnightly and weekly payments. Very few, less than 15 per cent of the plants make payments either dairy or on alternate days. In order to draw more milk from the producers to the organized sector, confidence should be built in them that they are getting a fair deal and better returns for their product. The mode of payment and revision of procurement price in a way helps to build up their confidence.

Bulk milk collection and transportation facility at village level, increasing density of chilling centres, better road and transportation will further add to the canalization of milk through organized sector. Use of vans, cycles, lorries, motorcycles, headloads and buses to transport the milk from collection centre to the chilling centre, not only increases the cost of transportation but also increases the handling losses. The mode of transportation from chilling centre to the plant is fairly better through insulated tankers and bulk milk cans in vans/trucks.

Table 23.2 shows procurement cost of milk incurred by different plants at different locations. In the lack of continuous study on procurement cost, data have been collected from post graduate students' thesis pertaining to different years. Though information in table does not fit to any statistical methods to draw conclusions but it certainly realizes the extent of procurement cost in marketing of milk and milk products in India where dairy is thinly spread across among rural households.

What clearly comes out of the figures in the table is that cost of procurement increase cost of milk from about a quarter to hundred per cent. In UP and Haryana, the cost of procurement was about 27 per cent of the farm gate prices. In AP, though the figures dates back to 1974-75 but the procurement cost was about 94 per cent of the farm gate price. If there is any truth in the data, it may be due to low amount of milk procurement at that point of time resulting into higher cost of procurement per unit. Another fact that is highlighted is that the procurement cost was only 7.93 per cent of the farm gate price of milk in case of private plant. Assuming that the plant

Table 23.2: Procurement Cost of Milk (Rs/lt) by different Plants in different States

Costs	*Dairy Plants*					
	AP	*UP*	*UP*	*Haryana*	*TN*	*TN*
	MPF, Vijaywada	*FBDP, Meerut*	*Private Dairy Plant*	*FBDP, Rohtak*	*Cooperative*	*Private*
Years	1974-75	1983-84	1986-87	1986-87	2006-07	2006-07
Farm gate price	0.35 (100.00)	2.00 (100.00)	2.90 (100.00)	2.90 (100.00)	8.60 (100.00)	8.93 (100.00)
Cost of collection	0.10 (28.57)	0.28 (14.00)	0.08 (2.76)	0.13 (4.48)	0.37 (4.30)	0.38 (6.72)
Cost of transportation	0.14 (40.00)	0.22 (11.00)	0.06 (2.07)	0.39 (13.45)	0.61 (7.09)	0.60 (4.48)
Cost of reception and sterilization	–	0.02 (1.00)	–	–	0.18 (2.09)	0.12 (1.34)
Cost of chilling	0.09 (25.71)	0.03 (1.50)	0.09 (3.10)	0.28 (9.65)	0.32 (3.72)	0.31 (3.47)
Cost of procurement	0.33 (94.28)	0.55 (27.50)	0.23 (7.93)	0.80 (27.59)	1.48 (17.21)	1.41 (15.79)

Source: Thesis, DESM, NDRI (Pundir, 1988; Chauhan, 1987; Khokar, 1985; Rao, 1976, Rangaswamy, 2007)

Note: Figures in the parenthesis are the percentage of farm gate price.

under reference represents the average picture, as mentioned above, the private plants adopt different methods of procurement to keep the cost low. It may be adoption the village/farmers, contract farming or supply through milk contractors. These plants generally concentrate in areas of high production of milk so that the scale of procurement is high. Further, it can be concluded from the figures of table that in general procurement cost is decreasing over the time regardless the locations of the study.

So far, the objective of cooperatives is the economic welfare of the small and marginal dairy farmers, reducing the cost of procurement is very difficult which resulting to inefficiencies. The goodwill the cooperatives enjoying may not last long in the new era of competition unless an alternative model of procurement is adopted.

Marketing Margin and Price Spread

The difference between the price paid by consumer and the price receives by the dairy farmer (producer) for a liter of milk is called as farm-retail spread or price spread or may be called as marketing margin. The marketing margin includes: (a) the cost involved in reaching the produce from producer to ultimate consumers and (b) the profit margin of market functionaries involved in the entire market chain. The study of marketing margin and price spread is important for the knowledge of nature, extent and genuineness of various marketing charges. It helps to develop appropriate pricing policy that aims to provide incentive prices to producers and assumes him to legitimate share in the consumer's prices which means higher marketing efficiency. It is also helpful in the development and evaluation of the market policies like regulation of market charges for different market functionaries and functions. The market margin may vary from channel to channel and market to market.

Fundamentals of Marketing Efficiency

A review of marketing efficiency fundamentals help in targeting improvement. Lot of research has gone into defining the marketing efficiency and determining the factors affecting it. In its simplest approach, marketing efficiency combines technical efficiency and the price efficiency of the marketing system. Technical efficiency alternatively called operational efficiency, deals with all aspects of raising productivity in all marketing functions such as storage, transportation and processing and thus determined by marketing costs. The evidence of pricing efficiency is efficient allocation of resources and maximization of economic output. Marketing input includes the resources (labour, packaging, machinery, energy and so forth) necessary to perform the marketing functions. Marketing output includes time, form, place and the possession utilities that provide satisfaction to consumers. Thus resources are the costs and utilities are the benefits of the marketing efficiency ratio. Alternatively, efficient marketing is the maximization of this input-output ratio. The input cost of marketing is simply the sum of all the prices of resources used in the marketing process. While output of the marketing process may best be measured by the price that consumers will pay in the marketplace for dairy products with different level of marketing utilities. If the consumer is willing to pay additional three rupees for per litre of packed and pasteurized milk than the raw milk, we may infer that the pasteurizing of milk adds three rupees worth of utility to raw milk. Using this concept, the marketing efficiency ratio can be increased in two ways. Firstly any marketing change that reduces the costs of performing the functions without altering the marketing utilities and secondly, enhances the utility-output of the marketing process without increasing marketing costs. The former situation refers operational efficiency where the costs of marketing are reduced without necessarily affecting the output side. An example may be a newly designed retail dairy case that reduces refrigeration energy costs. Technological innovations are not the only avenue leading to higher levels of operational efficiency. An organization that improves its raw material procurement practices, by say centralizing purchases, buying in larger quantities or taking advantage of unit freight rates, is likely to increase operating efficiency. In the same way, an organization that rearranges sales territories and distributes fewer but larger loads to each delivery point can improve its levels of operational efficiency. Reducing the physical losses is another method to improve operational efficiency. The higher the loss in marketing, the lower is the level of operational efficiency. The combination of improved cooling and handling equipments helps in reducing the milk losses during marketing.

The dilemma is that changes in the cost of marketing influence consumers' satisfaction, and efforts to increase the customer's utility often affect marketing costs. A new marketing practice that reduces costs but also reduces consumers' satisfaction may actually reduce the efficiency ratio. For instance, installation of bulk cooler at village level may reduce losses and increase consumer satisfaction by providing hygienic milk but at the same time it increases marketing cost and reduces marketing efficiency. The compromise which must be made between operational efficiency and customer satisfaction explains the difficulty of improving marketing efficiency. It is not difficult to reduce marketing costs by taking such measures as reducing the number of pack sizes or retail outlets but there may be a greater loss in customer satisfaction than is compensated for by the fall in marketing costs and retail prices.

Marketing firms, operating within a competitive environment, are especially well motivated in seeking to increase operational efficiency. Although their goal may be higher profits, often the benefits of improved operations accrue to customers in the form of lower prices. Competition acts as a brake on the extent to which profits increase and limits any tendency for customer service and satisfaction levels to fall.

Price efficiency is the second form of marketing efficiency. It is the capability of prices to allocate resources efficiently and in accordance with consumer preferences in the marketing process. It is concerned with the ability of the marketing system to allocate resources and coordinate the entire agricultural/food production and marketing process in accordance with consumer directives. The evidence of pricing efficiency is efficient resource allocation and maximum economic output. The pricing mechanism directly affects production, in this instance, by indicating that a certain amount of the available milk should be processed rather than sold as raw. Kriesberg (1974) says that the usefulness of pricing efficiency measures in evaluating any marketing system depends upon four conditions, namely (1) that customers have alternatives from which to choose in the marketplace and hence, the measure has little relevance to situations where there is an effective monopoly, (2) the prices of alternatives adequately reflect the costs of providing them provided there are no subsidies hidden or otherwise for competitive products, (3) organizations must be free to enter or leave the market, and (4) there must be competition between those in the marketplace avoiding any cartel-like behavior. With respect to these conditions, in order to improve pricing efficiency, there is a need to increase competition in marketing of milk and milk products both on procurement side and distribution of final products. The promulgation of MMP Order 1992 and its subsequent amendments in 1993 and 2003, abolished the restriction on entry of firms in milk processing and milk shed, which led to increase in establishment of large number of private and multi-national milk plants. Till 2006, there was an increase of 97 per cent in number of private dairy plants as compare to 27 per cent in cooperatives (Chandel and Chauhan, 2010). As a result, competition for procurement of milk enhanced which led to adoption of alternative and innovative ways of milk procurement like contract farming and the results of Chapter 4 have proved that it increased the milk price received by dairy farmers.

Still the private dairy plants do not enjoy the level playing field in marketing of milk and milk products because of various promotional schemes of the Government targeted at cooperative sectors like allocating prime milk booth locations to cooperatives, subsidized loan, tax holidays, etc. Yet the prices offered by the private plants are very competitive to prices of the cooperatives. In future business, the private sector has to establish the social credibility of the cooperative sector. In the marketing of milk and milk products, some how it is still important to make it a campaign of the people.

An attempt has been made to quantify the extent of marketing efficiency using the prominent methods. The commonly used measures of marketing efficiency are conventional output to input ratios. Other most important measures are the Shepherd's ratio of value (price) of goods marketed to the cost of marketing (Shepherd, 1965) and Acharya's modified marketing efficiency formula (Acharya and Agarwal, 2001). A brief description on marketing efficiency measures is given in Appendix I to remind readers of the approaches. Applying these measures, an attempt has been made to assess the marketing efficiency of milk marketing channels using the data from students' theses (Table 23.3).

Observed values in the table are not comparable due to difference of year and the region. Nevertheless, three measures of efficiency make them comparable. All these measures hint at that unorganized milk marketing systems specifically producer-vendor-consumer are more efficient one. Given the conditions of markets of milk and milk products, this may be one of the reasons that unorganized channels market major milk production in the country keeping in mind that non-price considerations of the product lack on the consumer side. The efficiency measures were observed moderately high for the cooperative marketing channel in Bhopal (MP). The values of conventional, Shepherd and Acharya's marketing efficiency were 1.56, 5.09 and 2.26, respectively. These values were higher than the marketing channels involving private dairy. All these measures of marketing efficiency ignore the non-price variables of the marketing channels which improve overall welfare of the producer and consumer. If these non-price variables like supply of inputs, veterinary facilities, fare practices,

Table 23.3: Price Spread and Marketing Efficiency of various Marketing Channels of Milk

Year and Author	Region / Marketing Channel	Producer Price Rs/lt	Marketing Cost Rs/lt	Marketing Margin Rs/lt	Retail Price Rs/lt	Price Spread Rs/lt	Marketing Efficiency Measures: Conventional	Shepherd	Acharya
Sujatha *et al.* (2003)	Chitoor, A.P.								
	Producer- Consumer	8.92	–	–	8.92	–	–	–	–
	Producer-Milk vendor-Urban consumer	7.17 (65.00)	1.5 (13.60)	2.36 (21.40)	11.03 (100.00)	3.86 (35.00)	2.57	7.35	1.86
	Producer-Private dairy/booth-Urban consumer	7.5 (57.69)	4.4 (33.85)	1.10 (8.46)	13.00 (100.00)	5.50 (42.31)	1.25	2.95	1.36
	Producer-Private dairy/Milk vendor-distributer/Urban consumer-distributer/Urban consumer	7.5 (57.69)	4.2 (32.31)	1.30 (10.00)	13.00 (100.00)	5.50 (42.31)	1.31	3.10	1.36
	Producer-Milk vendor-Private milk booth	7.75 (51.60)	1.25 (8.32)	6.02 (40.08)	15.02 (100.00)	7.27 (48.40)	5.81	12.01	1.07
Khare *et al.* (2003)	Bhopal, M.P.								
	Producer-Cooperative society-dairy plant-consumer	10.09 (69.30)	2.8 (19.64)	1.61 (11.06)	14.56 (100.00)	4.47 (30.70)	1.56	5.09	2.26
Vedamurthy (2004)	Shimoga, Karnataka								
	Producer-Consumer	10.32	0.68	–	11.00	0.68	1.00	16.18	15.18
	Producer-Milk vendor-Consumer	9.16 (81.42)	1.05 (9.33)	1.04 (9.24)	11.25 (100.00)	2.09 (18.57)	1.99	10.71	4.38
	Producer-Vendor processor-Consumer	8.88 (74.00)	1.04 (8.67)	2.08 (17.33)	12.00 (100.00)	3.12 (26.00)	3.00	11.54	2.85
Yogi (2006)	Jaipur, Rajasthan								
	Producer-Consumer	12.68	–	–	12.68	–	–	–	–
	Producer-Halwai-Consumer	11.40 (76.51)	1.58 (10.60)	1.92 (12.89)	14.90 (100.00)	3.50 (23.49)	2.22	9.43	3.26
	Producer-Milk vendor-Consumer	11.25 (75.96)	1.54 (10.40)	2.02 (13.64)	14.81 (100.00)	3.56 (24.04)	2.31	9.62	3.16
	Producer-Milk vendor-Contractor-Consumer	10.70 (67.51)	2.55 (16.09)	2.60 (16.40)	15.85 (100.00)	5.15 (32.49)	2.02	6.22	2.08

Figures in parentheses are the percentage of retail prices.

etc., of the cooperative marketing channels are taken into consideration perhaps it may be relatively efficient marketing system to the small and marginal dairy producers.

In some of the agricultural commodities like banana, the price paid by consumers of in the retail outlet of the co-operative society was less compared to the open market (wholesale channel). The modified efficiency ratio was higher in the co-operative channel than the open market mainly because of higher price realization by the farmer due to reduced marketing costs. The operational efficiency measured in terms of cost of performing marketing function was also higher in the cooperative channel due to lower marketing cost. As regards pricing efficiency, which the structural characteristics of marketing system, where the sellers were able to get the true value of their produce and the consumers received the true worth of their money, the co-operative channel was found more efficient (Murthy *et al.*). But the case of milk marketing is somewhat different as compared other agricultural commodities.

Producers' Share in Consumer's Rupee and Price Spread

The producer's share in consumer's rupee and the price spread are alternative ways of looking at the price efficiency of market in the sense that price signals are transmitted to production. The former is the share of the price paid by consumer that is received by the producer while the price spread is difference between prices paid by the consumer and received by the producer for a litre of milk. In other words, price spread is one *minus* share of producer in consumer's rupee. Higher is the price spread, lesser is the share of producer in consumer's rupee keeping all other things constant. Pricing efficiency of a marketing system is high if the share of producer in consumer's rupee is high and price spread is low. Table 23.3 also indicates in parenthesis the percentage share of producer and price spread as that of retail price. It can be observed that producer's share in consumer's rupee vary from about 52 per cent in Producer-Milk vendor-Private milk booth, Chittoor, AP to 81 per cent in Producer-Milk vendor-Consumer, Shimoga, Karnataka. The producer, higher the producer's share in consumer's rupee, higher is marketing efficiency. The share was 69 per cent in Producer-Cooperative society-dairy plant-consumer, Bhopal, MP.

The producer's share in milk and milk products is comparatively better than other highly perishable commodities of agriculture. As per data relating to vegetables and fruits, the share of farmers in consumer rupee for vegetables ranged from only 41.1 percent for onion to as high a s 69.3 percent for green pea, and for the selected fruits this share varied from only 25.5 percent for apple to 53.2 percent for sapota (Gandhi and Namboodri, 2004). Though, a part of the differential 60 -65 per cent can be attributed to cost of transportation, packaging, sorting, grading, taxes, loading/ unloading, branding, etc. still a large chunk is appropriated by the long chain of intermediaries ranging between 6 to 8 between farmer to end user. In US, the farmer gets roughly 60 to 65 per cent of the price paid by the end user. This implies that the process of agriculture marketing in US is 200 per cent more efficient compared to the Indian marketing system.

The price spread which is comprised of marketing cost and marketing margin, has been presented as percentage of consumer price (Retail price) along with producer's share in a stacked bar diagram. Large portion on the top of the bar is

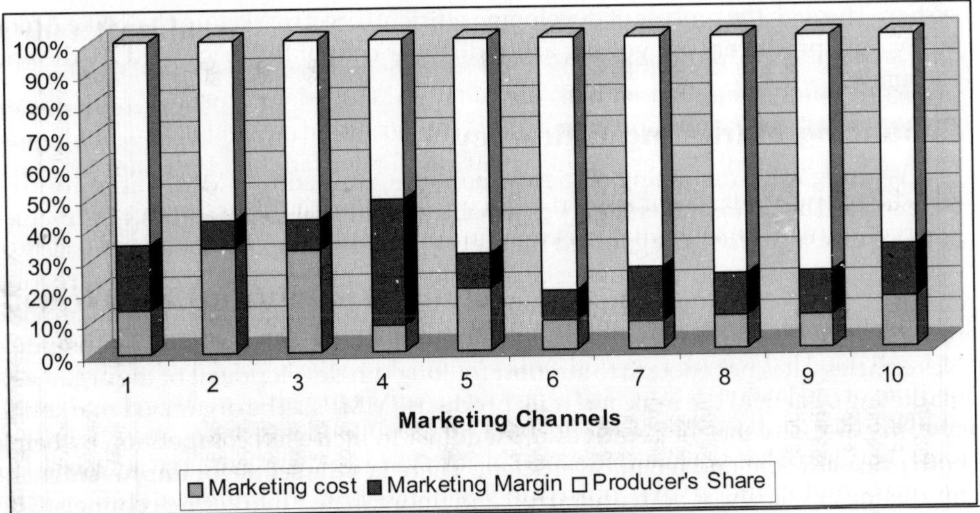

Note: Marketing Channels (1-10) are taken from Table 23.3 and are as following:

1. Producer-Milk vendor-urban consumer, Chittore (AP)
2. Producer-private dairy/private milk booth-urban consumer, Chittore (AP)
3. Producer-private dairy/milk vendor-distributer/urban consumar-distributer/urban consumer, Chittore (AP)
4. Producer-Milk vendor-private milk booth, Chittore (AP)
5. Producer-Cooperative society-Dairy plant-consumer, Bhopal (MP)
6. Producer-vendor-consumer, Shimoga (Karnataka)
7. Producer-vendor processor-consumer, Shimoga (Karnataka)
8. Producer-halwai-consumer, Jaipur (Rajasthan)
9. Producer-Milk vendor-consumer, Jaipur (Rajasthan)
10. Producer-Milk vendor-contractor-consumer, Jaipur (Rajasthan)

Figure 23.1: Marketing Cost, Marketing Margin and Producer's Share as Percentage of Retail Price in different Marketing Channels of MMPs.

indication of comparatively higher share of producer in consumer's rupee. Price spread has the indirect relationship with marketing efficiency and is represented together by middle and lower portion of the bar. It can be observed from the figure that larger is the price spread lower is the producer's share. In marketing channels second, third and fourth, the price spread is more than 40 per cent. These are the marketing channels involve private agencies. The component of marketing cost is high in the price spread when the agency was private dairy plant. Marketing margin in marketing of MMPs was, generally, low due to involvement of lesser number of intermediaries (3 to 4) as compare to 6-8 in other agricultural commodities.

It can be summarized that in the case of milk, there has been spectacular achievement through white revolution by developing cooperative dairy marketing system. Through the process of developing efficient direct marketing chain like Amul, today milk producers are getting around 70 per cent of the price paid by the end consumer.

Enhancing Marketing Efficiency

With regard to marketing efficiency measures, marketing of MMPs appears to be better placed as against other highly perishable agricultural commodities. High values of these measures for unorganized marketing channels explain their popularity to some extend among dairy farmer and consumers. Among organized marketing channels, the co-operative milk marketing system is adjudged better way to provide remunerative prices to producer farmers and reasonable prices to consumer. Nevertheless, it is not enough to support for long run development of the organized marketing system of the milk and milk products (MMPs). The organized marketing system especially the cooperatives must adjust to increased competition by being both price-efficient and technical efficient. There lies the scope for improvement of marketing efficiency of both organized and unorganized marketing channels. The chapter delineates key issues in marketing system to be efficient under different heads.

Reducing the Procurement Cost

Future of dairy lies in organized marketing. The biggest threat to the organized marketing is the procurement cost of milk. The procurement cost can be reduced by collecting large production from fewer collection centres. In order to do so, chilling and storage facilities at the village level or group of village level can prove to be helpful. A step in this direction has been initiated by installing the milk bulk coolers which have at least reduced cost by reducing collection from two times to one a day. Private dairy plants adopted innovative methods of contract farming and/or adopted village to increase the quantity of milk collection. The dairy cooperatives in the country have to go a long way in this direction. The comparative advantage of cooperatives lies among small and marginal producers which may reduce with specialized dairy production.

Pricing Policy of Milk

The government policy on fixing the producer and retail prices is a major detriment to the development of the dairy industry because prices are set under a climate of political influence with no relevance to general market conditions inside India or to border prices. The policy has set both the producer and retail price effectively constraining the dairy processing industry with margins that do not reflect general business cycles and the impact of rising costs, wages, utilities, taxes etc. The classified pricing system that is based on differentiation of markets and the demand characteristics of the products will provide a better return, a larger market, or both rather than a single price for milk for all uses.

Milk has different values in its various uses. Milk that is used for fluid products has a higher value than the milk used for processed dairy products. There are two reasons for this. First, the retail price of fluid products like whole milk, skimmed and low-fat milk, cream, flavored milk etc. is higher because it costs more to market highly perishable and bulky fluid milk than it does to market the processed forms of milk, such as cheese and butter. Second, the elasticity of demand is lower for fluid products than for processed dairy products. This provides an incentive for the industry to allocate some of the fluid milk to processing uses. This legal form of price discrimination results in a higher price for fluid products and a lower price for processed products than would be the case if milk supplies were divided equally between the two uses. Because of these milk-value differentials at the retail level, milk dealers can pay higher prices to farmers for milk unused in fluid products than for milk processed into cheese, butter, and other products.

Emerging Trends in Marketing of MMP

The dairy industry is undergoing substantial changes in market organization. Declining numbers and increasing sizes of dairy farms, changing consumer preferences for various dairy products, a tendency toward fewer but larger dairy product handlers and processors, chain stores integration into dairy processing, decentralized marketing and increased cooperative activity in assembly and processing markets have all altered the marketing patterns and competitive environment of the dairy industry. Product quality and price regulations make the dairy industry one of the most highly regulated food sectors. Farm milk prices and indirectly consumer prices are influenced by a complex system of competitive and regulatory arrangements, including federal price supports, import regulations and classified market order prices. Dairy product retail prices have not increased as fast as other fold products in recent years. Dairy farmer cooperatives have assumed most of the assembly market functions and also play in important role in negotiating farm milk prices.

Supply Chain Management

A supply chain is a network of facilities and distribution options that performs the functions of procurement of milk, transformation of milk into finished products, and the distribution of these finished products to customers. It is a total approach to marketing and akin to farm gate to consumer plate concept. The complexity of the chain may vary greatly from industry to industry and firm to firm. The supply chain of perishable products like MMPs is particularly important because they account for more than 50 per cent of supermarket sales, and the potential for extra profit from managing these items has been estimated at 15 per cent. The dairy industry is a vertically integrated industry with both parts of the supply chain (farm and manufacture) dependent on each other to ensure safe food.

Supply Chain Management in India is still in its nascent stage and going through a very interesting phase of learning through others' and own experience. Others' experience is in terms of know-how being acquired from professionals representing matured retail markets across the globe and the own experience supports the learning by considering the local aspects and realities of the country.

The biggest challenge in managing a supply-chain is to shorten the length of the chain by driving efficiencies across all the elements within the chain. Some of the key factors that contribute towards an efficient supply chain can be categorized into - physical Infrastructure, manpower talent, technology and quality of service providers.

Fostering Competition

Competition plays a key role in fostering marketing efficiency. Marketing firms compete for the consumer's favor by lowering marketing costs and increasing operational efficiency wherever possible. At the same time, there is competitive pressure on firms to add more utility to milk in order to gain an increased market share by catering to consumer preferences.

Cold Chain Management

Due to presence of important constituents, milk and milk products are magnificent mediums for growth of microorganisms. The growth is very fast in raw milk as compared to processed milk. Raw milk not chilled to 5-7°C within three hours after milking, the bacterial count just double. In Indian conditions, the first chilling point for the raw milk is the Chilling Centre and raw milk on an average takes 3-5 hours to reach the Chilling Centre. In some of the cases it may go up to 8-10 hours. The milk and milk products during transportation and after processing have to be stored at specified conditions till consumption. Therefore, a cold chain is essential right from the time of milking till is consumed either as fluid milk or as a dairy product. It will, further, increase the marketing cost but will, at the time, improve the consumer satisfaction when assessed considering reduced health expenditure.

Information and Communication Technology (ICT)

The historical information (market intelligence) on prevailed prices in the past, market arrival over time etc. and current information (market new) on prices, arrival and changes in market condition plays vital role in improving efficiency of agricultural marketing in general and milk marketing in particular which is very perishable. The analysis of past information helps to take the decision about future while the current information helps in operation marketing decisions. Further the availability of market intelligence and market news in time at with speed is of utmost important which is lacking and expensive in milk marketing. It is only by the introduction of ICT that information can provide time and upgraded information at the least cost. ICT will improve the efficiency in marketing activities by establishing a nation-wide information network, which will give details about market functionaries, sold and unsold stocks, as well as the sources of supply and destination. ICT can be used as decision support system, electronic auction system, e-catalogue of commodity profile national atlas of markets showing their features etc.

Market Integration

Market integration is an attempt at organizing and coordination the marketing process to increase operational efficiency and acquire greater power over the selling and/or buying process. In case of MMPs, processing plants can do the vertical integration by assuming the distribution function. Another for of integration *viz.* horizontal integration can be applied by forming the farmers groups for value addition and collective selling of MMPs.

Besides these measures, development of appropriate public-private partnerships and linkages also ensure necessary value chain to improve market linkages and efficiency for the output/s. Likewise, facilitating organization of self-help or other similar groups for initiating changes in the marketing system can also improve efficiency.

Appendix I: Measuring Marketing Efficiency

Conventional approach: it is the ratio of output to input where output is the value added by the marketing system and input is the real cost of marketing. The real cost of marketing is considered including some fair margins of the intermediaries only to the extent that margins are the wages to the services rendered by them. On the numerator side, measurement of value added is not easy. It is generally taken as the difference in the price received by the farmer and the price paid by the consumer. The practical problem with this measure is that it does not take market imperfection into consider. When marketing cost is take akin to value added, the former may be high due to lack of competition.

Shepherd Approach: Shepher (1965) suggested that the ratio of the total value of goods marketed to the marketing cost may be used as a measure of marketing efficiency. The higher the ratio, the higher is the efficiency and vice versa. This method eliminates the problem of measurement of value added in the conventional approach but only considers real marketing cost including some fair margin of the intermediaries and not the excessive margin retained by them.

Another important parameter of marketing ignored by both the approaches is the price received by the farmer. These limitations were taken care by the modified method suggested by Acharya (2001).

Acharya approach: The modified method of measuring the marketing efficiency (MME) suggested by Acharya is the ratio of price received by the farmer to the sum of marketing cost and marketing margins received by intermediaries. Since there is an exact relationship among marketing cost, marketing margin, price paid by consumer and price received by farmer, any three of these could be used to arrive at measuring the marketing efficiency. The method maintain its results if the price paid by consumer can be used as the numerator without any change in the conclusion. Acharya suggested that under Indian conditions, it is advisable to use farm harvest price in the numerator instead of price received by the farmers.

References

Acharya, S.S. and N.L. Agarwal (2001) *Agricultural Marketing in India*, Oxford and IBH Publishing Co., New Delhi.

Birthal, Pratap S. (2008), Linking Smallholder Livestock Producers to Markets: Issues and Approaches, *Indian Journal of Agricultural Economics*, Vol. 63(1): 19-37.

Chandel, B.S. and Chauhan, A.K. (2010) Present Situation and Future Potential for Milk Processing in India, *Indian Journal of Dairy Sciences*

Chauhan, Arun (1987) Economic Analysis of Different Milk Procurement Systems in Private Sector Dairy Plant in Western Uttar Pradesh. M. Sc Thesis, NDRI, Karnal.

Gandhi, Vasant P. and Namboodri, N. V. (2004) *Marketing of Fruits and Vegetables in India: A Study Covering the Ahmedabad, Chennai and Kolkata Markets*, Working Paper No. WP2004-06-09, Centre for Agribusiness Management, Indian Institute, Ahmedabad.

Gaski, John F. (1987), Toward Measurement of Consumer Market Efficiency, In Melanie Wallendorf and Paul Anderson (eds) *Advances in Consumer Research Volume 14*, Provo, UT : Association for Consumer Research, Pages: 314-318.

GOI (2003) *Annual Report 2003-04*, Ministry of Food Processing Industries, Government of India, New Delhi.

Houde S., Sonnad, J. S., Shivashankar, K. and Banakar, B. (2007) Processing and Marketing Management of Milk and Milk Products in North Karnataka. *Karnataka Journal of Agricultural Sciences* : 20 (2), 2007.

Khare, P., Sharma, H. O. and Singh, T.B. (2003), Marketing Analysis of Milk Production in Bhopal District of Madhya Pradesh. *Agricultural marketing*. Vol. XLVI (2): pp 9-14

Khokar, A.K. (1985) *Economics of Milk Procurement for Feeder Balancing Dairy, Partapur, Meerut (UP)*, M.Sc Thesis, NDRI, Karnal.

Kohls, R.L. (1956) Towards a More Meaningful Concept of Marketing Efficiency, *Journal of Farm Economics*, 38(1): 68-73

Kohls, Richard L. and Uhl, Joseph N. (1990) *Marketing of Agricultural Products* (Seventh Edition) Maxwell Macmillan International Editions, Macmillan Publishing company, New York. Milk and Dairy Product Marketing: 408-423

Kriesberg, M. (1974), *"Marketing Efficiency In Developing Countries"*, In: *Marketing Systems For Developing Countries*. INCOMAS Proceedings, Izraeli, D., and Dafna, pp. 18–29.

Murthy, D. S, Gajanana, T. M., Sidhu, M. and Dakshinamoorthy V. (2007), Marketing Lossess and Their Impact on Marketing Margins: A case study of Banana in Karnataka. *Agricultural Economics Research Review*. Vol. 20 (3)

Pundir, R.S. (1988) Economic Analysis of Milk Procurement in Public Sector Plants under Cooperative Set-up in Haryana. M.Sc Thesis, NDRI, Karnal.

Rangasamy, N. and Dhaka, J.P. (2007) Milk Procurement Cost for Cooperative and Private Dairy Plants in Tamil Nadu- A comparison, *Indian Journal of Agricultural Economics*, 62(4): 679- 693

Rangnekar D V. 2006. Livestock in the livelihoods of the underprivileged communities in India: A review. ILRI, Narobi, Kenya 71 pp.

Rao, C.J. (1976) *Cost of collection, transportation and chilling of milk*. M.Sc Thesis, NDRI, Karnal.

Shepherd, G.S. (1965) *Farm Products- Economic Analysis*, Iowa State University Press, USA.

Shobha, 1998, *Performance evaluation of fruit and vegetable processing units in North Karnataka*. M Sc. (Agri.) thesis, University of Agricultural Sciences, Dharwad.

Sujatha, R., Bhavani D. and Sastry T.V. N. (2003), Cost, Margins and Price Spread in Marketing of Milk in Chhitor District of Andra Pradesh. Indian Journal of Agricultural Marketing, Vol. 17(1): pp.27-33

Vedamurthy (2004), Economic Analysis of Milk Marketing in Shimoga district of Karnataka, M.Sc Thesis (unpublished), Division of Dairy Economics, Statistics and Management, National Dairy Research Institute (Deemed University), Karnal.

Waugh, F.V. (1954) *Readings in Agricultural Marketing*, Iowa State College Press, p. 195.

Yogi, Raj Kumar (2006) *Economics of Milk Marketing in Jaipur District of Rajasthan*, M.Sc Thesis (unpublished), Division of Dairy Economics, Statistics and Management, National Dairy Research Institute (Deemed University), Karnal.

2014, Sustainable Rural Development through Agriculture *Pages* ***338–344***
Editors: **Dr. Shobhana Gupta and Dr. S.S. Tomar**
Published by: **BIOTECH BOOKS, NEW DELHI**

Chapter 24

Accelerated Rural Finance for Rapid Agricultural Growth: Constraints and Solutions

Sarju Narain, Vikas Kumar, O.P. Maurya and R.R. Kushwaha

All the political parties, policy makers, bureaucrats, agricultural experts and corporate sector in India agree that the only way to develop the agriculture sector in the country is through the infusion of much needed credit in the agriculture. The Budget of 2010-2011 and 2011-2012 also advocated the above facts regarding more credit flow in rural India for socio–economic empowerment and agricultural development. The credit has a very important role to play in supporting agricultural production and investment activities. The per hectare investment in Indian agriculture is also very low as compared to China, Japan, America and other European countries. So, per hectare production and income is also low as compare to the developed countries. Except credit flow, all other factors also play an important role but credit is the basic necessity for eradicating all others causes. As one of the largest private enterprises of India, agriculture contributes about 14.2 percent to the national Gross Domestic Products (GDP), sustain livelihood of about 57 percent population and is the backbone of agro-based industry. But majority of the farmer's socio- economic condition are poor that results in very low investment in agriculture. It has been very rightly stated that if we all are concerned about increasing total agriculture output, we must provide credit first and foremost.

Demand of Credit

The demand of credit in agriculture is higher because of emergence of commercial agriculture especially in high value crops like horticultural, floricultural, medicinal, plantation etc. Technological advancement like green house, poly house, pack house, food processing, etc. have boost up the demand of credit. Today's farmer is different from yesterday's due to more market dependency for new seeds, irrigation equipment fertilizers, pesticides etc. Changing cropping and farming trends, need to earn more from smaller holding, increasing flow of information and communication, increasing socio-economic status, need for land and its improvement, increasing mechanization of agriculture and allied sector, purchasing and managing to livestock are the major causes of more credit demand. We hope in near future the demand of agricultural and rural credit would also increases at algebraic rate due to extension of education among the rural society especially in rural youth and women, increasing livelihood level, need for location specific technologies and their dissemination.

Supply of Credit

In India various agencies *viz.* Nationalized Banks, Lead Bank, Commercial Banks, Regional Rural Banks (RRBs). Land Development Banks (LDBs), Co-operative Banks, Co-operative societies, etc. are involved in providing agricultural credit with or without assistance of National Bank for Agriculture and Rural Development (NBARD). Indian rural financial institutions are unique in its reach and diversity considering the gigantic size of population involved in agriculture and the high credit requirements, catering to their needs in near impossible task. Nevertheless, the credit system has contributed significantly to the sustained growth of agriculture and social sector. However, the credit support to agriculture has mainly been confined in terms of the crop loan. About 80 percent credit flow is going only as crop loan which is not a good symptom for holistic agricultural growth. The non-crop agricultural activities and non-agricultural rural enterprises as well as area under adverse agro-climatic conditions where one or two crops are taken in a year have largely remained outside from 80 percent credit portion of crop loan. Only 20 percent credit of total credit is going for investment in agriculture which needs to be increased.

Institutional Credit Flow

In India, where large numbers of farmers are under small and marginal category with poor socio-economic conditions who feel tiredness from delayed documentation and loan sanctioning procedure for agricultural activities from institutional sources. There is unawareness of loan processes by poor farmers and tough documentation of banking system provide a chance to the involvement of middle n an or dalals. So, it is necessary to ensure easy credit flow to tenant farmers and women cultivators. The other issue is the seven percent rate of interest for agricultural activities which is not much cheaper for poor farmers. Farmers of some pockets like Vidarbh, Bundelkhand and Royalseema where rainfall is a limiting factor and agriculture always been the gamble of mansoon. So, the rate of interest should be less in these types of reagions. National Commission on Farmers (headed by Dr. M.S. Swaminathan, 2007) also advised the government to access credit for the farmers at 4 percent interest rate. In the last year, central government arranges Rs. 3,75,000 cr. for agricultural loan which is further increased in the budget of 2011-12 as Rs 4,75000 cr. The farmers who paid their loan timely they get additional 3 percent less on interest rate, *i.e.*, they pay only 4 percent rate of interest on their loan. This is good step but majority of farmers of Bundelkhand and Vidarbh like regions who could not repay their loan due to continue failure of monsoon, they become sufferer of 7 percent interest rate to the banks and have no repaying capacity. The informal tenancy arrangements restricted the bank loaning process. For the increasing credit flow to the farmers, government also increases Rs 1000 cr. as authorized capital share of National Bank for Agricultural and Rural Development (NABARD) to reach the Rs 3000 cr. The government also decided to enhance Rs 10,000 cr. as a credit to farmers for agricultural activities in the Budget of 2011-2012. All these steps has become a mile stone in the strengthening of rural credit, but they don't provide a 'safety-net' to the starved farmer from starvation death and suicide in the adverse monsoon pockets of India.

Non-Institutional Credit Flow

Availability of timely and adequate credit has been one of the major handicaps of Indian agriculture and rural sector since time immemorial. The majority of the credit to the farmers is from non-institutional sources like Bania, traders and commission agents, relative etc. For the eradication of the non- intuitional source of credit to the farmers, government has taken major steps during different five year's plans but the situation is not more improved. The findings of the National Sample Survey Organization (NSSO) 59th round (2003), reveal that only 27 per cent of the total number of cultivator households received credit from formal- institutional sources while 22 per cent received credit from informal non-institutional sources. The remaining household comprising mainly small and marginal farmers had no credit outstanding. In India, where about 20 percent land comes under informal tenancy system who can avail loan only from non-institutional sources at high rate of interest (report of the expert group on agriculture indebt ness, Ministry of Finance, Government of India). This critical situation increases the indebtedness of poor farmers. A sample survey of Kaimganj and Patiyali Tehseel of Farrukhabad and Etah respectively in U.P. in 2008 where, tobacco and potato crops are grown in larger area revealed that 43 percent tobacco growers and 39 percent of potato growers are depend on informal (non institutional) source of credit especially in the form of advances for very short period to grow up the crops on high rate of interest. After the harvesting of tobacco and potato crops, the farmers sale and pay. They also get the advances from Kisan Credit Card (KCC). The committee chaired by Prof. V.S. Vyas, agricultural economist and former Director of Indian Institute of Management (IIM), Ahmadabad observed that credit flow to the disadvantaged area and section has not shown much improvement over time.

Causes of indebtedness of rural farmers:

Commercial Agriculture needs new type of inputs for example new varieties of seeds especially hybrid, pesticides, fertilizers, Irrigation system, managerial practices,

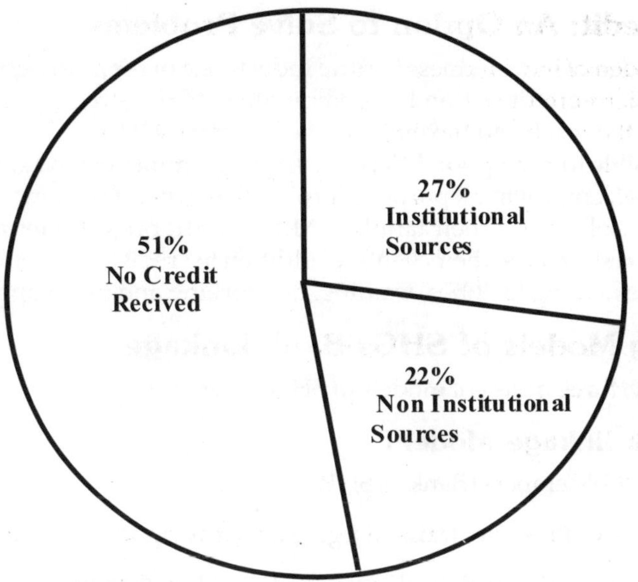

Figure 24.1: Credit Received by Farmers
***Source*: NSSO, 59th round (2003).**

etc. that demand timely and more credit. Vagaries of climate that cause endless miseries to the cultivator like per hectare low income, high rate of interest on loan, litigation, vicious moneylender, indebtedness etc. which encourage the farmers for suicide in some pockets of India. The poor farmers and agricultural labourers handle field with poor management and poor investment. In this case crop is failure is common and then farmers suffer from starvation.

Micro-Credit: An Option to Solve Problems

The burden of indebtedness in rural India is one of the most pervasive features of rural life. Micro-credit option has combination of the strength of formal banking system with the reach and flexibility of the informal self Help Group (SHG) to make credit accessible to rural poor. Micro credit programmes extend small loans to poor farmers for self employment purpose/project that generate income, allowing them to care for themselves and their families. Micro credit project offer a combination of services and resources to their clients in addition to credit for self employment. These often includes saving facilities, training, networking and peer support.

Emerging Models of SHGs-Bank Linkage

Broadly, three different models of SHGs have emerged.

SHG Bank linkage Model-I

Bank-SHG-Members (Bank as SHPI)

Banks ⟶ Promote, train and give credit support ⟶ SHGs

In this model, the bank itself acts as a Self–Help Group Promoting Institution (SHPI). It takes initiatives in forming the groups, nurtures them over a period of time and then provides credit to them after satisfying itself about their maturity to absorb credit.

SHG Bank-Linkage Model-II

Bank-Facilitator Agency- SHG Members

NGO ⟶ Promote, train and give credit ⟶ SHGs

↑ Bank

In this model, in most of the cases, the groups are formed by Non–Governmental Organizations (NGOs) or by government agencies. The groups are nurtured and trained by them. The bank, then, provides credit directly to the SHGs after observing their operations and maturity to absorb credit. While the bank provides loans to the groups directly, the facilitating agencies continue their interactions with SHGs. Most linkage experience begins with this model with NGOs playing a major role. This model has also been popular and more acceptable to banks, as some of the difficult functions of social dynamics are externalized

Model-III: Bank-NGO-MFI-SHG Members (NGOs as Financial Intermediaries)

NGOs as SHPIs

NGOs ⟶ Promote, training and helping in linkage with bank ⟶ SHGs

Bank ⟶ ↑ SHGs

Due to various reasons, banks in some areas are not in a position to even finance SHGs promoted and nurtured by other agencies. In such cases, the NGOs act as both facilitators and micro- finance intermediaries. First, they promote the groups, nurture and train them and then approach banks for bulk loans for lending to the SHGs. People also form SHGs, once the SHG movement starts in the same locality. Farmer's clubs are also motivating in formation of SHGs.

Opportunities and Constraints

Micro-credit is a viable and fast- growing system to support and empower rural poor and farmwomen. NGOs are the major players in promoting micro-finance through SHGs. SHG- Bank linkage is showing positive trends in cumulative growth. Southern region of the country shows tremendous growth in SHG- Bank linkage as compared to the other regions. Women SHGs are more sustainable as compared to male SHGs. As rural population is engaged in agriculture, agriculture shares maximum loaning followed by economic and domestic activities. Formation of SHGs is an approach for women empowerment. Women SHGs are inclined more towards economic activities. Farmers' Schools and SHG Federations are emerging to sustain the SHGs and related activities. Micro-insurance is the newer initiative, which needs strengthening. Micro enterprises are the mainstay for sustaining the SHGs and hence, micro-financing. Some of the major micro-enterprises are bee keeping, mushroom production, goat rearing, poultry, seed production, marketing of vegetable, etc.

But there are still some lacunae in the system as given here

1. Generation of viable economic activities is limited
2. Marketing of SHG products is still not up to the mark
3. Lack of sustenance measures for groups after withdrawal of the support from NGOs or the banks and high rate of interest taken by NGOs. The points pertaining to strength, weakness, opportunities and threats regarding NGOs led SHGs are observed as follows:

   The strength includes more NGO's penetration at village levels which help in community mobilization, perform micro financing work, help in SHG formation based on their needs and aim and provide need based services. The common weaknesses are majority of NGO's have not much strong network especially agriculture based NGO's are limited. They mostly focus on SHG's and their micro financing but their activities are not linked up to ground level. Most of them work due to self interest not for the rural people and monitoring of NGO's is lacking. The opportunities are need to initiate economic activities in agriculture, can help in development of social capital, norms of NGO's functions need to be standardized and strong monitoring enforced. At last the major threats are feel that miss utilization of founds and charging high rate of interest on loaning, belief of NGO's system may not remain longer among the people if they not perform actual worked at ground level.

Thus we can say that the rural market has a vast potential, which remains to be exploited. But the intervention of rural moneylenders and subsidy are major threats to the SHGs. There is further scope for support of financial institutions/Central or state govt./External agencies to SHG's. The creation of a massive rural micro financing infrastructure comprising SHGs, Federation of SHGs and other associated intermediation as a sub system of the formal banking structure will be of chief importance. The synergetic effect of this infrastructure combining with the already existing formal financial system will make the transformation a reality.

References

Economic survey- 2000- 01 to 2011-12

Agricultural finance and management: Reddy and Rao, 2006

Agricultural Prices: Analysis and Policy- Agarya and Agaarwal, 2005

Economics of farm production and management: V.T.Raju and Rao, 2010

2014, Sustainable Rural Development through Agriculture *Pages* ***345–349***
Editors: **Dr. Shobhana Gupta and Dr. S.S. Tomar**
Published by: **BIOTECH BOOKS, NEW DELHI**

Chapter 25

Alternate Use of Land for Better Livelihood in Western Rajasthan

Lokesh Kumar Jain, Neeraj Hada and S.S. Bhadauria

Drought is a common phenomenon in the western Rajasthan where this part faced 64 drought years from 1901 to 2010. The average annual rainfall in western Rajasthan is less than 300 mm but is markedly variable ranges from 200 mm in the west to about 370 mm in the East. Sand dunes and desert soils occupy major area in the zone (58 per cent). The climate of western Rajasthan is very fragile. Due to the uncertain and erratic weather conditions including very high temperature difference (-2° to 48°C), late onset of monsoon, frequent droughts, early cessation of monsoon, low rainfall (<300mm annually), less no. of rainy days (15-20) or hot winds (10-35 km/hour) during summer results in failure or very poor crop production is a common phenomenon. Drought and desertification are major problems of arid zone of India. Desertification mapping through remote sensing and field data has revealed that about 76 per cent area in western Rajasthan is affected by wind erosion/deposition alone, followed by water erosion (11 per cent), salinity/alkalinity (5 per cent) and water-logging (0.7 per cent). About 21 per cent area is affected by desertification of severe intensity, and 40 per cent by moderate intensity, especially due to high human and animal pressures. In western zone of Rajasthan 58 per cent of land is directly affected by movable sand dunes and vast cultivated land is directly or indirectly affected by these dunes. About 92 per cent area in arid Rajasthan is now affected by desertification. About 76 per cent area is affected by wind erosion of different intensities, and 13 per cent by water erosion. Also wind erosion creates greater health hazards as sand suspended in air causes asthmatic problem. Very high human and livestock demand for food, fodder and fuel wood are causing over-exploitation of fragile resources, resulting in wind and water erosion, water logging, salinity-alkalinity and vegetation degradation. There are aeolian soils, loamy fine to coarse and calcareous at places with poor organic matter content with low water holding capacity. Land degradation due to soil erosion is a more acute problem faced by the farmers in dryland regions and has social impact like shortage of water, loss of cultivable area, low productivity etc. Land degradation is the gradual conversion of productive land into less productive or unproductive ones. The problem of land degradation or desertification is a continuous one and creates due to excess of land abuse in any patch. Desertification is the process of land disturbance that ultimately leads to the transformation of productive land into these ecological wastelands. Situation often forces the region into a distress both in terms of quantity and quality aspects of the eco-system and collapse of the biotic and vegetative components lead to the disaster that always looms the area. Environmental degradation mostly affects the rural poor as their livelihoods largely depend on the available natural resources. Rainfed agriculture is complex, diverse and risk prone and is characterized by low levels of productivity and low input usage. Pearl millet is the predominant crop of the zone followed by cluster bean, moth bean, sesame and green gram in kharif while cumin, rapeseed and mustard, wheat and Isabgol are major Rabi crops in irrigated belts (10 per cent of cultivated area). The peoples are characterized by low literacy and awareness, poor socio economic status and low risk bearing capacity. According to the farmers, droughts seriously affected them both in the normally wet and dry seasons also as high rains causes flood situation due to impermeable layer beneath soil hinders infiltration. The water and soil quality also restricts the choice of crops. The livelihood of almost 60 per cent rainfed farmers depends on livestock especially small ruminants. The alternate source of income during droughts is byproducts of perennial trees. To fulfill the requirement of fodder for livestock is only possible through aforestation and stabilization of sand dunes by perennial trees and grasses and hedges. Now a days on fodder is available for 47 percent livestock in India and the condition is more severe in western Rajasthan where unavailability of green fodder causes imbalance nutrition. The pressure for limited resources in the desert area is tremendous and particularly worse in densely populated regions. In order to bring stability and increase productivity, several technologies for sand dune stabilization, shelterbelt plantation, soil and water conservation, improved agro-forestry systems, management of cropland, pasture and range areas, management of saline-sodic soils, disseminating the developed technologies to the stakeholders and contingency plans to combat drought, resource generation by women through value-addition to agricultural products, mushroom cultivation, dairying, etc have been developed.

Effect of Land Degradation on Livelihood

- ☆ Partial or complete crop failure (because of low soil moisture content and disease outbreaks).
- ☆ Choice of crops restricted.
- ☆ Livestock deaths.
- ☆ Productivity of pastures and grazing land reduced significantly.

- ☆ Wild life damaged the crops.
- ☆ Trees drying and dying.
- ☆ Boreholes, rivers, springs and other water points drying and caused scarcity of drinking water for both livestock and human.
- ☆ Transportation hinder due to sand deposition of raaods and tracks.
- ☆ Shortage of basic commodities in the local markets.
- ☆ High government expenditure because of food imports.
- ☆ Malnutrition, especially in children.
- ☆ Unemployment, coupled with rampant crime and robbery.
- ☆ Migration to urban areas for casual labour and herds to other state or districts for fodder.
- ☆ Loss of natural resources.
- ☆ Increased asthmatic problem due to increased suspension of sand in air.
- ☆ Mentally disturbance because of unemployment and loss of earnings.
- ☆ Conflicts increased for water and inability to repayment of credits.

Alternate Land Use

The western part of Rajasthan have lot of marginal or waste land not suited for agricultural production but may be put for cultivation with trees, shrubs, grasses, medicinal plants etc so farmers harvest additional income besides arable farming. Also it will protect the crop area by covering fine sand or removal of top layer from wind. To achieve greater efficiency in utilizing the resources in dryland and also to combat the "Energy crisis" and deficiency of animal feed, major approaches in this direction should be

- ☆ Putting the land (class IV and above) under suitable grasses, legumes and fodder crops and integrating with animal productive systems.
- ☆ Using the same land or a portion of it simultaneously or sequentially for food, fodder and fuel, due attention being given to interaction among various uses aiming to obtaining greater sustained production and securing both immediate benefits and long term environmental concerns.

The various alternate land use systems like agro-forestry, silvi-pastrol, agro-horticulture,silvi-horticulture, alley cropping system etc have been developed and for these systems, suitable crops, grasses, legumes and trees in western zone are given below.

Crops	*Grasses and Leguems*	*Trees*
Sorghum	*Cenchrus ciliaris*	*Zizyphus numularia*
Pearlmillet	*Cenchrus setigerus*	*Acacia tortilis*
Greengram	*Sehima nervosum*	*Calligonum polygonoides*

Contd...

Crops	Grasses and Leguems	Trees
Clusterbean	*Panicum antidotale*	*Prosopis juliflora*
Moth (Kidney bean)	*Lasiurus sindicus*	*Capparis Decidua*
Kulthi (Horsegram)	*Heteropogon controtus*	*Acacia Senegal*
Groundnut	*Dichanthium annulatum*	*Prosopis cineraria*
Mustard	*Atylosia scarabaeoides*	*Leucaena leucocephala*
Grain ameranth	*Stylosanthes humilies*	*Tecomella undulata*
Barley	*Clitoria ternatea*	*Salvadora persica*
Safflower	*Carchrus biflorus*	*Salvadora Oleoides*
Gram	*Aristida funiculata*	
Taramira	*Indigophora Cordifolia*	

In agri-hotriculture system, crops with erect nature and early maturing habit like greengram, moth and cluster bean may be intercropped with orchard of ber, datepalm, aonla, bael etc. or during early stage of establishment of orchards of pomegranate, castor may be intercropped to utilize the inter row space, enrichment with organic matter and irrigated with drip for water saving.

Silvi-pastoral system, several grass species like Sewan (*Lesiurus sindicus*), Anjan/ Buffelgrass (*Cenchrus ciliaris*), Dhaman/Birdwood grass(*Cenchrus setigarus*), Bharut (*Carchrus biflorus*), Lampra (*Aristida funiculata*), Bekaria(*Indigophora Cordifolia*) etc may be grown because some of them have mechanism to tolerate high salt concentration in the root zone soil. Also some of these highly tolerant grasses either exclude the absorption of salts from the soil and/or deposit the absorbed/ translocation salts at points within the plant system which do not allow them to interfere in metabolic processes. The performance of these grasses in association with salt tolerant trees like *Prosopis juliflora, Acacia nilotica, Prosopis cineraria* in a unified agroforestry system has vast potential to yield of green biomass per annum in saline soil. Agroforestry is recognized as one of the efficient land use systems to control desertification. Accordingly, most dwellers raise livestock as a subsidiary occupation and allow trees and shrubs to grow along with cultivated crops to mainly cover the risk and uncertainty of crop maturity. The khejri (*P. cineraria*) is an important multipurpose leguminous tree widely distributed in western Rajasthan. The value of every plant part of khejri is well known by the farmers of this region. Besides its value for human beings and animals it creates favorable microclimates for crop and forage species producing higher biomass under khejri tree canopy was due to high fertility status as it ameliorate harsh climate and add nutrients through leaf fall. The role of *Prosopis cineraria* on microbial C, N and total C, N under the canopy is appreciable. Therefore, it is possible to develop organic farm management practices in tillage with *P.cineraria* which can meet the crop nutrient requirements to some extent. The quantity of available nutrients held in microbial biomass is considerable and it constitutes a transformation matrix for all natural organic materials in the soil and acts as a source and sink of the nutrients. Therefore the purpose of synchronizing the release of nutrients from leaf litter with crop requirements (kharif and rabi cropping season) is to improve the ratio of nutrient uptake to losses by leaching. Plantation with Khejri will contribute to the efficiency of nutrient cycling in the nutrient 'hunger' soil. By doing so, not only will the crop productivity get a boost but also desertification will be taken care of in the extreme arid region. Other tree species grown on waste land *viz.*, Jaal/Peelu (Salvadora persica), Mithi jaal (Salvadora Oleoides), Kummat (*Acacia Senegal*), Rohida (*Tecomella undulata*), Kair (*Capparis Decidua*), Jharberi (*Zizyphus numularia*), Babool (*Acacia tortilis*), Vilayati babool (*Prosopis juliflora*), Fog (*Calligonum polygonoides*) etc. improve environment and pay sustainably for a long periods in western Rajasthan. Farm families also sell materials collected from perennial trees like gum, seed, fruit from these trees and sold in local market to earn income. The fruiting time do not match with Kharif season so family members collect these without affecting agriculture.

Medicinal and Aromatic Crops

The desert soils are good habitat for growing several valuable medicinal and aromatic plants. The wealth of medicinal plants of Thar Desert has been documented and reveled that some of these species are used as household remedies while some are used in traditional system of medicine. Approximate 15 species are commercially exploited. It will help in stabilization of sand dunes and alternate source of additional income during lean period. A number of medicinal and aromatic crops have been screened for salinity and sodicity tolerance in India. Crops like Isabgol (*Plantago ovata*) can be successfully cultivated in soils having pH of 9.5 and EC between 8-10 dS/m. Similarly, Salvadora, a non-edible oil tree can be grown in salt affected soils very successfully. Industrial species like *Euphorbia* and *mulethi* (*Glycyrrhiza glabra*) also have good scope for cultivation in salty environments. Other medicinal and aromatic crops showed potential yields are Guggal (*Commiphora whight*), *Aloe vera*, Sonamukhi (*Cassia angustifolia*), Tumba (*Citrullus colocynthis*), Shankhphusphi (Evolvulus alsinoides) etc. A variety of *Aloe* vera, which is a good source of gel and has high aloin content, has been identified in this zone. Farmers also collect and stored shankhpusphi a medicinal crop and collect guggal from tress to sold in market for purchase food grain also. These are principal source of cash income on western Rajasthan in rainfed conditions.

Other Policy Issues

A compressive national policy on "Forestation and desertification " may be prepared including both short term and long term measures should be in place specific to arid and semi arid areas. Also develop and strengthened linkage between research, disciplines and institutions involved in soil and water conservation and sand dune stabilization for efficient short term, medium term and long term planning. Forage production and research should be considered as a national agenda. Shift in investment from irrigated to rainfed areas and emphasis must be focus on aforestation and sand dune stabilization ultimately reflects national developments besides livelihood of human beings of western Rajasthan.

2014, Sustainable Rural Development through Agriculture *Pages* ***350–360***
Editors: **Dr. Shobhana Gupta and Dr. S.S. Tomar**
Published by: **BIOTECH BOOKS, NEW DELHI**

Chapter 26

Seed Production Technology of Wheat (*Triticum aestivum* L.)

***Vipin Chandra Joshi*[1], *M.K. Nautiyal*[2] *and Birendra Prasad*[2]**

Wheat is the staple food for millions of peoples in as many as 43 countries of the world. Wheat provides nourishment to 35 per cent of world population. Wheat cultivation has traditionally been dominated by the northern region of India. Today, India is exporting sufficient quantities of all types of wheat and extensive research efforts are underway for improving its cereals and grain output in the years to come. The major Wheat producing States are Uttar Pradesh, Punjab, Haryana, Madhya Pradesh, Rajasthan, Bihar, Maharashtra, Gujarat, Karnataka, West Bengal, Uttarakhand, Himachal Pradesh and Jammu and Kashmir. The share of wheat in total food grain production is around 35.5 per cent and share in area is about 21.8 per cent of the total area and their food grain.

Wheat is a crop of global significance. It is grown in diversified environments. Wheat is a staple food of millions of people worldwide. It supplies about 20 per cent of the food calories for the worlds growing population. In India wheat is the second most important food crop after rice, both in area and production. Overall India is the second largest producer of wheat after china. The success story of wheat in India is started from the introduction of semi-dwarf wheat, which leads on coming year's green revolution. The wheat production in India increases in many folds from 6.4 million tons in 1950 to 86.87 million tons in 2010-11.

Origin

The exact place and date of the origin of wheat plant that we recognize today is unknown. Hexaploid wheat is widely grown to day thought to have evolved before 7000 BC in an area from just south of the Caspian Sea in northern Iran eastward into northern Afghanistan. De Candolle believed – Valley of Euphrates and Tigris but Vavilov origin of durum wheat probably Abyssinia Soft wheat groups – In the region of Western Pakistan, SW Afghanistan.

Major Cultivated Species of Wheat

Durum - (*T. durum*) the only tetraploid form of wheat widely used today, and the second most widely cultivated wheat today. Einkorn - (*T. monococcum*) A diploid species with wild and cultivated variants. One of the earliest cultivated, but rarely planted today. Common Wheat or Bread wheat - (*T. aestivum*) a hexaploid species that is the most widely cultivated in the world. Emmer - (*T. dicocum*) a tetraploid species, cultivated in ancient times but no longer in widespread use. Spelta – (*T. spelta*).

Floral Biology

Main Culm flowers first and the tillers bloom later in order of their formation.Flowering starts at approximately 2/3 from the base and proceeds in both the directions.Blooming remains throughout the day and it takes 3-5 days for completion.Flower opening is usually during warmer part of the day *i.e.*, between 9 am to 2 pm and peak period between 10 am to 1 pm Anther dehiscence takes place simultaneously and hence the crop is highly self- pollinated (< 1 per cent cross pollination)

Seed

Monocot species like wheat have caryopsis (cereal grains) as propagation units. Caryopses are single-seeded fruits in which the testa (seed coat) is fused with the thin pericarp (fruit coat). Cereal grains have highly developed embryos and in cereal grains the triploid endosperm consists of the starchy endosperm (dead storage tissue) and the aleuronic layer (living cells). Organs of the cereal embryo are: coleoptile (shoot sheath), scutellum, the radicle and the coleorhizae (root sheath). Seed development Stages in Wheat Kernels at various stages during grain filling: a) kernel at watery ripe b) kernel at late milk c) kernel at soft dough d) kernel at hard dough showing loss of green color e) kernel ripe for harvest Physiological maturity: When the kernels have attained maximum dry weight it is physically matured. Note the green color is gone from the peduncle and head parts.

Land to be Used for Seed Production

For wheat seed production the land should be free of volunteer plants. The field should be well drained, Free of weeds. The soil neither too acidic nor to alkaline. Long interval of Crop rotation is desirable previous.

Recent Varieties Released by Central and State Variety Release Committee

Year	*Total*	*State Variety Release Committee*	*Central Variety Release Committee*
2001	6	HD 2733, HW 510	MACS 3125, NIAW 301, WH 711, WH 912
2002	20	DBW 14, GW 322, HD 2781, HI 1500, HS 375, HS 420, HW 2045, MP 4010, NW 2036, MACS 6145, VL 804, VL 829	DWR 225,VL 802, RAJ 3777, PDW 274, MPO 1106, K 9533, HI 1479
2003	9	VL 832, RAJ 4037, PBW 502, HD 2824	DWR 544, TL 2908, NIDW 295, HPW184, HD 2851
2004	11	SKW 196, PDW 291, HD 2864	UP 2565, PBW 509, NW 1076, NW 1067, NIDW 15, K 9423, K 9351, JW 3020
2005	14	DBW 16, DDK 1025, HD 2888, HI 1531, NIAW 917	UP 2572, UP2554, UP 2526, RAJ 6560, HPW 155, GW 1189, COW 1, AKAW 3722
2006	10	DBW 17, K 0307, AKDW 2997-16, GW 366, PBW 533, RAJ 483, TL 2942	UP 2584, MP 1142, DWR 1006
2007	14	HW 1021, PBW 550, VL 892, HS 490, HPW 251, HI 8663, HI 1544, HD 2932, DDK 1029	PBW 527, HD 2894, Birsa Gehun 2, Birsa Gehun 3, CG 511
2008	8	PBW 590, PBW 596, CBW 38, UAS 415, RAJ 4120, JW 3173, MP 1203, MACS 2971	
2009	9	VL 907, JW 1215, KRL 213, KRL 210	JW 1202, JW 3211, CG 516, WH 1025, UP 2628
2010	14	HD 2987, HD 2985, DBW 39, HI 1563, HS 507, MACS 6222, NIAW 1415, PBW 314, AKAW 4627, WHD 943	Wheat GW 11, MP 1201, COW 2, MP 4106

Cropping

The crop should be planted on a field with a known history to avoid contamination from volunteer plants, noxious weeds and soil-borne diseases that are potentially seed transmitted. A wheat seed crop should never immediately follow wheat, unless the wheat crop in the previous season was of the same variety.

Isolation Requirement

Normally a self-pollinated crop 1-4 per cent. It is sufficient to isolate seed fields with a strip of 3 meters all around which is planted with a non-cereal crop, or left uncroped. In cases where variety is susceptible to diseases treated with bavistin, vitavex @ 250g/kg. According to Indian minimum seed certification, standards In case of loose smut isolation required is 150 m from other wheat fields.

Time of Sowing

Long duration varieties like VL616, should be sown during the last week of October. Medeium duration varieties such as UP2003, UP2628, UP2572, WH711, and PBW17 have sown first week of November. Short duration varieties like UP2425,

Top Ten Varieties of Wheat According to the Demand

Sl.No.	2004-05	2005-06	2006-07	2007-08	2008-09	2009-10	2010-11	2011-12	2012-13
1.	PBW 343	LOK 1	PBW 343	LOK 1	LOK 1	LOK 1	PBW 502	PBW 550	PBW 550
2.	LOK 1	PBW 343	LOK 1	PBW 502	PBW 343	PBW 502	PBW 550	HD 2733	GW 273
3.	PBW 373	RAJ 3077	PBW 373	PBW 343	PBW 502	PBW 343	LOK 1	DBW 17	GW 322
4.	RAJ 3765	PBW 343	HD 2189	PBW 373	PBW 443	GW 273	GW 273	PBW 502	GW 366
5.	RAJ 3077	RAJ 3765	RAJ 3765	C 306	RAJ 3765	GW 332	PBW 343	GW 322	LOK 1
6.	HW 234	PBW 502	PBW 502	HW 2004	RAJ 3077	RAJ 3765	RAJ 3765	RAJ 4037	HD 2733
7.	UP 2338	HI 2189	RAJ 3077	UP 2425	PBW 373	WH 711	GW 322	GW 273	DBW 17
8.	JWS 17	HD 2687	HW 2004	RAJ 3765	PBW 550	DBW 17	WH 711	PBW 343	PBW 502
9.	HI 1418	WH 147	WH 147	RAJ 3077	WH 147	PBW 550	PBW 373	RAJ 3765	RAJ 3765
10.	WH 711	HW 2004	GW 273	WH 147	WH711	PBW373	RAJ 4037	WH 711	WH 711

Source: Kheti July 2013.

PBW116 sown during the last week of December. The optimum time of sowing for wheat is when the mean daily temperature is 23±3°C and for good tillering temperature should range between 16-20°C.

Preparation of Land

Deep ploughing must be done with a soil turning plough and Running a harrow before the pre-sowing irrigation.Give a light shallow ploughing or discing after pre-sowing irrigation. Levelling is an important part of seedbed preparation. Keep the seedbed free of weeds.Broadcast BHC, 10 per cent dust at 25 kg per hectare just before the last harrowing or ploughing. It may be added to the fertilizer.This will prevent white ant and Gujhiaattack

Source of Seed

Obtain breeder/foundation seed from a source approved by the certification agency.

Seed Rate

The recommended seed rate for seed crop is 100 kg per ha for timely sown and in late sown conditions 125 kg seed should be used. The seed should be treated with systemic fungicide to control loose smut.

Spacing

The row distance for seed crop should be kept at 22 to 23 cm to facilitate roguing and inspection work. For late sown wheat reduce the line spacing to 15-18 cm.

Crop Rotation

Wheat is mainly grown in rotation with Rice, Sugarcane, Arhar (pigeon pea) and Sorghum, Cotton, Pearl millet, Cluster bean, Sorghum, Groundnut

Method of Sowing

The seed crop is sown in rows with seed drill, or behind the plough in furrows. The depth of seeding should be 5 cm. Seed drill should be thoroughly cleaned and checked before use. Sowing of one variety should be completed before taking up another variety, to avoid mixture. If, for any reason, it has to be used for another variety, it should be thoroughly cleaned and checked so that not even a single seed of the previous variety is left.

Fertilizer

The recommended doses of fertilizers are 120 kg/ha nitrogen, 60 kg/ha phosphorus 40 kg/ha potash 20 kg/ha zinc may be given at the seeding time (in case of deficiency).Apply the whole of the phosphoric and Potassic fertilizers and half of nitrogenous fertilizers while sowing, or just before sowing. Apply the remaining half of nitrogenous fertilizer at first irrigation. In rainfed conditions, all the fertilizer should be applied at the time of sowing as basal.

Field Inspection

Genearlly two field inspections are conducted, first at the time of flowering and final at the time of maturity stage.

Irrigation

Depending on the soil, four to six irrigations may sufficient. The first irrigation should be given at crown root initiation stage, about 30-35days after sowing. Other irrigations should be given at late tillering, late jointing, and flowering, milk and dough stages. Two to three extra irrigations may be needed on light soils. In case of zero tillage, first irrigation should also be applied similar to conventional tillage. Crown root initiation and heading stages are the most critical to moisture stress.

Interculture

Timely weeding and intercultural operations are essential. Weed control by Periodic hoeing and weeding. For control of broad-leaved weeds, spray 2-4 D at@ 0.5kg active ingredient per hectare in 750 liters of water after 25 to 30 days of sowing. For control of Phalaris minor or wild oats make a pre-emergence application of Pendamethalin 2.3 liters per ha.

Rouging

Two or three rouging may be necessary First rouging: Just ahead of the flowering stage, or during flowering to remove any off-type plants which are obvious at this stage of growth.

Wheat Diseases

Flag Smut (*Urocystis agropyri*)

Masses of black teliospores are produced in narrow strips just beneath the epidermis of leaves, leaf sheaths and occasionally the culms. Seed treatment with Carboxin (75 WP @ 2.5 gm/kg seed) or Carbendazium (50 WP @ 2.5 gm/kg seed) or Tebuconazole (2DS @ 1.00 gm/kg seed),two days before sowing. **Loose Smut** (Ustilagotritici) The entire inflorescence, except the rachis, is replaced by masses of smut spores. The disease can occur wherever wheat is grown. Yield losses depend on the number of spikes affected by the disease; incidence is usually less than one percent and rarely exceeds thirty percent of the spikes in any given location. Control: Seed treatment with carboxin (75 WP @ 2.5 gm/kg seed) or carbendazium (50 WP @ 2.5 gm/kg seed) Seed treatment with fungicide should be done one or two days before sowing. **Cover Smut** (*Tilletia tritici*) Plants may be slightly shorter, and the heads are usually darker green than normal and remain green for a longer period. Control: seed treatment with Foliarflo-C, Maxiflo, Vitaflo C, or Vitavax Do not sow seed visibly infested with bunt (cover smut) Only sow disease free seed

Common and Dwarf Bunt (*Tilletia controversa*)

Bunt balls of Dwarf bunt is nearly spherical. When bunt balls are crushed, they give off a fetid or fishy odor. Infected spikes tend to be bluish green in color (or darker), and the glumes tend to spread apart slightly; the bunt balls often become visible after the soft dough stage.

Stem Rust or Black Rust (*Puccinia graminis* f.sp. Tritici)

Pustules (containing masses of urediospores) are dark reddish brown, and may occur on both sides of the leaves, on the stems, and on the spikes. If infection occurs during the early crop stages, the effects can be severe: reductions in tillering and losses in grain weight and quality. Under favorable conditions, complete crop loss can occur. In tarai region brown rust and yellow rust is also a serious problem. Control: Growing, varieties, like GW 322, HD 2781, HUW 510, MACS 2846, In Peninsular Zone CZ, PZ.

Powdery Mildew on Wheat (*Erysiphegraminis* f. sp. Tritici)

The first visible symptoms of this disease are white to pale gray, fuzzy or powdery colonies of mycelia, and conidia on the upper surfaces of leaves and leaf sheaths (especially on lower leaves),and sometimes on the spikes. Older fungal tissue is yellowish gray. Control: one spray of propiconazole (25 EC) @ 0.1 per cent at earhead emergence or appearance of disease (whichever is earlier) is recommended for the powdery mildew prone areas.

Karnal Bunt (Partial Bunt) *Tilletia indica* (syn. *Neovossia indica*)

Karnal bunt is not easily detected prior to harvest, since it is usual for only a few kernels per spike to be affected by the disease. Following harvest, diseased kernels can be easily detected by visual inspection: a mass of black teliospores replaces a portion of the endosperm, and the pericarp may be intact or ruptured. Diseased kernels give off a fetid or fishy odor when crushed. Karnal bunt is a relatively minor disease. Control: one spray of Propiconazole (25EC) @ 0.1 per cent may be given (in seed crop only) at ear head emergence stage.

Wheat Insect Pests

Russian aphid (*Diuraphis noxia*) Russian wheat aphids damage small grains by injecting toxic saliva into the leaves and by sucking sap from the leaves. Yield losses of 50 per cent or more have been attributed to Russian wheat aphids. The feeding of Schizaphisgraminin is especially damaging, resulting in the development of necrotic areas sometimes accompanied by purpling and rolling of the infested leaves. Control :. foliar spray of Imidacloprid 200SL @20g a.i./ha on border rows at the start of the aphid colonization be givenResistant varieties: Halt, Akron (Ankor), Lamar (Prowers99), TAM107 (Prairie Red),Yuma (Yumar) and Stanton.

Harvesting and Threshing

Soon after maturity, the seed crop should be harvested to avoid shattering and losses due to uncertain weather. Most suitable stage is grain moisture of 20-25 per cent.Mechanical harvesting is a common practice for seed production fields. Breeder and pre-basic seed are harvested by plot combine and do not constitute many problems. foundation and certified seeds have to be harvested with commercial combine harvesters.The most critical factors to be considered are: i) Seed moisture content, ii) Mechanical damage iii) Cleanliness of equipment. For seed crops, dry weather during ripening and harvesting is essential.Threshing or combine harvesting at 16 to 19 percent moisture content reduces mechanical damage. Harvesting and Threshing Harvesting may be done by sickle, Combine or reaper, and later the threshing with stationary thresher.

Threshing should be done promptly. Threshing equipment should be cleaned after threshing other wheat varieties. The threshing floor must be thoroughly cleaned to prevent mixtures. Care must be exercised to ensure that laborers do not mix the harvested certified seed with other wheat on the farm.

Processing

After a seed crop has been harvested, the seed, if necessary, has to be dried and cleaned. For wheat seed cleaning, mainly screens, indented cylinders and air screen cleaner are used Screens separate based on the width and thickness; a width separation is obtained by round screens, while for thickness separation oblong screens are used. Indented cylinders carry out length separation; the indents in the cylinder will, depending on their size, lift the seeds, which fit in the indents. Air separates seeds according to their behavior in an air stream. The most important characteristic is the weight; light particles will be lifted, whereas the heavier seed will fall down through the air stream.

Seed Standards

Factor	*Standards for each Class*	
	Foundation	*Certified*
Pure seed (minimum)	98.0 per cent	98.0 per cent
Inert matter (maximum)	2.0 per cent	2.0 per cent
Other crop seeds (maximum)	10/kg	20/kg
Total weed seeds (maximum)	10/kg	20/kg
*Objectionable weed seeds (maximum)	2/kg	5/kg
Seeds infested with Nematode galls of Ear-cockle (Anguina tritici Milne.) and Tundu (Corynebacterium michiganense pv. tritici and A. tritici Milne. complex) (maximum)	None	None
Seeds infected by karnal bunt (Neovossia indica (mitra) Mundkur) (Syn. Tilletia tritici (Bjerk)Wint) (maximum)	0.50 per cent (by number)	0.250 per cent
Germination (minimum)	85 per cent	85 per cent
Moisture (maximum)	12.0 per cent	12.0 per cent
For vapour-proof containers (maximum)	8.0 per cent	8.0 per cent

* Objectionable weeds shall be: wild morning glory (Hirankhuri) (Convolvulus arvensis L.) and Gulli danda (Phalaris minor Retz.)

Pre-Cleaner

It has one air channel to remove light material, one top scalping screen to remove large particles and one bottom grading screen to remove small particles. Dryer if wheat seed is above 11 to 12 percent moisture, it is dried before it goes into bulk storage or processing.

Air-Screen Cleaner

This is the basic cleaner, usually with two air channels and, preferably, four screens. The first air channel removes dust and light materials as the seed falls from the feed hopper. The second air channel removes light seed and materials after the seed passes through the last screen. Screen configurations vary considerably, one or two top or scalping screens remove particles larger than the good seed, and one or two bottom or grading screens remove particles smaller than the good seed. Because the average size of wheat seed varies according to the growing conditions, standard screen sizes cannot be recommended. In general size of Screen aperture for all wheat variety is: Top screen 6.40 mm(R); Bottom screen 2.10 mm(S)

Length Separator

A length separator is almost always used to clean wheat seed. By using the proper machine configuration, shorter or longer undesirable materials are removed. Broken grains and weed seeds, which are shorter than the good seed, are removed by using cylinders with smaller indents. Larger impurities can be removed by using a cylinder with indents that lift all good seed, but contaminants remain in the cylinder.

Gravity Separator

The gravity separator classifies a seed mixture mainly according to density or specific gravity. It can be used to remove unthreshed glumes and soil particles, which have similar sizes to wheat but different weights. Another application is the removal of weevil-infested grains from the seed lot and upgrading seed. Further-more, wild oats and some barley may be removed from the wheat seed lots.

Treater

Wheat seed should, if necessary, is treated with the appropriate fungicide to protect the seed and seedling after planting. Insecticides are sometimes applied to protect seed in storage and in the soil. Treatments may be applied to protect the seedlings or adult plants against pathogens carried on or in the seed. Dryer In humid and hot climates, seeds may be sealed in vapour-tight plastic bags to maintain viability over longer periods. In such cases, wheat seed moisture content must be below 9 percent, preferably not over 8.5 percent.

Seed Yield

The average seed yield varies from 30 to 40 q/ha.

Seed Standards

Contaminants	*Minimum Distance (meters)*	
	Foundation	*Certified*
Fields of other varieties	3	3
Fields of the same variety not conforming tovarietal purity requirements for certification	3	3
Fields of wheat, triticale and rye with infection of Loose smut (Ustilago tritici (Pers.) Jens.) disease in excess of 0.10 per cent and 0.50 per cent in case of Foundation and Certified seed, respectively	150	150
Off-types	0.050	0.20
**Inseparable other crop plants	0.010	0.050
***Plants affected by seed borne disease	0.10	0.50

* Standards for Off-types and inseparable other crops shall be met at the final inspection and for loose smut shall be met at any inspection conducted between ear emergences and harvesting.

** Inseparable othcr crops shall be barley, oats, triticale and gram.

*** Seed borne disease shall be loose smut (Ustilago tritici (Pers.) Jens.).

Seed Treatment

Seed health is an important attribute of quality, and seed used for planting should be free from pests. Seed infection may lead to low germination, reduced field establishment, severe yield loss or a total crop failureIn wheat, fungi (*Fusarium* spp., *Tilletia* spp., *Drechslera* spp., *Septoria* spp. and *Ustilago* spp.), bacteria (*Corynebacterium*, *Pseudomonas* and *Xanthomonas*) and nematodes (*Anguina tritici*) are the most important seed-borne diseases due to their worldwide distribution and losses they incur in crop production (Mamluk and van Leur, 1986; Diekmann, 1996a).Chemical seed treatment is one of the efficient and economic plant protection practices and can be used to control both external and internal seed infection.

Storage

Seed should be harvested when it reaches harvest maturity, dried to a safe moisture content and then stored under favorable conditions and protected from damage and pests until it can be planted. Immature or damaged seed cannot survive long storage periods. Mechanical injury to seed during harvest or handling makes it more susceptible to deterioration in storage. Fungi (*Aspergillus* and *Penicillium*) cause damage to stored seed if seed moisture is high. High storage temperature has a damaging effect on seed. The cleaned, bagged seed should be stored in a dry, insect and rodent proof warehouse. Effective rodent control (traps and poison) is essential in all seed stores. Insects should be controlled by a combination of insecticides and fumigants. Use safest fumigants (*e.g.* Phostoxin) because some fumigants (*e.g.* methyl bromide) will reduce germination. Lesser grain borer (*Rhizopertha dominica*) Angoumois grain moth (*Sitotroga cerealella*) Rice weevil (*Sitophilus oryzae*) Sawtoothed Grain beetle (*Oryzaephilus surinamensis*).

Germination Test

To obtain information with respect to the planting value of the seed and to provide results. This could be used to compare the value of different seed lots. Four replication of 100 seeds Substrata: TP, BP, Temperature: 20°C RH: 95±1 First Count: at 4th day Final Count: at 8th day Dormancy breaking treatment: Preheating (30-350 C)

Seed Moisture

Seed moisture content is one of the most important factor influencing seed quality and storability. Therefore, its estimation during seed quality determination is important. Seed moisture content can be expressed either on wet weight basis or on dry weight basis.

Wheat Varietal Identification

Phenol color reaction extensively used for identification wheat varieties It is easy, quick and reliable test Procedure: Soak 50 seeds in water for 16 hrs Place seeds in Petri dishes on 2 layers of filter paper soaked in 1 per cent phenol solution Petri dishes are immediately covered. Observe after 2 hrs and finally at 4 hrs. The varieties could be grouped into: Nil, no reaction, Light brown, Dark brown, Black.

Conclusion

Wheat is a high-volume, low-value crop and has been produced primarily by heavily subsidized government seed programmes. The private sector, however, may not focus on wheat seed due to its characteristics (self-pollinating, high-volume and low-profit). To meet the demand for improved seeds of wheat, new improved varieties developed by National Agricultural Research Systems (NARSs) should be multiplied and made available to farmers in the shortest possible time. Appropriate seed production techniques coupled with strict quality control measures ensure that varietal purity and identity is maintained, which is the key foundation of the entire quality seed program.

References

Agarwal, R.L. 1993. Seed Technology - Oxford and IBH Publishing Co., New Delhi.

Agarwal, P.K. 1994. Principles of Seed Technology, ICAR Publication, New Delhi.

Tunwar, N.S. and S.V. Singh. 1988. Indian Minimum Seed Certification Standards Published by Central Seed Certification Board, New Delhi.

2014, Sustainable Rural Development through Agriculture *Pages* ***361–377***
Editors: **Dr. Shobhana Gupta and Dr. S.S. Tomar**
Published by: **BIOTECH BOOKS, NEW DELHI**

Chapter 27

Women and Livestock: A Sustainable Management

Sudhir Kumar Rawat, S.C. Singh, M.K. Awasthi and Sarju Narain

Introduction

Without women involvement in livestock, it is not possible to manage the livestock in a proper way. Women in livestock play a vital role in wide range of activities, thereby contributing to livestock development. To achieve inclusive agricultural growth, empowering women by having comprehensive understanding about work participation, gender issues, drudgery and health and nutritional status is necessary. Further, these issues are to be addressed through gender friendly technology assessment, refinement and extension methodologies. Women worldwide play important roles in livestock keeping and provision of livestock services. However, a large number of challenges face the livestock sector, including ensuring food and feed resources, and livelihood security for poor small holder producers and processors. Many countries still face challenges in translating legislation related to women's access to and control of resources into action at the community and household level, impacting women's capacity to control and benefit from livestock. Women and men have different needs and constraints related to livestock production systems. Women often have a predominant role in managing poultry, dairy and other animals that are housed and fed within the homestead. Men are more likely to be involved in constructing housing and herding of grazing animals and in the marketing of products if women's mobility is constrained. Women strongly know the use of eggs, milk and poultry meat for home consumption, and often have control over marketing and the income from these products. Ownership of livestock is particularly attractive and important to women in societies where, due to cultural norms, women's access to land and mobility are restricted (FAO, 2011 and 2012).

Major constraints faced by farm women in adopting improved packages of practices were lack of grazing resources (86.7 per cent), lack of awareness about vaccination and deworming (83.3 per cent), inadequate availability of veterinary services (73.3 per cent), and inadequate knowledge and poor appreciation for AI services (66.6 per cent). Women played major role in care and management of animals; care during pregnancy (36.7 per cent), care during and after parturition (45 per cent), feeding animals (25 per cent), watering animals (48.3 per cent), care of new born (50 per cent), churning of milk (56.7 per cent), making dung cake (68.3 per cent), cleaning shed (41.7 per cent) etc. Similarly, the participation of women in goat rearing overweighed men. However, women were not exposed to training and use of scientific rearing and management of livestock (Annual Report, 2011–12).

Women: The First Livestock Keepers

The beginnings of livestock keeping date back some nine to ten thousand years, and there is good reason to believe that women played a key role in the process of domesticating the major livestock's. Before the commencement of farming, in the hunter gatherer stage of human evolution, humans practiced an archetypal division of labour, men were hunters, while women gathered plants and fruits. The exact cause of the transition to farming remains disputed among scientists, but there is widespread agreement, where sheep and goats were first domesticated, the cultivation of cereals preceded the keeping of domestic animals (Clutton. B.,1999).

Sheep and goats can be easily domesticated due to their small body size and bond easily when young; they certainly fit the requirements for being pre adapted to domestication. Based on ethnographic observations about women nursing young animals, scientists believe that women played a major role in the taming of young stock and in bonding between humans and young animals during the early phases of domestication (Serpell, 1989 and Uerpmann, 1996). The transition from hunting to herding had important connotations and implications for the stratification of human society, while hunted animals usually have no particular owner, domesticated animals are private property; their ownership was the first step towards wealth differentials within society. Considering the role that women, as a home and hearth bound segment of society, may have played in taming, nursing and raising the first livestock, it is logical to assume that they were also the world's first livestock owners (Kohler.R. and Rollefson, 2002).

Why Women Need Livestock

The role of livestock for the rural people is crucial and complex; it goes far beyond just providing marketable products. For the question at hand, it is noteworthy that small scale livestock keepers generally pursue a diverse range of livelihood activities. Instead of specializing in any one activity, such as dairying or fattening, their livelihood portfolio consists of a number of different activities. Natural resource based activities are supplemented by wage labour, trade and crafts to provide for the livestock keepers' various needs and buffer against risks (Waters. B. and Bayer, 1992).

Women's Roles and Responsibilities in Livestock Production

In India, women generally have more control over livestock ownership than elsewhere and have total responsibility for the care and use of small animals, poultry etc. Women also take care of large animals (buffalo, cattle and horses) and their produce, but men have overall responsibility for their husbandry and marketing. In India, the majority of small holders are marginal or landless farmers with small herds of cattle and buffalo whose size and productivity are restricted by the availability of fodder. Jobs performed mostly by women are the milking, feeding and cleaning of animals. Jobs generally shared by men and women are the caring for sick animals, calving and fodder and hay collection. Jobs carried out mostly by men are fodder production, the purchase of feed and medicines, etc., attending difficult births, taking the animals for breeding and the care and handling of male animals. In terms of decision making, the handling and marketing of milk is mostly done by women. Men make decisions about male animals. Decisions on the sale of female animals are generally taken by both men and women. Often the major objective of most intensive systems is commercial dairy production. In northern India, women look after lactating animals, milking them and taking the milk to collection centers run by private traders or dairy cooperatives. The raising of dairy cows is the sole responsibility of women, but men will help in the cow shed if there is an expensive, high yielding breed or if the shed has comparatively modern technology. Women are involved in fodder production, and they do use kitchen wastes to feed animals. Women are heavily involved in livestock production, cleaning out animal sheds, keeping hens, milking animals and preparing manure and butter. Girls and boys collect fodder. Men are responsible for breeding and curing cattle. Men and women are equally responsible for collecting fodder and feeding cattle. Livestock production takes up 98 per cent of the women's non- domestic labour, and crop production the remainder. Small landless farmers rared most of the goat. Women gather their goats on roadsides or wastelands or feed them on cut grass and crop residues.

Women do much of the daily work with livestock, meaning that their roles and responsibilities often are not immediately obvious to people coming from outside the community. In settled mixed farming systems, women and girls usually carry out most of the work related to collecting and cutting feed, bringing water and cleaning pens. If interventions are aimed at intensifying livestock production, such as by shifting from grazing to stall feeding systems or by keeping potentially higher yielding but also higher demanding breeds. In many livestock systems, women customarily care for sick and very young animals kept near the home. If only men are trained to be community animal health workers or "paravets", women's role in animal healthcare is undermined and their knowledge assets are underused. Similarly, when livestock research is conducted in realms where women normally do most of the work, a major part of relevant local knowledge is foregone if the researchers interact only with male household heads rather than including the female members of the household. The different roles of women in the given below:

Women and Land Rights

Women are very much at a disadvantage with respect to land ownership. Worldwide, less than 20 per cent of landholders are women. The situation is especially skewed in India where women constitute less than 10 per cent of landholders. The situation is slightly better in other country (FAO, 2010). Because women rarely own private land, access to common property resources is especially important for them. Locally adapted livestock breeds that have evolved in a particular ecosystem are an ideal means of accessing common property resources and of converting waste and crop by-products into food and fiber and manure. The ability to access common property resources is especially crucial for women, given the enormous gender inequalities that exist with respect to land ownership.

Rural Livelihoods

The role of livestock in rural livelihoods as inferior in terms of productivity is facing serious threats from habitat degradation and cross-breeding. In many rural communities, the exchange of livestock is a way of maintaining social relationships and invoking mutual responsibilities and commitments within the extended kin system. Life cycle events are marked by gifts of livestock, and animals are necessary for giving dowry and paying bride price. Scholars applying the sustainable livelihood framework have concluded that poor households are best served by locally adapted livestock breeds and that a commodity focus is inappropriate for poor livestock development. India, in which a large number of livestock keeping by women. The majority of women were opposed to increasing the number of their animals, whether buffaloes, cows or goats, because of their limited time and resources (Herath, S., 2007).

Leadership

According to Kilavuka (2003) it may be difficult to find out skilled and confident leaders among the women. Some of the challenges women in particular face in group leadership. Further a study of women's groups in Kenya showed that failures of women as leaders have due to them not linked each other, not being unified but selfish; the mismanagement of funds and resources, and problems due to illiteracy and the fact that women like to gossip which in many cases has torn the group apart.

Animal Genetic Resource Management

The roles of women in animal genetic resource management is a serious shortcoming, as ignoring this angle will negatively affect economic and growth aimed at the conservation and sustainable use of animal genetic resources. In traditional societies, the management of livestock, as well as the ownership and control of animals and their products, is very often the domain either of men or women. It is reasonable to assume that a similar division pertains to breeding management. Only if we understand the respective roles of women and men in making breeding decisions and in deciding which type of breed to keep, can we design and implement appropriate intervention strategies and overall policies for the sustainable management of livestock diversity.

Women's Decision Making Power

It is considered that men generally control decision-making, although in many situations, decisions are often made jointly. Women likewise make important decisions when the household is female headed or when the enterprise is one in which women predominate, such as milking, feed and fodder, making and sale milk products, manure, poultry, and kitchen garden produce etc. Thus, any effort to increase the resources and skills available to women in income generating activities should improve their decision making power in the household. Although rural women are heavily involved in almost all aspects of agricultural production. Women are heavily involved in livestock production, with the exception of herding and marketing, decisions related to when to buy or sell, when to vaccinate, feed or water, like most often with the male members of the household. Decisions regarding field crops such as purchase of inputs, and practices such as when to sow, fertilize, irrigation, weeding, harvesting and what machinery to use, mainly rest with the men, although women may be physically involved in many of these activities. Since women are often solely responsible for poultry, dairy products and handicrafts, they are the primary decision-makers in issues regarding these enterprises.

Ownership and Control of Livestock and Livestock Product

In general, men and women tend to own different animal species. In many societies, cattle and larger animals are owned by men, while smaller animals, such as goats and backyard poultry that are kept near the house, are under the control of women. However, there are many exceptions to this general rule, and the widespread transition from agricultural to industrial economies is also changing the equation. In addition, livestock ownership patterns are very complex and layered. Women and girls play a key role in managing the animals. Ownership patterns and control are determined by the type of production system, species, religion and other factors. Often it is men who control the income generated from livestock, but not always (Bravo B., 2000). In Indian cultures, women and all their belongings, including livestock that they may have received from their parents or purchased themselves, are the property of men.

However, women are rarely livestock owners; they often have the right to allocate meat and milk to household members and to sell livestock products. Furthermore, they also have the opportunity to influence breeding decisions. In India, small holder dairy production is often performed by women. They have learned to keep their own personal accounts, and the pattern of income management in women managed households is quite different from that of men (Bravo, B., 2000). In many societies, men do the herding, but selling animals is a joint decision by husband and wife. Among the cattle breeders of men do not sell animals without the permission of the women. Women often handle the sale of sheep and goats, as the men are out herding. Keeping livestock also requires access to grazing resources, and here women usually have entitlements based on their husbands' access. In some societies, women may "own" some animals but have little say about selling or slaughtering them (Talle, 1988). However, in other societies, women may have a say, even though they do not "own" the animals. Women often have rights to use the milk, but there are big differences to extent to which they control the proceeds from selling it. In some societies, the proceeds go to the husband, while in others; the husband has no idea how much his wife earns through milk sales. If interventions demand additional work by women who have little control over the products, then their motivation to participate is likely to be lower as will the level of improvement in livestock production (Waters, B, 1988).

Access to Livestock Services and Markets

Livestock extension, input delivery and financial services staff are usually dominated by men who are most likely to talk with male family members about, for example, how to improve livestock feeding and housing. The women and girls who carry out the actual work receive the relevant information only indirectly, if at all. Information days are often held for existing groups, such as livestock associations or committees, which tend to be composed mainly or purely of men. In some parts of the world, particularly in Muslim areas, there are cultural barriers to direct communication between male advisors and rural women, and as a result, women do not have the same access as men to information that could help increase their work efficiency and productivity. Time consuming training sessions held far from the women's homes may not fit into their busy daily work schedules including care of livestock, and some men may forbid their wives to attend such training. Women are usually less mobile than men and find it more difficult to access services and obtain relevant information. In many countries, because of changing economic circumstances, women are taking on responsibilities for types of livestock that had traditionally been the realm of men, such as cattle in central and southern area. However, livestock service providers are often oblivious to women's changing role and do not give them enough technical, organizational and capacity building support. Many stories are told about how originally women used to own all the livestock but due to their mismanagement their rights were taken away. However today, in most societies, women exercise substantial and recognized rights over livestock which may vary according to the category of livestock, its sources and the purpose of its disposal. However women own only the livestock that they brought with them as dowry.

Women as Income Generators

The success of livestock keepers as a production strategy has been heavily dependent on women's diverse economic roles as traders (Hodgson 2000a). As pressure has increased on societies and economies to become more diversified as livelihoods based on livestock become ever more challenging, women in particular have taken up more income generating activities describe for the women themselves are desperately seeking ways to earn their own income, including dairy projects, goat businesses, piggery, poultry and producing beadwork crafts for the tourist market (Beaman, 1983 and Sikar and Hodgson, 2006).

Credit and Savings

For women savings and credit institutions can have two major benefits. They stabilize income and consumption, not only minimizing sale of livestock during scarcity and drought when prices are low but also savings allow livestock keeper to have more regular income and consumption patterns. They enable people to diversify income sources and reduce vulnerability to future shocks provides an example of credit provision to a cheese making business in India, Credit saved the business, local jobs and stimulated the local economy (Smith *et al.*, 2001; Gamba 2005 and Chakravarty-Kaul, 2008). In the case of livestock loans the repayments were exceptionally low; while in the case of the women's programmed, loan repayments have been exemplary. 84 per cent of those paid to individuals have been paid back, and 68 per cent of those provided to women's cooperatives. The loans were mostly used for dairy processing including making ghee, cheese, butter, curd, flavored milk, srikhand and ice cream.

Several factors were thought to explain the high repayment rates:-

- ✰ Activities were of interest to beneficiaries
- ✰ Adequate pre training on loan management was provided by a Rural Women's Unit who effectively 'adopted' the women who had taken loans
- ✰ A supportive environment was built based on trust between the beneficiaries and the extension agents
- ✰ There was a constant monitoring of beneficiaries' activities by the extension agents
- ✰ There were healthy profits made from the activities. The greatest problem was with marketing which the women had to do themselves.

Women as Natural Resource Users

Women depend heavily on natural resources in many areas for food, water, agriculture, animal fodder and the like. Women tend to collect natural resources closer to home, often whilst carrying out other activities, opportunistically, and can be considered to be 'generalists'. Women can have an intimate relationship with natural resources. For example, Samburu pastoralists make milking bowls out of wood or gourds. "Once made these containers, are regarded almost as human, and their health and vitality is strongly associated with the person who regularly drinks from them" (Straight, 2007).

Many natural resources, particularly plants, are used as medicines and in rituals. Species of plants that can be used as medicine for both humans and livestock, as insecticides and fumigants, for shades and pens. Cuts are treated with the sap of certain plants and soups are prepared with various roots and barks depending on the ailment of the patient. Sheep fat mixed with herbs is given to expectant mothers while babies three months old are fed on cow milk with herbs and root extracts to control colic and provide roughage. When a child is about four years old, the mother teaches him/her about poisonous and edible plants (UNCCD, 2007). Perfume is made from plant extracts and widely used by female pastoralists. Women are utilizing the aloe plant and with assistance sustainably harvesting and processing for commercialization. Women play a central role in natural resource management. Women in particular are intricately interlinked to the environment and are mindful of its needs and variability's. Woman avoids cutting down live trees and before she cuts a branch from a live tree. She has to make a request and give an explanation. Women more often have rights of renewable use for example harvesting leaves from trees, while men have rights of consumptive use harvesting the tree itself.

Gender Issues around Livestock

Many aspects of livestock keeping, including knowledge, labour, ownership, and user rights are gendered, that is men and women have different knowledge about livestock, are in charge of different livestock related tasks, own different types of livestock, and have different rights to the products of livestock. This pertains especially to pastoralist societies, with their long traditions of livestock keeping. However, while gender roles may be deeply embedded in a community's social fabric, they are not written in stone (FAO, 2003). They can and do adapt when the social context of livestock keeping changes. In particular, women tend to take over male tasks if there is no suitable male available to perform urgent work such as taking animals for grazing. Men are said to be more reluctant to take up tasks that are traditionally performed by women.

Rural Women Preferred Local Breeds of Livestock and Poultry

Locally adapted livestock's have additional advantages that are rarely made explicit. It is a self reproducing asset, with an inbuilt self replicating mechanism. The animals generate interest like money in the bank. In fact, the concept of "interest" is assumed to be based on livestock, which multiply if given on loan (Ferguson, 2008). This is again a stark contrast to cross-bred and improved animals. Cross-bred cattle regularly suffer from fertility problems. The broilers and layer hens that are promoted in development interventions are hybrids and therefore do not lend themselves to breeding by farmers, as they lose their hybrid vigor in subsequent generations (Anderson, S., 2003). Moreover, as their brooding instinct has disappeared, they can no longer reproduce on their own. Therefore, the transition to commercial poultry obliges the farmer to keep purchasing new batches of chickens. Commercially bred pigs are hybrids; natural reproduction has often become problematic for them because of the enormous development of their hind quarters.

Reasons given by women livestock keepers in India for preferring indigenous breeds (Rangnekar, 2002):

- ☆ Low external input feed, fodder, medicine, advice
- ☆ Well adapted to local conditions and have fewer health problems
- ☆ Thrive on local feed, fodder and coarse roughage
- ☆ Easy to handle and manage
- ☆ Replacements are easily available
- ☆ Backyard poultry
- ☆ Well adapted to local conditions– no disease problems once they are adult
- ☆ Low external input – thrive on waste, insects, weeds, etc.
- ☆ Can protect themselves from predators

- ✰ Good market demand for their produce, which sells at a premium
- ✰ Replacements are easily available.
- ✰ Management of livestock near the household

However it is more often the case that women play a greater role in the management of livestock kept around the homestead.Women care for newborn and young animals, which are not old enough to go to pasture with the herd, together with ill animals kept away from the other animals. Women also remove ectoparasites such as ticks, collect manure and restrain animals when necessary as well as help with cutting up of fodder. Women may be responsible for caring for and counting the grazing animals as they come home for the night and for signaling any problems (sickness, birthing, poor health, and missing animals).

Milking by Women

In most societies, women are responsible for milking the livestock, whether sheep, goat, cattle or camels in India (Geerlings, 2004). However though in most circumstances women, it is suggested that women are prevented from decision making about the herd (Nduma, *ct al.*, 2000). As confirm they may not have any control over decisions about which animals are sent out grazing and which remain at the house. It is said that they will give first priority to satisfying the milk needs of their children while men put the needs of calves and by implication the herd first (Joekes and Pointing 1991). Milk selling tends to be the domain of women, once the women have been given the milk, they will decide on how much they will allocate for home consumption and how much they will sell. Similarly sheep and goat products, most especially dairy products, are shown to be tied directly to them, and the roles they play and the power they wield in the community. This also helps to explain female ownership of part of the herd, and the special responsibilities women share in the feeding of the young, and their eventual weaning from their mothers.

Women have acquired milk separators for cheese preparation, not only reducing the drudgery of cheese making but hiring them out for a fee to other women. The cheese is hauled to town in a cart and sold by women. The main products are nano (sour milk) and butter which can last for several days without refrigeration. Women deposit their dairy products at the depot for storage and marketing, and come back to collect their receipts afterwards. Amongst other things they have provided training on basic milk hygiene, milk collection, handling and quality control, built a Milk Centre which acts as a focus for villagers' activities including testing of milk quality, training etc., improved market access and strengthened milk collection networks through simple cooling facilities and introducing appropriate processing technology for production of storable and marketable milk products such as oriental sweets, condensed milk, flavored milk, srikhand, cheese, butter and ghee etc.

Breeding Aspect

Though it is often assumed that men know more about breeding livestock than women, this need not be the case in reality. Women are often more knowledgeable when it comes to assessing the mothering abilities of calf, ewes and issues relating to milk production (Geerlings 2004). Because they care for new born lambs they are also very knowledgeable about the character, vitality and health of lambs. Ramdas (2007) suggest that women in southern India recognize the value of indigenous breeds of cattle which have qualities to adapt as against new breeds which may perish.

Fodder Management

Women tend to be responsible for collection of fodder in order to supplement the feed of the livestock kept close to the homestead. Married women traditionally supplement the calf diet with cut and carried native grasses and water hauled form wells and springs. Over 90 per cent of dry season fodder reserves for animals and other less mobile stock. However, again it should not be assumed that it is only women who collect fodder in India (Ramdas, S., 1999). For example it tends to be the men who are responsible for fodder collection. However, it is often the case that women too are experts, particularly those who are involved in livestock herding and grazing (Geerlings, 2004). For example, shepherdesses have good traditional knowledge of the best grazing grounds, on which they base daily decisions about animal grazing and for the optimal use of forage, and of pasture rotation systems, in order to prevent overgrazing.

Manure Utilization

One of the most highly valued animal products in India is manure. Manure collected by women at the homestead is used for fertilizer, fuel and building houses. It is so valuable that old animals are kept even when they no longer produce milk or are strong enough to pull a plough. In addition, there is a widespread market in manure, often used to pay for services and labour and exchanged as a gift among relatives and friends. In all these operations, it is the women who are responsible for procuring and processing the manure (Mc Corckle, *et al.*, 1987).

Hides and Wool

There is a dearth of information on the use of and sale of other animal products such as hides, wool, horn etc. Only two examples were found both on the development of successful businesses based on wool and woolen products. This is despite the fact that women can have a dominant role in the curing and selling of hides and wool processing. Again, when there are large numbers of hides to sell, men may take over believing that women cannot handle business deals or large amounts of money (Wangui, 2003). Indeed in much women process livestock fiber products from sheep, goats and camels into articles such as carpets and clothing for home and for sale. The women were given processing equipment including spinning wheels and taught how to spin and dye Merino wool that is produced by the Merino sheep. Since women are recognized within the rural society as having specialized knowledge and abilities in handling animal fibers, the development of wool enterprises will enable women to raise their status by yielding more income for their families (GL-CRSP 2006).

Purchasing, Sale, Livestock, Products and Disposal of Livestock

The women of highly involved in informal milk marketing, the women revealed that about one third of the total cash income of the households came from milk sales. This was a regular source of income with which the women could meet the family's daily needs. Interventions intended to increase milk production by encouraging men to grow improved pastures and to feed supplements to cows did not lead to significantly higher milk off take/yield, because the men did the milking and thus controlled the off take, but the women controlled the income from milk sales and decided how this was used. The women seldom invested the milk income in the inputs needed for pasture and feeding, this being the role of men. The men controlled the income from animal sales, so targeted the use of the inputs and the intensity of milking with a view to reducing animal mortality and increasing livestock off take rates, *i.e.* meat rather than milk production. While men may have management control over livestock, they can not freely dispose of animals in which women or children have rights. In most cases purchase, disposal or sale of livestock is discussed between husband and wife prior to action, and often a wife's approval is needed. Women negotiate dung prices with farmers, while men negotiate wool prices and as amongst the woman will make the decisions concerning her own herd there can be conflicts between men and women and their different priorities in raising livestock for different purposes. Women fatten rams for home consumption or for sale (Geerlings, 2004).

Veterinary Works

In the early 1990s the trained only men despite the obvious knowledge that women had of livestock due to their livestock related roles and the consent of both male and female community members that women could and should be trained. Not only were the men trained to deal with illnesses in livestock that was related to their own roles, but also to treat illnesses that normally a woman would take care of. As such men gained increased access to women's domains and interests and this risked their taking over such as milking and thus access to milk (Catley, A. S., *et al.*, 2002).

It is said that men tend to have the most veterinary knowledge. While men typically deal with diagnosis and the choice of treatment, women collect and prepare various herbs used in traditional remedies. If modern medicines are used, then it is usually the man who procures and administers them. However, recent veterinary research not only shows that the women play a greater role in the care of livestock, but suggests that they also know as much, and sometimes more, about livestock health and disease than the men (Davis, 2005). As described above women have a close relationship with many of the herd. It is suggested that the differences in labour responsibilities for veterinary work relates to the different veterinary knowledge of women and men. For example women have more knowledge of external and internal parasites, as well as such as mastitis. Because they do the milking, women are often the first to notice behavioral changes and other initials signs of disease (Kohler. R. and Rathore, 2000). Women are responsible for delicate tasks such as giving injections and children look after calves and lambs under the supervision of women. However extension workers are invariably men and often barred by custom from addressing women. Not surprisingly, advice on the treatment of such as veterinary mastitis has not reached the majority of women. Indeed, veterinary services and extension programmed, and advisory services are mainly designed by men for men (Bravo, B. 2000). Extension personnel are often not trained to teach technical subjects to women or to react to their specific questions. Due to limited time and resources primarily, attention is given to men's animals. Extension work with women often requires special didactic knowledge and communication skills because women often only speak the local language or dialect and illiteracy is high. Few if any women are trained as animal health workers despite them often having better knowledge of small ruminants, milking cows but also risked removing women's access to such as milk, because men were the ones trained in milk related illnesses (Riviere. C. and Eregae, 2003). This is despite the many benefits of including women as animal health workers. One reason often given for not including women in trainings for animal health workers is that they are illiterate.

Women in Fisheries and Agro Forestry

Compared to crop and livestock production, fisheries and agro forestry are not major sources of employment in the India. The proportion of total land area grown to forests is less than 22 per cent in India. Indicates these sub sectors are generally male dominated, and that women play very limited roles in fisheries and agro forestry. In fisheries, some women may be involved in net making, net maintenance, net repair, marketing of fish and fish products. Some women are involved to a limited extent in processing fish, and medical purposes. A very small number of women in southern area participate in fish catching and feeding. Women's participation in agro forestry is also limited in the region, with the exception of the Sudan, where women are solely responsible for agro forestry, where labour is shared with men. In the many regions, women are involved in some production and transplanting of seedlings.

Need to Strengthen Local Women's Organizations

Particularly good results in empowering women have been gained through encouraging women to organize themselves around production and processing of livestock products. It is usually easier for groups of women rather than individuals to access resources for production, also through credit, and to achieve economies of scale in marketing the products. Existing informal groupings, whether traditional or more recently developed by the women themselves can provide good starting points for enhancing women's managerial and leadership skills. This can eventually lead to women becoming more active in community based organizations involving both men and women (Haramata, 2006).

Need of Education and Training for Improvement in Women's and Girls'

Women and girls need better access to general education as well as to specific training and information related to livestock keeping. To improve livestock husbandry and value addition to animal products, women need to be trained directly, not through second hand information via male family members. They need training in literacy and numeracy, small enterprise management, and group management and leadership. Extension agents for crop and livestock husbandry are usually male, whereas those for home economics in countries where such advisory services are offered are usually for female. If female agricultural extension and home agents are trained in livestock production, marketing and participatory experimentation for local adaptation of technologies, they will be able to give relevant support to rural women (Nori, *et al.*, 2006; Pantuliano, 2002 and Waters. B., 1988). Extension staff both male and female should also be capacitated to facilitate community discussion on gender issues that affect family welfare, such as property and inheritance rights to livestock, land and other resources. Rural women can more easily take part in training and other extension activities that take place in or near their villages rather than in district. Ways need to be explored further to improve women's access to livestock information, such as through radio and village based information and communication technology (ICT), and their access to livestock services such as veterinary care, *e.g.* by training both men and women as paravets. In addition, where transmission of knowledge and skills to younger generations is being ruptured by livestock, it is vital to offer orphaned girls and boys possibilities to develop their livestock knowledge and skills.

Recognize Dynamism and Openings for Positive Change

Livestock plays multifunctional and changing roles in poor households, especially those that are confronted by rapid changes in their livelihood possibilities. In efforts to survive despite these changes, local people develop their own coping mechanisms and adaptations and explore alternative ways of making livelihoods from livestock. Some women have developed new ways of organizing themselves and collaborating with men so as to gain better access to more lucrative markets for livestock products. Many of these innovations, including changes in women's roles and activities, serve to maintain or enhance the multiple functions of livestock. This has been confirmed by more recent work in India, where ingenious women are using locally available resources to improve the husbandry of goats and chickens. By growing supplementary feed for the chickens to encourage them to stay at home rather than wandering to other households. Giving this woman the opportunity to share her knowledge with other farmers has further strengthened her self confidence as well as the gender sensitivity. Women in crop farming and livestock keeper societies often lack confidence and undervalue their own achievements. Therefore, important steps to empower women are to raise awareness of how they contribute to livestock development through their own innovation and to support this innovation process.

Need to Maintain Gender Equality in Livestock Services and Organizations

In addition to recognizing the situation and seeking gender equality at grassroots level, it is also necessary to sensitize people in organizations working with livestock keepers (research, extension, education, private sector) about gender issues at the grassroots level and to seek gender equality in these very organizations (Book, A., 2013 and Panjwani, A., 2005). Beintema, *et al.* (2010) found that, on average, women made up less than 22 percent of professional staff trained in animal sciences in 15

African countries Although their numbers are growing in relative and absolute terms, particularly in the industrialized countries, women are still in the minority among graduates in animal sciences, range and pasture science and veterinary medicine, as well as in research, development and education institutions concerned with livestock production.

Conclusion

There is huge information about gender aspects of animal genetic resources management. Women play a vital role in sustainable animal genetic resources management than men. In a global scenario in which the livestock sector is undergoing rapid and dramatic change due to demand for meat, milk, and eggs. It is mainly women that act as guardians of the remaining locally adapted livestock breeds. This is due to women's responsibility for shouldering the reproductive economy women's roles and responsibilities in animal production are recognized, if women have more rights of ownership over livestock, if women have better access to livestock services and markets, if women have more say in decision making about inputs and outputs of animal production and have more control over the income from this, then family welfare can be improved and poverty and hunger can be reduced up to the mark. In addition to meeting the basic needs of women and their families, enhancement of the role of women in the livestock value chain helps address their strategic needs. A major contribution to focusing attention and action on empowering women through livestock can be made by spreading powerful images of women who use livestock to meet family and community needs and to address their strategic interests. Such high profile documentation would give strong messages to women and men at all levels about women's actual and potential contributions to livestock production, and help to change perceptions and attitudes at all levels. It is obviously an uphill struggle to change the perceptions of many agricultural R&D professionals about the contribution that women can make to livestock development and the contribution that livestock can make to enhance the economic and socio political status of women.

References

Anderson, S. (2003). Animal genetic resources and sustainable livelihoods. Ecological Economics, 45(3): 331–339.

Annual Report. (2011-12). Empowering Women in Agriculture, Dare/ICAR, pp 93-95

Beaman, A. (1983). "Women's participation in pastoral economy: income maximization among the Rendille" Nomadic Peoples Vol. 12: 2-25.

Beintema, N.M., and Di Marcantonio, F. (2010). Female participation in African agricultural research and higher education: New insights. IFPRI Discussion.

Book, A. (2013). Gender and Livestock Production, Deutsche Gesellschaft für Internationale Zusammenarbeit (GIZ) GmbH.

Bravo-Baumann, H. (2000) Gender and Livestock. Capitalization of Experiences on Livestock Projects and Gender. Working document. Berne: SDC.

Catley, A. S. Blakeway and Leyland, T. (2002). Community-based Animal Healthcare. UK: ITDG.

Chakravarty-Kaul, M. (2008). WISP Good Practice Study on Gender and Pastoralism / Women's Empowerment – Asia.

Clutton-Brock, J. (1999). A natural history of domesticated mammals. Cambridge, UK, Cambridge University Press

Davis, D. (2005). "A Space of her Own. Women, Work and Desire in an Afghan Nomad Community" in G. W. Falah and C. Nagel,

FAO, (2003). Livestock and gender: the Tanzanian experience in different livestock production systems, by C. Hill. A Glance at LinKS: LinKS Project Case Study No.3. Rome

FAO, (2010). Gender and land rights. Understanding complexities; adjusting policies. Economic and Social Perspectives Policy Brief No. 8. Rome

FAO, (2011). The role of women in agriculture. ESA Working paper No. 11-02

FAO, (2012). Invisible Guardians - Women manage livestock diversity. Animal Production and Health Paper No. 174. Rome, Italy.

Ferguson, N. (2008). The ascent of money: a financial history of the world. London, Allen Lane.

Gamba, P. (2005). Policy Analysis Study. Improving Market Access for Drylands Commodities Project. Unpublished report for the EU, Nairobi, Kenya.

Geerlings, E. (2004). "The Black Sheep of Rajasthan" Seedling October 2004.Geographies of Muslim Women. Gender, Development and Religion. US: Guildford Press

GL-CRSP. (2006). Developing Institutions and Capacity for Sheep and Fiber Marketing in Central Asia (WOOL). Annual Report 2006. US: GL-CRSP.

Haramata, (2006). "Women and leadership: lessons from the Sahel" Haramata No. 49: 8 -11.

Herath, S. (2007). Women in livestock development in Asia. Journal of Commonwealth Veterinary Association, 24(1): 29–37.

Hodgson, D. (2000a). "Gender, Culture and Myth of the Patriarchal Pastoralist." *In*: D. Hodgson (ed.), Rethinking Pastoralism in Africa. London: James Currey.

Joekes, S. and Pointing, J. (1991). Women in Pastoral Societies in East and West Africa. Dryland Issues Paper No 28. London: IIED.

Kiluva, J. M. (2003). A Comparative Study of the Socio-Economic Implications of Rural Women, Men, and Mixed Self-Help

Kohler-Rollefson, I. and Rathore H. S. (2000). " Building on pastoralists' cosmovision" Compas Newsletter July: 19-20.

Kohler-Rollefson, I. and Rollefson, G. (2002). Brooding about breeding: social implications for the process of animal domestication. *In*: R.T.J. Cappers and S.

Bottema, eds. The dawn of farming in the Near East. Studies in Early Near Eastern Production, Subsistence, and Environment, 6. pp. 177–182. Berlin, Ex Oriente.

McCorckle, C. (1987). Highlights from sociological (crsp) research on small ruminants. Pastoral Development Network Paper 24d, London : ODI.

Nduma, I. P., Kristjanson and McPeak, J. (2000). Diversity in income generating activities for sedentarized pastoral women in Northern Kenya Submitted to Human Organization, Nov 2000.

Nori, M., Kenyanjui, M.B., Yussef, M.A., and Mohammed, F.H. (2006). Milking drylands: The emergence of camel milk markets in stateless Somali areas. Nomadic Peoples, 10 (1): 9–28.

Panjwani, A. (2005). Energy as a key variable in promoting gender equality and empowering women: A gender and energy perspective on MDG No. 3.Discussion paper.

Pantuliano, S. (2002). Sustaining livelihoods across the rural-urban divide: Changes and challenges facing the Beja pastoralists of north eastern Sudan. Pastoral Land Tenure Studies 14. International Institute of Environment and Development (IIED), London, UK.

Rangnekar, S. (2002). Perception of wcmen about selection/breeding of local livestock. *In*: Local livestock breeds for sustainable rural livelihoods. Towards community-based approaches for animal genetic resources conservation. Proceedings of a workshop held on 1-4 November, 2000 in Udaipur and Sadri, Rajasthan, India, pp 69-74. Sadri, India, Lokhit Pashu-Palak Sansthan.

Ramdas, S. (1999). Between the green pasture and beyond. an analytical study of gender issue in the livestock sector, Orissa.Technical Report No. 21. India: Indo-Swiss Natural Resources Management Programmes.

Ramdas, S. and Ghotge, N. (2007). "Whose Rights? Women in Pastoralists and Shifting Cultivation Communities.

Riviere-Cinnamond, A. and Eregae, M. (2003). Community Animal Health Workers (CAHWs) in Pastoralist Areas of Kenya: A Study on Selection Processes, Impact and Sustainability. Kenya: AU/IBAR.

Serpell, J. (1989). Pet keeping and animal domestication: a reappraisal. *In*: Journal of Clutton-Brock, ed. The walking larder: patterns of domestication, pastoralism and predation, pp. 10–21. London, Unwin Hyman.

Sikar, N. K. and Hodgson, D. L. (2006). "In the Shadow of the MDGs: Pastoralist Women and Children in Tanzania" in Indigenous Affairs Vol. 01/06.

Smith, D., Gordon, A., Meadows, K. and Zwick, K. (2001). "Livelihood Diversification in Uganda : Patterns and Determinants of Change Across Two Rural Districts" Food Policy Vol 26 : 421-35.

Straight, B. (2007). "Development Ideologies and Local Knowledge among Samburu Women in Northern Kenya". *In:* D. Hodgson (ed). Rethinking Pastoralism: Gender, Culture and the Myth of the Patriarchal Pastoralist. London: James Currey.

Talle, A. (1988). Women at a loss: Changes in Maasai pastoralism and their effects on gender relations. Thesis, Dept of Social Anthropology, University of Stockholm, Stockholm, Sweden.

Uerpmann, H.P. (1996). Animal domestication – accident or intention? *In*: D.R. Harris, ed. The origin and spread of agriculture and pastoralism in Eurasia, pp. 227–237. London, UCL Press.

UNCCD. (2007). Women Pastoralists. Preserving Traditional Knowledge. Facing Modern Challenges. Bonn: UNCCD.

Wangui, E. (2003). Links between Gendered Division of Labour and Land Use in Kajiado District. Kenya. The Land Use Change, Impacts and Dynamics Project Working Paper No. 23.

Waters-Bayer, A. (1988). Dairying by settled Fulani agropastoralists in central Nigeria: The role of women and implications for dairy development. Vauk Wissenschaftsverlag, Kiel, Germany.

Waters-Bayer, A. and Bayer, W. (1992). The role of livestock in the rural economy. Nomadic Peoples, 31: 3–18

Index

A

Acacia nilotica 348
Acacia senegal 131, 142
Acacia tortilis 131, 142, 347
Advertising 278
AESA 252
Aestivum 38
Affordability 13
Agri-biodiversity 144
Agri-business 133, 241, 276, 281
Agri-business education 278
Agri-business system 277
Agri-clinics 133
Agri-horticultural system 77
Agricultural biodiversity 25, 141
Agricultural crops 91
Agricultural development 93
Agricultural emission 115
Agricultural extension 234
Agricultural growth 62, 338
Agricultural input sector 278
Agricultural insurance 134
Agricultural marketing-distribution sector 278
Agricultural processing-manufacturing sector 278
Agricultural product price insecurity 19
Agricultural production 23, 94
Agricultural production growth rate 16
Agricultural production sector 279
Agricultural risk management practices 149
Agriculture 1, 8, 17, 33
Agro forestry 372
Agro-based Industries 262
Agro-climatic zones 283
Agro-forestry Industry 272
Agro-forestry system 138
Agro-horticulture system 138
Agroforestry 131
Agronomic approaches 72
Agronomy 24
AICRPAM 66
AICRPDA 66
Air-screen cleaner 358

Alley cropping 76
Alley cropping system 138
Aloe vera 134, 349
Alphonso 190
Alternate use 345
Animal feeding 310, 313, 314
Animal genetic resource management 364
Animal production 307
Annapoorna 14
APEDA 271
Apiculture industry 274
APL 21
Apple pomace 312
Arid agro-ecosystem 144
Aristida funiculata 131, 348
Aromatic crops 134, 349
ATIC 236
ATICs 234
ATMA 236, 238
ATP 198
Aurumani 193
AVRDC 47, 55
Azotobacter 138

B

Bajra 137
Bangalora 190
Banganpalli 191
BBC 4
Bed preparation 172
Bedding system 70
Biennial bearing 204
Bio-fertilizers 283
Biodiversity 16, 117, 123, 143, 145
Biotechnology 46, 52
Black Rust 356
Bombai 191
Bombay green 191
Bottle gourd 300
Breeding aspect 369
Brinjal 289
Bulb-to-seed method 293

C

Cabbage 303
CAG 12
Calligonum polygonoides 131, 142, 347
Capparis decidua 131, 142
Carbon dioxide 120
Carbon ratio 205
Carrot 294
Cassia angustifolia 134, 349
Castor 137
Cauliflower 305
Cenchrus ciliaris 347
Cenchrus setigarus 131, 347
CFC 115
CH_4 29, 113, 115
Chausa 192
Chilli 290
Chilli cultivation 225
CIDA 90
CIIFAD 169
Citrullus colocynthis 134, 349
Cleaning 290, 291, 301
Climate 104, 195, 305
Climate change 18, 29, 30, 33, 42, 46, 54, 113, 116, 119
Climatological factors 205
Clitoria ternatea 348
Clusterbean 137
CO_2 29, 30, 113, 115
Cold chain management 334
Commiphora whight 134, 349
Communication 161

Communication network 94
Composition 189
Composting 130
Contour bunding 136
Contract farming 323
Cooperation 134
Cooperative system 323
Cost components 321
Cottonseed hulls 310
Cow pea 74
CPR 132
Credit 134, 161, 339, 366
Credit flow 339
CRIDA 65
Crop diversification 142, 147, 218
Crop insurance 25, 123
Crop management 164
Crop production 67
Crop rotation 137, 354
Cropping 352
Cropping patterns 75
Cropping season 122
Cropping systems 74, 137
Crops 104
Cross pollinated crops 185
Cruciferous 302
CSP 200
Cucumber 299
Cucurbits 295
Cultivation 214
Cultivation practices 85
Cultural practices 48
Curd pruning 305
Cyclone 117
Cyclones 150
Cynodon dactylon 308

D

Dairy industry 268
DAP 177
Dashehari 191
Desertification 349
Disaster preparedness 123
Diseases 117, 160, 176, 177, 183
Distribution 13
Diversification 147
Drought 45, 117, 128, 160
Drought management 134
Drought mitigation 127
Drought tolerance 51
Droughts 150
Dry farming 77
Dry farming zones 67
Dry matter 206
Drying 292, 301
Dryland farming 62
Durum 38

E

e-chaupal 18
Earthquakes 150
Eco-development 91
Education 93
EIU 4
Employment programmes 14
Empowering women 239
Engineering stress tolerance 53
Environmental quality 176
Extension approaches 234, 242
Extension service 235
Extension system 231
Extension workers 251
Extraction 301

F

FACs 238
FAO 4, 43, 141, 248
Farm credits 5
Farm forestry system 138
Farm school 242
Farm yard manure 172
Farmer field school 247
Farmers feedback 36
Farmers' indebtedness 17
FCI 12, 20
Feeding programmes 14
Fermentation method 288
Fernnadin 191
Fertilization 198
Fertilizer 354
Fertilizer consumption 93
Fertilizer management 74
Fertilizer placement 200
FFS 242, 247
Field crops 31
Field inspection 355
Field preparation 196
FIG 242
Financial risks 153
Financial sustainability 240
Fiscal deficit 21
Fisheries 267, 372
Fisheries industry 271
Fishery 94
Flood 117, 150
Flooding 46
Floral biology 351
Floriculture industry 271
Flowering 202
Fodder 118, 283
Fodder banks 131
Fodder management 370
Food availability 13
Food grain production 10
Food processing industry 267
Food production 3
Food safety 14
Food schemes 14
Food security 3, 6, 7, 13, 20, 22
Food security act 14
Food self-sufficiency 17
Food subsidy programmes 14
Food utilization 14
Forest trees 32
Forests 16
Fruit growth 204
Fruit harvest 300
Fruit set 202, 296
Fruit setting 300
Fruits 120

G

GCMMF 321
GDP 17, 19, 276, 338
Genetic approaches 71
Genetic purity 182
Genetic resources conservation 162
Genomics 46, 52
Genotype 121
Germination test 360
GHGs 42, 114
GIS 239
Glycyrrhiza glabra 134, 349
Grading 207
Grafting 49
Grain processing 267
Granaries 129
Granite formation 68
Grassland 91

Gravity separator 358
Green gram 74
Green manure crop 172
Green manuring 138
Green revolution 18, 23, 145
Greengram 137
Groundnut 137

H

Hand pollination 186, 297
Harmful effects 88
Harvest 300
Harvesting 206, 288, 289, 290, 291, 292, 293, 294, 295, 297, 299, 301, 304, 306, 356
Harvesting process 86
Heat Use efficiency 39
Hedging 206
Hides 370
Hilly states 81
Himsagar 191
Hormonal balance 205
Human resource risks 154
Hybrid varieties 193

I

ICAR 213, 231
ICEF 90
ICT 18, 234, 334, 373
IFFCO 239, 242
IIM 279
Improved technology 93
Income generators 366
Indian economy 263
Indigenous technical knowledge 103, 123
Indigophora cordifolia 131
Indo-Gangetic plains 68
Innovated shifting agriculture 87
Insects 104
Insurance 278
Integrated approach 285
Integrated nutrient 64
Integrated nutrient management 138
Integrated pest management 247
Intercropping 201
Interculture 202, 355
IPCC 31, 117
IRMA 279
Irrigation 173, 197, 355
Irrigation management 47
Irrigation water 51
Irrigations 178
Isolation 288, 289, 290, 291, 292, 293, 294, 295, 296, 298, 300, 305
Isolation requirement 352
ITC 271
IVLP 234, 237

J

Jhum 81

K

Karnal bunt 356
KCC 134, 239
Kesar 191
Kharif onion 222
Knowledge management 149
KRIBHCO 242
KVK 122, 213, 214, 234, 235

L

Labour saving technology 161
Lac culture industry 275
Land degradation 346
Land leveling 136
Land rights 364
Land use 76
Land use management 122

Langra 192
LDBs 339
Leadership 364
Leather industry 270
Length separator 358
Lesiurus sindicus 131
Livelihood 345
Livestock 94, 120, 129, 361, 368, 371
Livestock production 363
Livestock services 373

M

MAFC 162
Mahatma Gandhi National Rural Employment Guarantee Act 14, 238
MANAGE 279
Mangifera indica 188
Mangoes 188, 208
Manjeera 194
Manure utilization 370
Manuring 198
Market demand 21
Market integration 334
Market risks 152
Marketing 161
Marketing efficiency 326
MAS 53
Mass media 239
Match industry 273
Maturity testing 206
Meat industry 271
Medicinal properties 208
Methane emission 120
Methodology 150
Micro credit 243
Milk marketing 318, 321
Milk marketing channels 320
Milk procurement 322
Milk products 317
Milking 369
MMPs 333
Modified shifting agriculture 87
Moisture conservation 136
Monsoon 117
Monsoon grasses 308
Moth 137
Mulgoa 192
Mutations 182

N

N_2O 115
NAARM 279
NABARD 340
NAIP 234, 238
NAIS 134
National Food Security Mission (NFSM) 25, 238
National rural health mission 15
NATP 234
Natural crossing 182
Natural resource 16, 367
Natural resource management 24
NBARD 339
NEPED 90
NERICA 162
NFE 255
NGOs 243, 343
NHB 271
NIAM 279
Nitrogen 164
Nitrous oxide 120
NPCC 66
NPGRC 162
NPK 177
Nutritional value 14

O

Okra 292
Onion 292
Orchard management 188
Orchard site 195

P

PACBs 153
Packing 206
Panicum antidotale 348
Participatory monitoring 254
PDS 11, 20
Pea 291
Pests 117, 160, 176, 183
Pests management 64, 123
Phenological development 36
Phenothermal index 39
Photosynthesis 32
Physiological approaches 70
Pits 196
Plant breeder 183
Plant protection 76
Plant respiration processes 32
Planting 196
Plywood industry 273
Policy reforms 234
Policy risks 155
Pollination 202, 296, 301
Post-harvest losses 18
Post-harvest processing technologies 161
Poultry 368
Poultry processing 267
Poverty 8, 16
Pre-cleaner 358
Preparation 354
Preventing sap burn 207
Pricing policy 333
Processing 165, 294, 298, 357
Procurement cost 332
Products 371
Prosopis cineraria 131, 142, 348
Prosopis juliflora 131, 142, 348
Pruning 201, 202
PSC 92
PTD 255
Public private partnership 241
Public-private sector cooperation 270
Pulp and paper industry 273
Pulses 119
Pumpkin 298

Q

QTLs 53

R

Radish 295
Rajiv Gandhi Drinking Water Mission 15
RAPD 53
Ratna 193
RCT 123
Restorer period 86
Rice cultivation 178
Rice darmers 150, 156
Rice farming 149
Rice intensification 168
Ridges 69
Ripening 207
RKC 235
RNA 53
Roguing 288, 289, 290, 291, 293, 296, 298, 300, 301, 304, 306
Root-to-seed 294, 295
Rouging 355
RRBs 134, 153, 339
Rural development 262

Rural finance 338
Rural livelihoods 364
Rural transport 94
Rural women 368

S

S. chilense 51
S. peruvianum 51
Saline soils 51
Salinity 45
Salvadora oleoides 131
Salvadora persica 131
SAMETI 238
Sargassum wettai 314
Savings 129, 366
Science 103
Scooping 69, 305
Sea level 117
Sea weeds 313
Seed 129, 285, 351, 354
Seed banks 130
Seed drying 298, 299
Seed extraction 288, 290, 297, 298, 299, 300
Seed industry 269
Seed moisture 360
Seed production 291, 295, 304, 305, 351
Seed production technology 287, 350
Seed rate 354
Seed storage 299
Seed treatment 359
Seed-to-seed 293, 294
Sehima nervosum 347
Select 289
Selection 289, 290, 291, 292, 293, 294, 295, 297
Self-pollinated crops 185
Sericulture 274
Sesame 74, 137
Sex expression 296
Sharbati 38
SHGs 243, 342
Shifting cultivation 81
Shoots 205
Shrinking 16
Silvi-horticulture system 138
Silvi-pastoral System 77, 138
Skin browning 207
Social safety net programmes 14
Socio-economic constraints 95
Sodium carbonate 288
Soil 136, 195
Soil composition 87
Soil fertility 142, 163, 171
Soil health 163
Soils 104
Solid wastes 25
Sorghum 137
Sowing 352, 354
Spacing 354
Spice industry 272
Spinach 291
Storage 207, 278, 295, 304, 359
Strategies 90
Stress tolerance 49
Stresses 53
Stylosanthes humilies 348
Subsidies 21
Sugar industry 266
Sugarcane byproducts 309
Sulfur 164
Supply chain management 333
Sustainable agriculture 23
Sustainable crop production 135
Sustainable management 361
Sweet corn 220

T

TAR 235
Technological constraints 94
Technological interventions 36
Tecomella undulata 131, 142, 348
Temperatures 44, 50, 162
Textiles industry 265
Threshing 290, 291, 292, 293, 295, 304, 306, 356
Tilletia tritici 355
TOF 251
Tomato 287
Tomato cultivation 227
Traditional shifting agriculture 87
Training 112, 201
Transplantation 173
Transport 278
Transportation 161
Trapian plateau 68
Treater 358
Triticum aestivum 350

U

Ultra high temperature 317
UNCCD 367
UNFCCC 42
Universities 280
USDA 7
Uses 189

V

Vanraj 192
Varietal maintenance 184
Vegetable crop production 42
Vegetable crops 287
Vegetables production 43, 285
Vegetative stage 296, 299
Vernalization 303
Veterinary works 371
Vigna acontifolia 130

W

Warehouses 278
Washing 299
Water 116
Water harvesting 129, 131, 136
Water harvesting system 70
Water management 173
Water use efficiency 51
Watermelon 301
Weed control 75
Weed management 173
Weeds 104, 117
Wheat 119, 137, 356
Wheat diseases 355
Wheat insect pests 356
Wool 370
Woolen textile industry 266
Work scheme 22
World resources institute 4

Z

Zizyphus numularia 131, 347